Advances in Intelligent Systems and Computing

Volume 976

The series "Advances in Intelligent Systems and Computing" contains publications on theory, applications, and design methods of Intelligent Systems and Intelligent Computing. Virtually all disciplines such as engineering, natural sciences, computer and information science, ICT, economics, business, e-commerce, environment, healthcare, life science are covered. The list of topics spans all the areas of modern intelligent systems and computing such as: computational intelligence, soft computing including neural networks, fuzzy systems, evolutionary computing and the fusion of these paradigms, social intelligence, ambient intelligence, computational neuroscience, artificial life, virtual worlds and society, cognitive science and systems, Perception and Vision, DNA and immune based systems, self-organizing and adaptive systems, e-Learning and teaching, human-centered and human-centric computing, recommender systems, intelligent control, robotics and mechatronics including human-machine teaming, knowledge-based paradigms, learning paradigms, machine ethics, intelligent data analysis, knowledge management, intelligent agents, intelligent decision making and support, intelligent network security, trust management, interactive entertainment, Web intelligence and multimedia.

The publications within "Advances in Intelligent Systems and Computing" are primarily proceedings of important conferences, symposia and congresses. They cover significant recent developments in the field, both of a foundational and applicable character. An important characteristic feature of the series is the short publication time and world-wide distribution. This permits a rapid and broad dissemination of research results.

** Indexing: The books of this series are submitted to ISI Proceedings, EI-Compendex, DBLP, SCOPUS, Google Scholar and Springerlink **

More information about this series at http://www.springer.com/series/11156

Alfredo Vellido · Karina Gibert ·
Cecilio Angulo · José David Martín Guerrero
Editors

Advances in Self-Organizing Maps, Learning Vector Quantization, Clustering and Data Visualization

Proceedings of the 13th International Workshop, WSOM+ 2019, Barcelona, Spain, June 26–28, 2019

Editors
Alfredo Vellido
Department of Computer Science
UPC BarcelonaTech
Barcelona, Spain

Cecilio Angulo
Department of Automatic Control
UPC BarcelonaTech
Barcelona, Spain

Karina Gibert
Knowledge Engineering and Machine
Learning Group (KEMLG) at Intelligent
Data Science and Artificial Intelligence
Research Center
UPC BarcelonaTech
Barcelona, Spain

José David Martín Guerrero
Departament d'Enginyeria Electrònica
Universitat de València
Burjassot, Valencia, Spain

ISSN 2194-5357   ISSN 2194-5365   (electronic)
Advances in Intelligent Systems and Computing
ISBN 978-3-030-19641-7   ISBN 978-3-030-19642-4   (eBook)
https://doi.org/10.1007/978-3-030-19642-4

This Springer imprint is published by the registered company Springer Nature Switzerland AG
The registered company address is: Gewerbestrasse 11, 6330 Cham, Switzerland

Preface

The Association for Computing Machinery (ACM) has named Yoshua Bengio, Geoffrey Hinton, and Yann LeCun as recipients of the 2018 Turing Award for their major contribution to the development of deep neural networks as a critical component of computing. This is a timely reminder of the renewed vitality of the machine learning field, in which self-organizing systems have played a major role from the 1980s not only from the perspective of data analysis, but also as in silico models in computational neuroscience.

This book contains the peer-reviewed and accepted contributions presented at the 13th International Workshop on Self-Organizing Maps (WSOM+2019) held at Universitat Politècnica de Catalunya (UPC BarcelonaTech), Barcelona (Spain), during June 26–28, 2019. WSOM+2019 is the latest in a series of biennial international conferences that started with WSOM'97 in Helsinki, Finland, with Prof. Teuvo Kohonen as General Chairman. We would like to express our gratitude to Prof. Kohonen for serving as Honorary Chair of WSOM+2019.

The reader will find here a varied collection of studies that testify to the vitality of the field of self-organizing systems for data analysis. Most of them relate to the core models in the field, namely self-organizing maps (SOMs) and learning vector quantization (LVQ), but the workshop also catered for research in the broader spectrum of unsupervised learning, clustering, and multivariate data visualization problems. It is also worth highlighting that the book includes a balanced mix of theoretical studies and applied research, covering a wide array of fields that vary from business and engineering to the life sciences. As a result, the book should be of interest to machine learning researchers and practitioners in general and, more specifically, to those interested in keeping up with developments in self-organization, unsupervised learning, and data visualization.

The book collects the work of more than 90 researchers from 18 countries, and it is the result of a collective effort. It would not have been possible without the advice and guidance of the international WSOM Steering Committee, and the quality of the final selection of papers is the result of the selfless reviewing work performed by the Program Committee members and the anonymous additional reviewers, which enhanced the sterling work of the authors themselves. We are truly indebted to the international researchers who agreed to participate as plenary speakers in WSOM+2019: Prof. Paulo J. G. Lisboa (Liverpool John Moores University, UK), Prof. Tobias Schreck (Graz University of Technology, TU Graz, Austria), Dr. Aïda Valls (Universitat Rovira i Virgili, Spain), and Prof. Alessandro Sperduti (Universita degli Studi di Padova, Italy). The Local Organizing Committee would like to acknowledge the support of the UPC BarcelonaTech, the Intelligent Data Science and Artificial Intelligence (IDEAI) Research Center at UPC and the RDLab at the Department of Computer Science of the UPC. We also truly appreciate the support of the sponsoring companies: Amalfi Analytics, LumenAI, and Predict Assistance and the invaluable help provided by our postgraduate students from the master's and the Ph.D. in artificial intelligence programs of the UPC in the organization of the workshop.

June 2019

Organization

Steering Committee

Teuvo Kohonen (Honorary Chairman), Finland
Marie Cottrell, France
Pablo Estévez, Chile
Timo Honkela, Finland
Jean Charles Lamirel, France
Thomas Martinetz, Germany
Erzsébet Merényi, USA
Madalina Olteanu, France
Michel Verleysen, Belgium
Thomas Villmann, Germany
Takeshi Yamakawa, Japan
Hujun Yin, UK

Program Committee

Cecilio Angulo	Universitat Politècnica de Catalunya, Spain
Guilherme Barreto	Universidade Federal do Ceará, Brasil
Abdel Belaid	Université de Lorraine, France
Michael Biehl	University of Groningen, Netherlands
Hubert Cecotti	California State University, Fresno, USA
Cyril de Bodt	Université catholique de Louvain, Belgium
Pablo Estévez	Universidad de Chile, Chile
Jan Faigl	Czech Technical University in Prague, Czech Republic
Jérémy Fix	CentraleSupélec, France
Hervé Frezza-Buet	CentraleSupélec, France

Contents

Self-organizing Maps: Theoretical Developments

Look and Feel What and How Recurrent Self-Organizing Maps Learn

Jérémy Fix[✉] and Hervé Frezza-Buet

LORIA, CNRS, CentraleSupélec, Université Paris-Saclay, 57000 Metz, France
{jeremy.fix,herve.frezza-buet}@centralesupelec.fr

Abstract. This paper introduces representations and measurements for revealing the inner self-organization that occurs in a 1D recurrent self-organizing map. Experiments show the incredible richness and robustness of an extremely simple architecture when it extracts hidden states of the HMM that feeds it with ambiguous and noisy inputs.

Keywords: Recurrent self-organizing map · Sequence processing · Hidden Markov Models

1 Introduction

Self-organizing maps (SOMs) or Kohonen maps, introduced in [9], is a particular topographically organized vector quantization algorithm. It computes a mapping from a high dimensional space to a usually one or two dimensional regular grid with the specificity that close positions in the regular grid are associated with close positions in the original high dimensional space. We have a pretty good understanding of what a SOM is doing. Even if there is no energy function associated with the Kohonen learning rule which could formally state what Kohonen maps do actually capture (some authors actually suggested some alternative formulations derived from an energy function, see for example [6]), we can still pretty much see Kohonen maps as a K-means with a topology i.e. capturing the distribution of input samples in a topographically organized fashion. As soon as we experiment with, for example, 2D-Kohonen maps with two dimensional input samples, we quickly face the nice unfolding of the map sometimes trapped in some kind of local minima where there remains some twist in the map. While our understanding of Kohonen SOMs is pretty clear, the things become more complicated when we turn to recurrent SOMs.

Recurrent SOMs are a natural extension of SOMs when dealing with serial inputs in order to "find structure in time" (J. Elman). This extension follows the same principle introduced for supervised multi-layer perceptrons by [3,7] of feeding back a context computed from the previous time step. These recurrent SOMs are built by extending the prototype vector with an extra component which encapsulates some information about the past. There are indeed various proposals about the information memorized from the past, e.g. keeping only the

A. Vellido et al. (Eds.): WSOM 2019, AISC 976, pp. 3–12, 2020.
https://doi.org/10.1007/978-3-030-19642-4_1

location of the previous best matching unit [4] or the matching over the whole map [11]. An overview of recurrent SOMs is provided in [5]. Cellular and biologically inspired architectures have been proposed as well [8]. When the question of understanding how recurrent SOMs work comes to the front, there are some theoretical results that bring some answers. However, as any theoretical study, they are necessarily limited in the questions they can address. For example, [10] studied the behavior of recurrent SOMs by analyzing its dynamics in the absence of inputs. As for usual SOMs for which mathematical investigations do not cover the whole field yet [2], these theoretical results bring only a partial answer and there is still room for experimental investigation. Despite numerous works, how the recurrent SOMs deal with serial inputs and what they actually learn is not obvious: "The internal model representation of structures is unclear" [5]. We indeed lack the clear representations that we possess for understanding SOMs.

In order to tackle this issue, we focus in this paper on the simplest recurrent SOM where the temporal context is only the position of the best matching unit (BMU) within the map at the previous iteration (which bears resemblance to the SOM-SD of [4]). This simplicity comes with the ability to design specific visualizations to investigate the behavior of the map. As we shall see in the experiments, despite this simplicity, there is still an interesting richness of dynamics. In particular, we will investigate and visualize the behavior of this simple recurrent SOM when inputs are provided sequentially by different hidden Markov models. These will illustrate the behavior of the recurrent SOM in the presence of ambiguous observations, long-term dependencies, changing dynamics, noise in the observations and noise in the transitions.

2   Methods

2.1   Algorithm

Let us consider a stream of inputs $\xi^t \in X$ available at each successive time step t. Here, let us use $X = [0, 1]$. Let us consider a topological set $\mathcal{M}$ where unit (sometimes called neuron) positions lie. We use a 1D map in this paper, thus $\mathcal{M} = [0, 1]$ is considered with a topology induced by the Euclidean distance. The map is made of N units, each unit is denoted by an index $i \in \mathcal{I} \subset \mathcal{M}$. Indexes are equally spread over $\mathcal{M}$, i.e. $\mathcal{I} = \{0, 1/(N-1), 2/(N-1) \cdots, 1\}$.

For the sake of introducing notations for our recurrent algorithm, let us start by rephrasing the self-organizing map algorithm (SOM). Each unit i is equipped with an input weight $w_i \in X$ also referred to as a *prototype*. When ξ^t is presented to the map, all the units compute a matching value $\mu_i = \mu(w_i - \xi^t)$, where $\mu(d) = \max(1 - |d|/\rho_\mu, 0)$ is a linear decreasing function that reaches 0 for the distance value $\rho_\mu{}^1$. The best matching unit (BMU) $i_\star$ can be computed here from μ_i as $i_\star = \mathrm{argmax}_i \, \mu_i$. It could have been computed directly as the unit for which $|w_i - \xi^t|$ is minimal, as usual SOM formulation do, but the current formulation involving μ_i allows for the forthcoming extension to recurrent SOM.

---

[1] A more classical Gaussian function could have been used as well.

Once $i_\star$ is determined consecutively to the submission of ξ^t, the prototypes in the neighborhood of $i_\star$ have to be updated. The strength of that update for any w_i is $\alpha\mathrm{h}\left(i - i_\star\right)$, with $\alpha \in [0,1]$ and $\mathrm{h}\left(d\right) = \max\left(1 - |d|/\rho_\mathrm{h}, 0\right)$. Let us stress here that the width ρ_h of the learning kernel is kept constant, as opposed to usual SOM and recursive SOM implementations where it continuously decreases.

The recurrence is added to our formulation of SOM by using *context* weights $c_i \in \mathcal{M}$. They are trained in the same way as w_i, except that they are fed with $i_\star^{t-1}$ instead of ξ^t. A context matching distribution $\mu_i' = \mu\left(c_i - i_\star^{t-1}\right)$ is computed as we did for μ_i. The BMU needs to be determined from both matchings. To do so, each unit computes a global matching $\mu_i'' = \frac{\mu_i + \mu_i'}{2}$, such as the BMU is determined as $i_\star = \mathrm{argmax}_i\, \overline{\mu}''_i$. The overline of $\overline{\mu}''$ indicates a low pass spatial filtering with a gaussian kernel of standard deviation σ_k. Moreover, the selection of the BMU is done by randomly sampling in the set of possible BMUs. These two elements improve the algorithm in our settings where observations are drawn from a discrete set. The whole process studied in this paper can then be formalized into Algorithm 1. For all the experiments, we used $N = 500$ units, a neighbour kernel width $\rho_\mathrm{h} = 0.05$, a learning rate $\alpha = 0.1$, a matching sensitivity $\rho_\mu = 0.4$ and a Gaussian convolution kernel standard deviation $\sigma_k = 0.0125$.

Algorithm 1. Architecture update at time t.

1: Get ξ^t, compute $\forall i \in \mathcal{I}$, $\mu_i = \mu\left(w_i - \xi^t\right)$, $\mu_i' = \mu\left(c_i - i_\star^{t-1}\right)$
2: $\forall i \in \mathcal{I}$, $\mu_i'' = (\mu_i + \mu_i')/2$
3: $i_\star^t \in \mathrm{argmax}_i\, \overline{\mu}''_i$ // $\overline{\mu}'' = \mu'' * k$, $i_\star$ is taken randomly in argmax.
4: $\forall i \in \mathcal{I} \begin{cases} w_i^t = w_i^{t-1} + \alpha\mathrm{h}\left(i - i_\star\right).\left(w_i^{t-1} - \xi^t\right) \\ c_i^t = c_i^{t-1} + \alpha\mathrm{h}\left(i - i_\star\right).\left(c_i^{t-1} - i_\star^{t-1}\right) \end{cases}$

In our experiments, the inputs in X that are provided at each time step are generated from a Hidden Markov Model (HMM). The HMM has a finite set $S = \{s_0, s_1, \cdots\}$ of states. Each state is an integer (i.e. $S \subset \mathbb{N}$). At each time step, a state transition is performed according to a transition matrix. In the current state s^t, the observation is sampled from the conditional probability $\mathbb{P}\left(\xi \mid s^t\right)$, defined by the observation matrix of the HMM. Different states of the HMM may provide a similar observation. In this case, the recursive architecture is expected to make the difference between such states in spite of the observation ambiguity. In other words, the current BMU $i_\star^t$ value is expected to represent the actual s^t even if several other states could have provided the current input ξ^t.

2.2 Representations

Algorithm 1 can be executed with any dimension for $\mathcal{M}$ without loss of generality. Nevertheless, we use 1D maps ($\mathcal{M} = [0,1]$) for the sake of visualization. Weights w_i are in $X = [0,1]$ as well. They can be represented as a gray scaled value, from black (0) to white (1). In the bottom left part of Fig. 1, the background of the chart is made of w_i^t, with t in abscissa and i in ordinate. On this chart, red curves are also plotted. This is done when the HMM is deterministic (and thus cycling through its states, visiting $s_0, s_1, \cdots, s_{p-1}, s_0, s_1, \cdots$). If the state sequence that is repeated throughout the experiment has a length p ($p = 10$ in experiment of Fig. 1), p red curves are plotted on the chart. For $0 \le k < p$, the kth red curve links the points $\{(t, i_\star^t) \mid t \bmod p = k\}$. The curves show the evolution of the BMU position corresponding to each of the p states throughout learning. From left to right in that chart in Fig. 1, some red curves are initially overlaid before getting progressively distant. Such red curves splits show a bifurcation since the map allocates a new place on $\mathcal{M}$ for representing a newly detected HMM state. This allocation has a topography since the evolution is a split and then a progressive spatial differentiation of the state positions.

Let us take another benefit from using 1D maps and introduce an original representation of both w and c weights. This representation is referred to as a *protograph* in this paper. It consists of a circular chart (see three of them on top left in Fig. 1). The gray almost-closed circle represents $\mathcal{M} = [0, 1]$. At time step t, one can plot on the circle the two weights related to $i_\star^t$. First weight, related to the input, is $w(i_\star^t)$, which is a value in X to which a gray level is associated. This is plotted as a gray dot with the corresponding gray value, placed on the circle at position $i_\star^t$. The second weight to be represented for $i_\star^t$ is $c(i_\star^t)$, related to the recurrent context, which is a position in $\mathcal{M}$ and thus a position on the circle. $c(i_\star^t)$ is represented with an arrow, starting from position $c(i_\star^t)$ on the circle and pointing at $i_\star^t$ on the circle, where the dot representing $w(i_\star^t)$ is actually located. This makes a dot-arrow pair for $i_\star^t$. The full *protograph* at time t plots the dot-arrow pairs $(w(i_\star), c(i_\star))$ for the 50 last steps. The third protograph in Fig. 1 seems to contain only 10 dot-arrow pairs since many of the 50 ones are identical to others. This last protograph corresponds to an organized map, it reveals the number of states visited by the HMM (number of dots), where they are encoded in the map (dot positions), which observation each state provides (dot colors), and the state sequence driven by the HMM transitions (follow the arrows from one state to another). Making movies from the succession of such protographs unveils the dynamics of the organization of spatio-temporal representations in the map. The splits and separation mentioned for the red curves is then visible as a split of one dot into two dots that slide afterwards away one from the other. Movies of the experiments are available online[2].

2.3   Evaluation

The representations presented so far enables to unveil the inner dynamics of a single run. Nevertheless, the ability of the architecture to encode the hidden states of the HMM providing the inputs needs to be measured quantitatively from several runs (a thousand in our experiments). At time step t, let us store the dataset $D^t = \{(i_\star^{t-99}, s^{t-99}), \cdots, (i_\star^{t-1}, s^{t-1}), (i_\star^t, s^t)\}$ that is a 100-sized sliding window containing the last observed BMU position/HMM state pairs.

---

[2] http://www.metz.supelec.fr/~fix_jer/recsom1D.

If the map encodes the HMM states with a dedicated BMU position, each observed BMU position must be paired with a single state. In this case, D^t can be viewed as a set of samples of a *function* from $\mathcal{M}$ to S. To check this property for the map at time t, a supervised learning process is performed from D^t, that is viewed here as an input/output pairs container. As $S \subset \mathbb{N}$, this is a multi-class learning problem. A basic bi-class decision stump is used in this paper (i.e. a threshold on map position values makes the decision), adapted to the multi-class problem thanks to a one-versus-one scheme. Let us denote by χ^t the classification error rate obtained on D^t (i.e. the empirical risk). The value χ^t is null when one can recover the state of the HMM from the position of the BMUs collected during the 100 steps. It is higher when a small contiguous region of the map is associated with several HMM states.

In our experiment, χ^t is computed every 100 steps in a run. As previously said, 1000 runs are performed in order to compute statistics about the evolution of χ^t as the map gets organized. At each time step t, only the best 90% of the 1000 χ^t are kept. The evolution curve, as reported in the right of Fig. 1, plots the upper and lower bounds of these 900 values, as well as their average. There are indeed less than 10% of the runs for which the map does not properly self-organize. A deeper investigation of this phenomenon is required, but it is out of the scope of the present paper, which is focused on the dynamic of the self-organization when it occurs. This is why the corresponding runs are removed from the performance computation.

3   Results

As mentioned in Sect. 2.1, the serial inputs ξ^t are observations provided by the successive states of a HMM. Let us use a comprehensive notation for the HMMs used in our experiments. Observations are in $X = [0, 1]$ as previously stated and 6 specific input values are represented by a letter (A $= 0$, B $= 0.2$, C $= 0.4$, D $= 0.6$, E $= 0.8$, F $= 1$). The HMM denoted by **AABFCFE** is then a 7-state HMM for which s_0 provides observation A, s_1 provides A as well, s_2 provides B, ... s_6 provides E. The states are visited from s_0 to s_6 periodically. In this particular HMM, (s_0, s_1), as well as (s_3, s_5) are ambiguous since they provide the same observation (A and F) as an input to the recurrent SOM. When a state provides an observation uniformly sampled in $[0, 1]$, it is denoted by $*$. The notation $\mathsf{AB}\overline{\mathsf{CD}}^\sigma\mathsf{EF}$ means that values for both s_2 and s_3 are altered by an additive normal noise with standard deviation σ. Last, the notation $\mathsf{ABC}|_q^p\mathsf{DEF}$ means that the HMM is made of two periodical HMMs **ABC** and **DEF**, with random transitions from any of the state of **ABC** to any of the state of **DEF** with a probability p. Random transitions from **DEF** to **ABC** occurs similarly with a probability q.

3.1   Ambiguous Observations

In order to test the ability of the recurrent SOM to deal with ambiguous observations, we consider the HMM **ABCEFEDCB**, i.e. a HMM with 10 states and 6 observations. There are 8 states which provide an observation that is ambiguous (the state cannot be identified given only the current observation). The recurrent SOM receives observations during 5000 time steps, i.e. 500 presentations of the full sequence. A single run is depicted on the left of Fig. 1. The SOM initially captures the individual observations, irrespective of their context (at $t = 200$, approximately 6 units are the BMUs), and then, ambiguous observation begin electing their own BMUs and the recurrent prototypes begin to reflect the context of the observations. This splitting of the winning positions ultimately leads to one distinct BMU for each state of the HMM and, at $t = 5000$, the structure of the HMM can be uncovered from the weights $(w(i_\star), c(i_\star))$ as displayed in the respective protograph. Red curves show how the 6 areas in the map insensitive to the context split to build the expected appropriate 10 areas. Running the same experiment for 1000 runs indicates that the ability of the algorithm to identify the structure of the HMM is statistically significant (right of Fig. 1). The observations produced by this HMM could be easily disambiguated, taking into account the observation of the previous time step. In the next experiment, we study longer term dependencies.

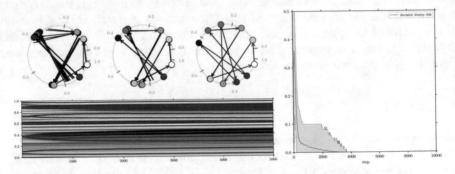

Fig. 1. Observations from ABCDEFEDCB. The photographs are recorded at $t = 200, 500, 5000$.

3.2   Long Term Dependencies

In the second experiment, the algorithm receives observations from the HMM AAAAAAAF. This experiment seeks to test if the algorithm can capture long-term dependencies. Indeed, this HMM produces exactly the same observation A for a fixed number of steps before outputting F which seeds the ambiguity of A. The experiment is run for 5000 steps, i.e. 625 repetitions of the full sequence. The results of a single run are displayed on the left of Fig. 2. Initially, the SOM captures the two observations A and F independently of the context (see the plot of the prototypes on the top left of the figure). Then, we observe several units specializing to the observation of A in a context dependent manner. The logic of the propagation of the context can be appreciated from the split of the red curves. The first state producing a A to be clearly identified is the one associated with the unit with the smallest position (the node shown in back on the first protograph around position 0.3). This is the simplest to be identified because it is the state just after the one outputting F. Then, the dependence on the context propagates through all the previous steps. The $c(i_\star)$ weights are continuous over $\mathcal{M}$, as for usual weights in SOMs. In the specific run of Fig. 2, when we consider the black dots counterclockwise, the arrow origins $c(i_\star)$ progress clockwise, i.e. $c(i_\star)$ is a monotonously decreasing function. Running this experiment on 1000 independent trials reveals that the algorithm is able to capture the structure of this HMM (see Fig. 2, right).

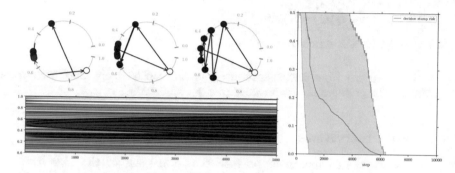

Fig. 2. Observations from AAAAAAAF. The protographs are recorded at $t = 40, 1000, 5000$.

3.3   Adapting to a Changing Dynamics

In this third experiment, the algorithm receives observations from the HMM ABCDEFEDCB for the 10000 first steps and then from the HMM ABCBAFEDEF for the last 10000 steps. The prototypes obtained at $t = 10000$ and $t = 20000$ as well as the evolution of the observation weights and winner locations are displayed on the left of Fig. 3 for a single run. The algorithm successfully recovers the structure of the two HMM. Analyzing the red curves at the time the second HMM is presented is illuminating. One can note there is a reuse of the previously learned prototypes and some adaptation of the prototypes. Indeed, there was a single BMU responsive for a F (white node on the first protograph) for the first sequence which splits and two BMUs are now responsive for a F for the second sequence, which makes sense given the second HMM has two different states producing the observation F. The same comment holds for the BMUs when the observation A is produced by the HMM. On the contrary, while two BMUs had observation prototypes w close to a C and D during the first training period, only one BMU is remaining for each C and D after learning with the second HMM.

The performance of the algorithm ran for 1000 independent trials is shown on the right of Fig. 3. Similarly to the first experiment, it takes around 5000 steps to learn the sequence. At the time the HMM is changed, there is a degradation in the performances that quickly drops.

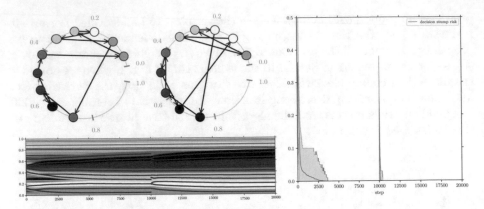

Fig. 3. Observations from ABCDEFEDCB for the first 10000 steps and then from ABCBAFEDEF. The protographs are recorded at $t = 10000, 20000$.

3.4 Noisy Observations

We now perform an experiment to test the robustness of the algorithm in the presence of noise in the observations. This experiment involves the HMM $\overline{\text{BCDEDC}}^{0.05}$, i.e. each observation is perturbed with a normal noise of standard deviation 0.05. Given that the unperturbed observations are separated by 0.2, a normal noise of standard deviation 0.05 leads to slightly overlapping observations. On the left of Fig. 4, one can recognize the 6 states identified by the algorithm. The blur in the representation of the prototypes comes from the fact that the circles arranged along the ring are displayed with a transparency proportional to the winning frequency of the displayed BMU. Given the observations are noisy, there is jitter in the location of the BMUs but still, the elected BMUs for each state of the HMM remains in a compact set. The sequence B, C, D, E, D, C tends to elicit BMUs in positions around 0.5, 0.65, 0.1, 0.35, 0.2 and 0.8. Running the experiment on 1000 runs, the performance χ^t decreases almost down to 0.0 as shown on the right of Fig. 4. This confirms that the locations that are BMUs for a given HMM state are indeed compact sets. Finally, while the absence of noise and the linear neighborhood function kept the BMUs confined within the map in the previous experiments, the presence of noise in this experiment ultimately leads to populate all the map; all the positions within the map tend to be recruited to encode the HMM.

3.5 Perturbed by a Noise State

In the last experiment, we consider a challenging HMM ABCDEFEDCB$|_q^p*$, with $p = 0.03$ and $q = 0.1$. This HMM is based on the sequence ABCDEFEDCB. There is a probability p for the HMM to jump from any of the states ABCDEFEDCB to the state we denote $*$ which emits a uniformly distributed observation. There is also a probability q to jump from the state $*$ back to one of the states in

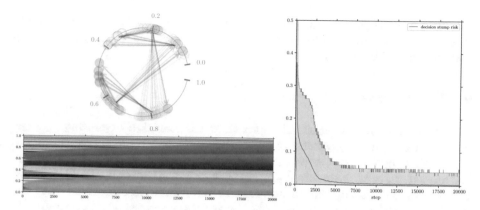

Fig. 4. Observations from $\overline{\text{BCDEDC}}^{0.05}$. The protograph is recorded at $t = 20000$.

ABCDEFEDCB. The clean sequence of observations from **ABCDEFEDCB** is therefore regularly corrupted with a uniform noise. This challenging HMM can mimic for example a temporal disruption of sensors. With $p = 0.03$, there is a probability $(1 - p)^9 = 0.76$ to completely unroll the sequence **ABCDEFEDCB** when starting from **A**. When the HMM is in the state $*$, it stays in this state for $\frac{1}{q} = 10$ steps in average. The results of a single run are displayed on the left of Fig. 5. The structure of the HMM, without the noisy state, can be recognized from the plot of the prototypes on the top of the figure (if we omit the BMUs just before $i = 0.2$ and just after $i = 0.8$). It should be noted that the structure of the algorithm does not allow it to capture the noisy state and the latter is therefore filtered by the algorithm. Running the experiment for 1000 runs indicates that the ability of the algorithm to capture the structure of the clean HMM is statistically significant, as shown on the right of Fig. 5. For computing the statistics on Fig. 5, the samples labelled with $*$ are removed from the dataset D^t. However, this still does not

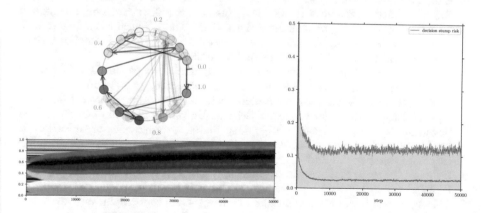

Fig. 5. Observations from $\text{ABCDEFEDCB}\big|_q^p *$. The protograph is recorded at $t = 20000$.

lead to a perfect classification. Indeed, when the HMM is back from the noisy state $*$, it sometimes requires two successive observations to identify in which state the HMM is. This explains why χ^t is not null.

4 Conclusion

This paper presents an empirical approach of recurrent self-organizing maps by introducing original representations and performance measurements. The experiments show how spatio-temporal structure gets organized internally to retrieve the hidden states of the external process that provides the observations. An area of the map associated with an observation splits into close areas when observation ambiguity is detected, and then areas get progressively separated onto the map. Unveiling the emergence of such a complex and continuous behavior, from both the SOM-like nature of the process and a simple re-entrance, is the main result of this paper. Such a simple architecture also shows robustness to temporal and spatial damages in the input series, as well as the ability to deal with deep time dependencies while the recurrence only propagates previous step context. Forthcoming work will consist in using such recurrent maps in more integrated multi-map architecture, as started in [1].

Acknowledgement. This work is supported by the European Interreg Grande Région/Région Grand-Est project GRONE.

References

1. Baheux D, Fix J, Frezza-Buet H (2014) Towards an effective multi-map self organizing recurrent neural network. In: ESANN, pp 201–206
2. Cottrell M, Fort J, Pags G (1998) Theoretical aspects of the SOM algorithm. Neurocomputing 21(1):119–138
3. Elman JL (1990) Finding structure in time. Cognit Sci 14:179–211
4. Hagenbuchner M, Sperduti R, Tsoi AC, Member S (2003) A self-organizing map for adaptive processing of structured data. IEEE Trans Neural Netw 14:491–505
5. Hammer B, Micheli A, Sperduti A, Strickert M (2004) Recursive self-organizing network models. Neural Netw 17(8–9):1061–1085
6. Heskes T (1999) Energy functions for self-organizing maps. In: Oja E, Kaski S (eds) Kohonen maps. Elsevier Science B.V, Amsterdam, pp 303–315
7. Jordan MI (1996) Serial order: a parallel distributed processing approach. Technical reporty 8604, Institute for Cognitive Science, University of California, San Diego
8. Khouzam B, Frezza-Buet H (2013) Distributed recurrent self-organization for tracking the state of non-stationary partially observable dynamical systems. Biol Inspired Cognit Archit 3:87–104
9. Kohonen T (1982) Self-organized formation of topologically correct feature maps. Biol Cybern 43(1):59–69
10. Tiňo P, Farkaš I, van Mourik J (2006) Dynamics and topographic organization of recursive self-organizing maps. Neural Comput 18(10):2529–2567
11. Voegtlin T (2002) Recursive self-organizing maps. Neural Netw 15(8–9):979–991

Self-Organizing Mappings
on the Flag Manifold

Xiaofeng Ma, Michael Kirby$^{(\boxtimes)}$, and Chris Peterson

Colorado State University, Fort Collins, CO 80523, USA
`xiaofeng.ma@rams.colostate.edu`,
`{michael.kirby,christopher2.peterson}@colostate.edu`

Abstract. A flag is a nested sequence of vector spaces. The type of the flag is determined by the sequence of dimensions of the vector spaces making up the flag. A flag manifold is a manifold whose points parameterize all flags of a particular type in a fixed vector space. This paper provides the mathematical framework necessary for implementing self-organizing mappings on flag manifolds. Flags arise implicitly in many data analysis techniques for instance in wavelet, Fourier, and singular value decompositions. The proposed geometric framework in this paper enables the computation of distances between flags, the computation of geodesics between flags, and the ability to move one flag a prescribed distance in the direction of another flag. Using these operations as building blocks, we implement the SOM algorithm on a flag manifold. The basic algorithm is applied to the problem of parameterizing a set of flags of a fixed type.

Keywords: Self-Organizing Mappings · SOM · Flag manifolds · Geodesic · Visualization

1 Introduction

Self-Organizing Mappings (SOMs) were introduced as a means to *see* data in high-dimensions [7–10]. This competitive learning algorithm effectively transports the notion of proximity in the data space to proximity in the index space; this may in turn be endowed with its own geometry. This tool has now been widely applied and extended [4]. The goal of the SOM algorithm is to produce a topology preserving mapping in the sense that points that are neighbors in high-dimensional space are also represented as neighbors in the low-dimensional index space.

The geometric framework of the vanilla version of the SOM algorithm is Euclidean space. In this setting, the distance between points is simply the standard 2-norm of the vector difference. The movement of a center towards a pattern takes place on a line segment in the ambient space. The only additional ingredient to the algorithm is a metric on the index space.

© Springer Nature Switzerland AG 2020
A. Vellido et al. (Eds.): WSOM 2019, AISC 976, pp. 13–22, 2020.
https://doi.org/10.1007/978-3-030-19642-4_2

Motivated by the subspace approach to data analytics we proposed a version of SOM using the geometric framework of the Grassmannian [2,14–16]. This subspace approach has proven to be effective in settings where you have a collection of subspaces built up from a set of patterns drawn from a given family. Given one can compute distances between points on a Grassmannian, and move one point in the direction of another, it is possible to transport the SOM algorithm on Euclidean space to an SOM algorithm on a Grassmannian [6,11].

An interesting structure that generalizes Grassmannians and encodes additional geometry in data is known as the *flag manifold*. Intuitively, a point on a flag manifold is a set of nested subspaces. So, for example, given a data vector, a wavelet transform produces a set of approximations that live in nested scaling subspaces [5]. The nested sequence of scaling subspaces is a flag and corresponds to a single point on an appropriate flag manifold. Alternatively, an ordered basis, $v_1, v_2, \ldots, v_k$ for a set of data produced by principal component analysis induces the flag $S_1 \subset S_2 \subset \cdots \subset S_k$ where S_i is the span of $v_1, \ldots, v_i$. In this paper we extend SOM to perform a topology preserving mapping on points that correspond to nested subspaces such as those arising, for instance, from ordered bases or wavelet scaling spaces. To accomplish this we show how to compute the distance between two points on a flag manifold, and demonstrate how to move a flag a prescribed distance in the direction of another. Given these building blocks, we illustrate how one may extend SOM to the geometric framework of a flag manifold.

This paper is outlined as follows: In Sect. 2 we provide a formal definition of the flag manifold and illustrate with concrete examples. In Sect. 3 we introduce the numerical representation of flag manifolds. Here we indicate explicitly how distances can be computed between flags, and further, how a flag can be moved in the direction of another flag. In Sect. 4 we put the pieces together to realize the SOM algorithm on flag manifolds. We demonstrate the algorithm with a preliminary computational example. Finally, in Sect. 5 we summarize the results of the paper and point towards future directions of research.

2 Introduction to Flag Manifold with Data Analysis Examples

Let us first introduce the flag manifold. A *flag* of type $(n_1, n_2, \ldots, n_d; n)$ is a nested sequence of subspaces in $\mathbb{R}^n$ where $\{0\} \subsetneq V_1 \subsetneq V_2 \subsetneq \cdots \subsetneq V_d = \mathbb{R}^n$, $dim\, V_j = \Sigma_{i=1}^{j} n_i$ and $n_1 + n_2 + \cdots + n_d = n$. We let $FL(n_1, n_2, \ldots, n_d; n)$ denote the *flag manifold* whose points parameterize all flags of type $(n_1, n_2, \ldots, n_d; n)$. As a special case, the flag of type $(1, 1, \cdots, 1; n)$ is referred to as a full flag and $FL(1, 1, \cdots, 1; n)$ is the full flag manifold in $\mathbb{R}^n$. Figure 1 illustrates the nested structure of the first three low-dimensional elements comprising a full flag in $\mathbb{R}^n$. A flag of type $(k, n - k; n)$ is simply a k–dimensional subspace of $\mathbb{R}^n$ (which can be considered as a point on the Grassmann manifold $Gr(k, n)$). Hence $FL(k, n - k; n) = Gr(k, n)$. The Grassmannian-SOM algorithm is developed in [6,11]. The idea that the flag manifold is a generalization of the Grassmann

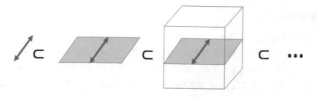

Fig. 1. Illustration of a nested sequence of subspaces corresponding to a point on the flag manifold $FL(1, 1, \cdots, 1; n)$.

manifold will be utilized later to introduce the geodesic formula on the flag manifold. The nested structure inherent in a flag shows up naturally in the context of data analysis.

1. Wavelet analysis: Wavelet analysis and its associated multiresolution representation produces a nested sequence of vector spaces that approximate data with increasing resolution [1,12,13]. Each *scaling* subspace V_j is a dilation of its adjacent neighbor V_{j+1} in the sense that if $f(x) \in V_j$ then a reduced resolution copy $f(x/2) \in V_{j+1}$. The scaling subspaces are nested

$$\cdots \subset V_2 \subset V_1 \subset V_0 \subset V_{-1} \subset \cdots$$

and in the finite dimensional setting can be considered as a point on a flag manifold. The flag SOM algorithm provides a means to visualize relationships in a collection of discrete wavelet transforms and organize the corresponding sequences of nested subspaces in a coherent manner via a low-dimensional grid.

2. SVD basis of a real data matrix: Let $X \in \mathbb{R}^{n \times k}$ be a real data matrix consisting of k samples in $\mathbb{R}^n$. Let $U \Sigma V^T = X$ be the thin SVD of X. The columns of the n-by-d orthonormal matrix U is an ordered basis for the column span of X. This basis is ordered by the magnitude of the singular values of X. This order provides a straightforward way to associate to U a point on a flag manifold. If $U = [u_1|u_2|\ldots|u_d]$ then the nested subspaces $span([u_1]) \subsetneq span([u_1|u_2]) \subsetneq \cdots \subsetneq span([u_1|\cdots|u_d]) \subsetneq \mathbb{R}^n$ is a flag of type $(1, 1, \ldots, 1, n - d; n)$ in $\mathbb{R}^n$. After we introduce the distance metric on the flag manifold in Sect. 3.2, one could consider computing the distance between two flags, perhaps derived from a thin SVD of two different data sets, which takes the order of the basis into consideration.

3   Numerical Representation and Geodesics

A point in the vector space $\mathbb{R}^n$ can be naturally represented by an $n \times 1$ vector. For a more abstract object like a Grassmann or flag manifold, we need a way to represent points in such a way that we can do computations. In this section, we describe how we can represent points and we describe how to determine and express geodesic paths between points. Note that in this paper we are using exp and log to denote the matrix exponential and the matrix logarithm.

3.1   Flag Manifold

The flag manifold $FL(n_1, n_2, \ldots, n_d; n)$ consists of the set of all flags of type $(n_1, n_2, \ldots, n_d; n)$. The presentation in [3] describes how to view the Grassmann manifold $Gr(k, n)$ as the quotient manifold $SO(n)/S(O(k) \times O(n-k))$. Similarly, we can view $FL(n_1, n_2, \ldots, n_d; n)$ as the quotient manifold $SO(n)/S(O(n_1) \times O(n_2) \times \cdots \times O(n_d))$ where $n_1 + n_2 + \cdots + n_d = n$. Let $P \in SO(n)$ be an n-by-n orthogonal matrix, the equivalence class $[P]$, representing a point on the flag manifold, is the set of orthogonal matrices

$$[P] = \left\{ P \begin{pmatrix} P_1 & 0 & \cdots & 0 \\ 0 & P_2 & \cdots & 0 \\ \vdots & & \ddots & \vdots \\ 0 & \cdots & & P_d \end{pmatrix} : P_i \in O(n_i), \, n_1 + n_2 + \cdots + n_d = n \right\}.$$

It is well known that the geodesic paths on $SO(n)$ are given by exponential flows $Q(t) = Q \exp(t\mathbf{A})$ where $\mathbf{A} \in \mathbb{R}^{\mathbf{n} \times \mathbf{n}}$ is any skew symmetric matrix and $Q(0) = Q$. Viewing $FL(n_1, n_2, \ldots, n_d; n)$ as a quotient manifold of $SO(n)$, one can show that geodesics on $SO(n)$ continue to be geodesics on $FL(n_1, n_2, \ldots, n_d; n)$ as long as they are perpendicular to the orbits generated by $S(O(n_1) \times O(n_2) \times \cdots \times O(n_d))$ (for a derivation on a Grassmann manifold, see [11]). This leads one to conclude that the geodesic paths on $FL(n_1, n_2, \ldots, n_d; n)$ are exponential flows:

$$P(t) = P \exp(t\tilde{\mathbf{C}}) \tag{1}$$

where $\tilde{\mathbf{C}}$ is any skew symmetric matrix of the form

$$\tilde{\mathbf{C}} = \begin{pmatrix} \mathbf{0}_{n_1} & & & * \\ & \mathbf{0}_{n_2} & & \\ & & \ddots & \\ -*^T & & & \mathbf{0}_{n_d} \end{pmatrix}, \, \mathbf{0}_{n_i} = \mathbf{0}^{n_i \times n_i}.$$

3.2   Geodesic and Distance Between Two Points on Flag Manifold

By Eq. (1), one may trace out the geodesic path on a flag manifold emanating from P in the direction of $\tilde{\mathbf{C}}$. In this section we utilize Eq. (1) to solve the inverse problem:

Given two points $Q_1, Q_2 \in SO(n)$, whose equivalence classes $[Q_1], [Q_2]$ represent flags of type $(n_1, n_2, \ldots, n_d; n)$, obtain a factorization

$$Q_2 = Q_1 \cdot \exp(H) \cdot M \tag{2}$$

for H and M where H and M are constrained to be of the form

$$H = \begin{pmatrix} \mathbf{0}_{n_1} & & & * \\ & \mathbf{0}_{n_2} & & \\ & & \ddots & \\ -*^T & & & \mathbf{0}_{n_d} \end{pmatrix} \quad \text{and} \quad M = \begin{pmatrix} M_1 & 0 & \cdots & 0 \\ 0 & M_2 & \cdots & 0 \\ \vdots & & \ddots & \vdots \\ 0 & \cdots & & M_d \end{pmatrix}$$

where H is skew symmetric, $M_i \in O(n_i)$, and $M \in SO(n)$. The **distance** between $[Q_1]$ and $[Q_2]$ along the geodesic given by H is

$$d([Q_1], [Q_2]) = \sqrt{\Sigma_{i=1}^{l} \lambda_i^2} \tag{3}$$

where the λ_i's are the distinct singular values of H.

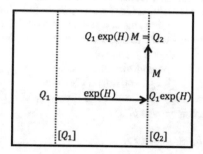

Fig. 2. Illustration of Eq. (2). The vertical lines represents the equivalence classes $[Q_1]$ and $[Q_2]$ respectively. Q_1 is mapped to an element in $[Q_2]$ by right multiplication with $\exp(H)$ which is then sent to Q_2 by multiplying with M.

Equation (2) can be interpreted in the following way. First, we map Q_1 to a representative in $[Q_2]$ via the geodesic determined by the velocity matrix H. Second, we map this element in $[Q_2]$ to Q_2 via the matrix M. Figure 2 is a pictorial illustration of the idea behind Eq. (2). For $FL(k, n-k; n)$ i.e. the Grassmannian $Gr(k, n)$, one can solve for H analytically. Please see [3] for details. For the more general case, we will present an iterative algorithm to obtain a numerical approximation of H and M in Sect. 3.3. Before we proceed to the algorithm, let us further simplify Eq. (2) by letting $Q = Q_1^T Q_2$. This allows us to rewrite (1) as

$$Q = \exp(H) \cdot M \tag{4}$$

Here we define $\mathcal{W}$ as the vector space of all n-by-n skew symmetric matrices. Let $\mathbf{p} = (n_1, n_2, \ldots, n_d; n)$. We define $\mathcal{W}_{\mathbf{p}}$ to be the set of all block diagonal skew symmetric matrices of type $\mathbf{p}$ and its orthogonal complement $\mathcal{W}_{\mathbf{p}}^{\perp}$ in $\mathcal{W}$, i.e.

$$\mathcal{W}_{\mathbf{p}} = \left\{ G \in \mathcal{W} \mid G = \begin{pmatrix} G_1 & \cdots & 0 \\ \vdots & \ddots & \vdots \\ 0 & \cdots & G_d \end{pmatrix} \right\}, \quad \mathcal{W}_{\mathbf{p}}^{\perp} = \left\{ H \in \mathcal{W} \mid H = \begin{pmatrix} \mathbf{0}_{n_1} & & * \\ & \ddots & \\ -*^T & & \mathbf{0}_{n_d} \end{pmatrix} \right\}.$$

where, by definition, $G_i \in \mathbb{R}^{n_i \times n_i}$ is skew symmetric for all i. Instead of solving Eq. (4) directly, we propose to solve the following alternative equation:

$$Q = \exp(H) \cdot \exp(G) \tag{5}$$

where $G \in \mathcal{W}_{\mathbf{p}}$ and $H \in \mathcal{W}_{\mathbf{p}}^{\perp}$. It is important to note that in these computations, we are implicitly working on the fully oriented flag manifold $SO(n)/SO(n_1) \times SO(n_2) \times \cdots \times SO(n_d)$. There is a natural 2^{d-1} to 1 map from the fully oriented flag manifold to the flag manifold. As the output of the algorithm, we must pick the "optimal" H with the shortest distance arising from this map.

Algorithm 1. Iterative Alternating Exp-Log Algorithm

Input Data: Load matrices $Q_1, Q_2 \in SO(n)$ from the flag manifold, desired
     flag type $\mathbf{p} = (n_1, n_2, \ldots, n_d; n)$, initial $G_0 \in \mathcal{W}_{\mathbf{p}}$
Output Data: Optimal skew symmetric matrices $H \in \mathcal{W}_{\mathbf{p}}^{\perp}, G \in \mathcal{W}_{\mathbf{p}}$ such that
     $Q_1^T Q_2 = exp(H) \cdot exp(G)$
Result: Geodesic path and geodesic distance between $[Q_1]$ and $[Q_2]$
Initialization: Let $Q = Q_1^T Q_2$, set initial $d_H = \infty$
Define: Geodesic distance associated to H: $d_H = \sqrt{\Sigma_{i=1}^{l} \lambda_i^2}$ where λ_i's are
    distinct singular values of H.

1 **for** $i = 1, \cdots, m$ **do**
2   Generate initial randomized G_0
3   **for** $k = 1, \cdots, l$ **do**
      Step 1: $H_k = \text{Proj}_H(\log(Q \cdot \exp(G_{k-1})^T))$
      Step 2: $G_k = \text{Proj}_G(\log(\exp(H_k)^T Q))$
4   **end**
5   **if** *current H is associated to a smaller d_H* **then**
6   Update d_H
7   Set current H and G as our output
8   **else**
9   continue
10   **end**
11 **end**

3.3 Iterative Alternating Algorithm

The idea of the Iterative Alternating algorithm is straightforward. Given an initial guess $G_0 \in \mathcal{W}_{\mathbf{p}}$, since Q and G_0 are known, we can solve for H numerically. Let $\hat{H} = \log(Q \cdot \exp(G_0)^T)$, since $\hat{H}$ is generally not of the desired form ($\hat{H} \notin \mathcal{W}_{\mathbf{p}}^{\perp}$), we project $\hat{H}$ onto $\mathcal{W}_{\mathbf{p}}^{\perp}$ to obtain the updated H. This projection zeros out certain select entries in $\hat{H}$, which is denoted by $H_1 = \text{Proj}_{\mathcal{W}_{\mathbf{p}}^{\perp}}(\hat{H})$. Then we start updating G. Let $\hat{G} = \log(\exp(H_1)^T Q)$ we project $\hat{G}$ onto $\mathcal{W}_{\mathbf{p}}$ which zeros out other select entries, i.e. $G_1 = \text{Proj}_G(\hat{G})$. Now then iterate this process until it converges. The pseudo code of our Iterative Alternating algorithm is presented in Algorithm 1.

  We walk through two examples as an illustration of the numerical computation of the geodesic formula and distance between two points from the flag manifold. Here two types of flag manifold are utilized to illustrate the

---

Algorithm 2. Flag Manifold Self-Organizing Mapping

Input Data: Load class labeled orthonormal data matrices $\{X_i\} \in R^{n \times k}$ for
$i = 1, \cdots, P$ such that $X_i^T X_i = I_k$ where X_i is the nested
subspace of interest, k is the dimension of the most outer
subspace of interest, n is the dimension of the data, i is the matrix
index. Let $\{Y_i\}$ for $i = 1, \cdots, P$ be the corresponding label set.

Output Data: Final centers and indices of each data subspace.

Result: Representation of points on $FL(n_1, n_2, \cdots, n_d; n)$ as indices of SOM
centers.

Initialization: Complete the orthogonal complement of X_i by computing the
QR-decomposition, i.e., $X_i = Q_i R_i$, such that $Q_i \in SO(n)$.
Select number of centers and the structure of low dimensional
index set. Initialize centers $\{C_i\}$ so that $C_i \in SO(n)$

Define: Geodesic distance d_g on the flag manifold

Step 1: Present a random point(nested sequence of subspaces) to the network.

Step 2: Move all the centers C_i proportionally towards the presented nested
sequence of subspaces along the appropriate geodesic.

Step 3: Repeat until convergence.

---

different geometry between a Grassmann and a flag manifold. Let $X = \begin{pmatrix} 1 & 0 \\ 0 & 1 \\ 0 & 0 \\ 0 & 0 \end{pmatrix}$

and $Y = \begin{pmatrix} \frac{1}{\sqrt{2}} & \frac{1}{\sqrt{3}} \\ 0 & \frac{1}{\sqrt{3}} \\ 0 & \frac{1}{\sqrt{3}} \\ -\frac{1}{\sqrt{2}} & 0 \end{pmatrix}$ be two data matrices of interest. Let $X = Q_1 R_1$ and

$Y = Q_2 R_2$ be the full QR-decomposition of X and Y. Here we look at two different flag structures:

1. Flag manifold of type $\mathbf{p} = (2, 2; 4)$: Let $Q = Q_1^T Q_2$ and the initial G_0(or any

    G_i in the iterative procedure) should be of the form $G_i = \begin{pmatrix} 0 & g_1 & 0 & 0 \\ -g_1 & 0 & 0 & 0 \\ 0 & 0 & 0 & g_2 \\ 0 & 0 & -g_2 & 0 \end{pmatrix}$.

The output velocity matrix H(or any H_i) should be of the form $H_i = \begin{pmatrix} 0 & 0 & h_1 & h_2 \\ 0 & 0 & h_3 & h_4 \\ -h_1 & -h_3 & 0 & 0 \\ -h_2 & -h_4 & 0 & 0 \end{pmatrix}$. The unique singular values of output H are $\lambda_1 = 1.0172$,

$\lambda_2 = 0.5536$ and the geodesic distance is therefore $d([Q_1], [Q_2]) = \sqrt{\lambda_1^2 + \lambda_2^2} = 1.1581$. One thing to note is that $FL(2, 2; 4)$ is equivalent to $Gr(2, 4)$. It is easy to verify that λ_1, λ_2 are exactly the principal angles between X and Y.

2. Flag manifold of type $\mathbf{p} = (1, 1, 2; 4)$: For this example, the G_i's and H_i's should be of the form $G_i = \begin{pmatrix} 0 & 0 & 0 & 0 \\ 0 & 0 & 0 & 0 \\ 0 & 0 & 0 & g_1 \\ 0 & 0 & -g_1 & 0 \end{pmatrix}$ and $H_i = \begin{pmatrix} 0 & h_1 & h_2 & h_4 \\ -h_1 & 0 & h_3 & h_5 \\ -h_2 & -h_3 & 0 & 0 \\ -h_4 & -h_5 & 0 & 0 \end{pmatrix}$ respectively. The unique singular values of output H are $\lambda_1 = 1.0469$, $\lambda_2 = 0.5404$ and the geodesic distance is therefore $d([Q_1], [Q_2]) = 1.1782$. The geodesic distance is larger than the previous example since we have imposed more structure in this example.

4   SOM on Flag Manifolds

In this section we extend the SOM algorithm to the setting of flag manifolds. The general setting of SOM starts with a set of training data $x^{(\mu)}$ $\mu = 1, \cdots, p$ and an initial set of randomized centers $\{C_i\}$ where the subscript i is associated to the label of the low dimensional index a_i. The standard SOM center update equation is given by,

$$C_i^{m+1} = C_i^m + \epsilon_m h(d(a_i, a_{i^*}))(X - C_i^m).$$

The superscript m is indicating the m-th iteration in the SOM algorithm. Here i^* is the winning center of data point X, i.e.

$$i^* = \arg \min \|X - C_i\|_2.$$

We also set the localization function as the standard

$$h(s) = e^{-s^2/\sigma^2}$$

and d is the metric which induces the geometry on the index set. Here we mainly focus on the simple one,

$$d(a_i, a_j) = \|a_i - a_j\|_2$$

where the indices are enumerated by subscript, i.e. the index set contains $a_1, a_2, \cdots, a_N$. In the following example, we use $a_1 = (1, 7), a_2 = (1, 6), a_3 = (1, 5), \cdots, a_{49} = (7, 1)$. On the flag manifold, points are no longer living in a Euclidean space thus cannot be moved using the standard update equation. For a given data point X from a flag manifold of type $\mathbf{p} = (n_1, n_2, \cdots, n_d; n)$, we identify the winning center, from the set of all nested subspaces of type $\mathbf{p}$ which represent centers $\{C_i\}$, that is closest via

$$i^* = \arg \min_i d_g(X, C_i)$$

where d_g is defined in Eq. (3). To move the centers towards the nested subspace pattern X according to the SOM update we compute the geodesic, using the

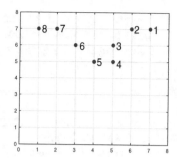

Fig. 3. 8 data points are sampled uniformly along the geodesic curve in $FL(1,1,2;4)$ and labelled in the ascending order from 1 to 8. We observe that 7×7 2D lattice index sorts the label/order of the data points and preserves the geometry of this geodesic line.

Iterative Alternating algorithm described in Algorithm 1, between each center C_i and nested subspace pattern X.

Our localization term now becomes

$$t = \epsilon_n h(d(a_i, a_{i*})).$$

We now take

$$h(s) = \exp(-s^2/\sigma^2)$$

where $\epsilon_n = \epsilon_0(1/\ln(e + n - 1))$. The centers thus change along the geodesic by moving from $C_i(0)$ to $C_i(t)$ where t is adjusted for the step size. The algorithm for SOM on a flag manifold is summarized in Algorithm 2.

4.1   Numerical Experiment

In this section we present an illustrative example concerning a straight line on a flag manifold. We select an initial position $Q \in SO(n)$ from a flag manifold of type $\mathbf{p} = (1, 1, 2; 4)$ and a velocity matrix $H \in \mathcal{W}_{\mathbf{p}}^{\perp}$. Then a set of 8 points are uniformly sampled along the geodesic path emanating from P in the direction H, i.e.,

$$X_i = Q \cdot \exp(t_i H), \, i = 0, 1, 2, \cdots, 7$$

where $t_i = 0.1i$. We employ the 7×7 2D grid index set described above for this example. In Fig. 3, we see that the square lattice index set captures the geometry of the points living on a geodesic in the high dimensional flag manifold and sorts the labels (subscripts of data points X_i's) in the right order.

5   Conclusions and Future Work

We have presented algorithms for Self-Organizing Mappings on flag manifolds. Techniques for computing the key ingredients of the SOM on flags are determining distances between flags and moving one flag a prescribed distance in the direction of another flag. The algorithm was tested on a sample problem that involves computing an ordering of points on a flag manifold. In future work we will explore the application of this flag SOM algorithm to real-world data sets.

References

1. Daubechies I (1992) Ten lectures on wavelets. CBMS-NSF regional conference series in applied mathematics. SIAM, Philadelphia
2. Draper B, Kirby M, Marks J, Marrinan T, Peterson C (2014) A flag representation for finite collections of subspaces of mixed dimensions. Linear Algebra Appl 451:15–32
3. Alan E, Arias TA, Smith ST (1998) The geometry of algorithms with orthogonality constraints. SIAM J Matrix Anal Appl 20(2):303–353
4. Kaski S, Kangas J, Kohonen T (1998) Bibliography of self-organizing map (SOM) papers: 1981–1997. Neural Comput Surv 1(3&4):1–176
5. Kirby M (2001) Geometric data analysis: an empirical approach to dimensionality reduction and the study of patterns. Wiley, New York
6. Kirby, M, Peterson, C (2017) Visualizing data sets on the Grassmannian using self-organizing mappings. In: 2017 12th international workshop on self-organizing maps and learning vector quantization, clustering and data visualization (WSOM), June 2017, pp 1–6
7. Kohonen T (1982) Self-organized formation of topologically correct feature maps. Biol Cybern 43:59
8. Kohonen T (1990) The self-organizing map. Proc IEEE 78(9):1464–1480
9. Kohonen T (1998) The self-organizing map. Neurocomputing 21(1):1–6
10. Kohonen T (2013) Essentials of the self-organizing map. Neural Netw 37:52–65
11. Ma, X, Kirby, M, Peterson, C, Scharf, L (2018) Self-organizing mappings on the Grassmannian with applications to data analysis in high-dimensions. Neural Comput Appl
12. Mallat S (1989) Multiresolution approximations and wavelet orthonormal bases of l^2. Trans Am Math Soc 315:69–87
13. Mallat S (1989) A theory of multiresolution signal decomposition: the wavelet representation. IEEE Trans Pattern Anal Mach Intell 315:69–87
14. Tim, M, Beveridge, JR, Draper, B, Kirby, M, Peterson, C (2015) Flag manifolds for the characterization of geometric structure in large data sets. In: Numerical mathematics and advanced applications-ENUMATH. Springer, pp 457–465
15. Tim, M, Bruce, D, Beveridge, JR, Kirby, M, Peterson, C (2014) Finding the subspace mean or median to fit your need. In: Computer Vision and Pattern Recognition (CVPR). IEEE, pp 1082–1089
16. Marrinan, T, Beveridge, JR, Draper, B, Kirby, M, Peterson, C (2016) Flag-based detection of weak gas signatures in long-wave infrared hyperspectral image sequences. In: Proceedings SPIE Defense + Security. International Society for Optics and Photonics, p 98401N

Self-Organizing Maps with Convolutional Layers

Lars Elend[(✉)] and Oliver Kramer

Computational Intelligence Group, Carl von Ossietzky University of Oldenburg,
26111 Oldenburg, Germany
{lars.elend,oliver.kramer}@uni-oldenburg.de

Abstract. Self-organizing maps (SOMs) are well appropriate for visualizing high-dimensional data sets. Training SOMs on raw high-
dimensional data with classic metrics often leads to problems arising
from the curse-of-dimensionality effect. To achieve more valuable semantic maps of high-dimensional data sets, we assume that higher-level features are necessary. We propose to gather such higher-level features from
pre-trained convolutional layers, i.e., filter banks of convolutional neural
networks (CNNs). Appropriately pre-trained CNNs are required, e.g.,
from the same or related domains, or in semi-supervised scenarios. We
introduce SOM quality measures and analyze the new approach on two
benchmark image data sets considering different convolutional network
levels.

Keywords: Self-Organizing Maps · Convolutional neural networks ·
Dimensionality reduction · Quality measures · Data visualization

1 Introduction

Visualization of high-dimensional data has an important part to play in many
data science domains. SOMs, introduced by Teuvo Kohonen [5], are established
techniques to visualize high-dimensional data sets on 2- or 3-dimensional maps.
However, the SOM learning results significantly depend on the features of the
input data. Weak learning results are often observed when raw input features are
used, e.g., high-dimensional pixel vectors from raw images. We argue that maps
that represent semantic neighborhoods between patterns are more valuable.

Convolutional neural networks are successful image recognizers computing
higher-level representations of features with convolutional layers [6,9]. The approach introduced in this paper unifies both worlds by using the outputs of convolutional layers from pre-trained convolutional networks as high-order features
for the SOM training process. The resulting richer features are supposed to yield
more efficient and robust SOM learning results.

This paper is structured as follows. Section 2 introduces the basic SOM algorithm, while Sect. 3 gives an overview of related work. Section 4 presents the
foundations of convolutional layers. Section 5 introduces the convolutional SOM,

© Springer Nature Switzerland AG 2020
A. Vellido et al. (Eds.): WSOM 2019, AISC 976, pp. 23–32, 2020.
https://doi.org/10.1007/978-3-030-19642-4_3

while Sect. 6 presents relevant quality measures for dimensionality reduction. The convolutional SOM is experimentally analyzed in Sect. 7. Findings are discussed in Sect. 8.

2 Self-Organizing Maps

A self-organizing map (SOM) is a biologically inspired neural model and well appropriate for visualizing the structure of high-dimensional data sets. It has a set of neurons n_i that is usually arranged on a 2- or 3-dimensional map. Each neuron n_i has a position $\mathbf{p}_i \in \mathbb{R}^q$ on the map. A common setting is $q = 2$, i.e., the arrangement of neurons on a 2-dimensional grid. The SOM receives a training set $\{\mathbf{x}_1, \ldots, \mathbf{x}_N\}$ of N data patterns $\mathbf{x}_i \in \mathbb{R}^d$ as input. Each neuron has a weight vector $\mathbf{w}_i \in \mathbb{R}^d$ of the same dimension as the data input.

At the beginning of the SOM learning algorithm, all weight vectors $\mathbf{w}_i$ are initialized with uniformly drawn random values. Given input pattern $\mathbf{x}$, the neuron with the closest weight vector

$$\mathbf{w}^*(\mathbf{x}) := \mathbf{w}_j \text{ with } j = \arg \min_j \|\mathbf{x} - \mathbf{w}_j\|_2 \tag{1}$$

is called winner neuron. Accordingly, $\mathbf{p}^*(\mathbf{x})$ is the winner neuron's position. As the algorithm proceeds, the weights are pulled into the direction of the data patterns based on the winner neuron, a neighborhood function, and a learning rate. The neighborhood function $h(\mathbf{p}_i, \mathbf{p}_j, \sigma) \in [0, 1]$ measures the link between neurons n_i and n_j. Its value is high for small distances, while it should approach zero if the distance is outside the range of σ, i.e., $\|\mathbf{p}_i - \mathbf{p}_j\|_2 > \sigma$. We define the neighborhood function as:

$$h(\mathbf{p}_i, \mathbf{p}_j, \sigma) := e^{-\frac{\|\mathbf{p}_i - \mathbf{p}_j\|_2}{\sigma}}. \tag{2}$$

Each weight $\mathbf{w}_j$ is updated in every iteration of the SOM training process with:

$$\mathbf{w}'_j := \mathbf{w}_j + \alpha \cdot h(\mathbf{p}^*(\mathbf{x}), \mathbf{p}_j, \sigma) \cdot (\mathbf{x} - \mathbf{w}_j) \tag{3}$$

with learning rate $\alpha \in \mathbb{R}^+$. Usually, the parameters σ and α are decreased in the course of the learning process.

Instead of the standard SOM's pattern-wise training, it is also possible to present a batch of patterns to the network leading to more efficient implementations. An advantage of the batch-SOM variant is a decrease in runtime due to parallel calculations. The weight adaptation from Eq. (3) is modified for a batch of patterns $\mathcal{X}$ as follows:

$$\mathbf{w}'_j := \mathbf{w}_j + \alpha \cdot \frac{\sum_{\mathbf{x} \in \mathcal{X}} h(\mathbf{p}^*(\mathbf{x}), \mathbf{p}_j, \sigma) \cdot (\mathbf{x} - \mathbf{w}_j)}{\sum_{\mathbf{x} \in \mathcal{X}} h(\mathbf{p}^*(\mathbf{x}), \mathbf{p}_j, \sigma)}. \tag{4}$$

The sum of all neighborhood-based differences is divided by overall neighborhood influence for neuron n_j. This product is weighted with learning rate α and added to the previous weight $\mathbf{w}_j$.

3 Related Work

The original SOM we described in the previous section was presented by Teuvo Kohonen at the beginning of the 1980s [5]. It usually works on normalized input patterns. Feature transformations can be employed, e.g., the feature transformation of kernel SOMs, which map the patterns to some feature space employing kernel functions [1, 10]. An example for the successful application of kernel SOMs detects interturn short-circuit faults in a three-phase converter-fed induction motor [2].

Kutics *et al.* [8] introduce the layered SOM for the segment-based image classification. They select the most relevant segment and pass it to different image descriptors, which insert their results into different interconnected layers. Transformed Fourier features are combined with kernelized matrix learning vector quantization in [14]. In [12] hierarchical feature maps are used for script recognition.

A conceptual paper exploring the combination of multilayer feedforward networks and learning vector quantizers is presented in [15] using the multilayer networks as adaptive filters. The approach does not use convolutional layers and SOMs. In [3] the combination of convolutional layers and SOMs was considered, but the algorithm is only vaguely described with a lack of experimental evidence. Contrary to our approach, the output of the SOM is fed to a convolutional layer. A related approach uses a SOM as preprocessing for supervised learning and enables the discovery of new classes [13]. In [16], a combination of CNN and SOM is used for monitored quantization to solve the approximate nearest neighbor search problem. A combination with a denoising autoencoder is proposed in [4].

4 Convolutional Layers

In CNNs [6, 9], convolutional layers are added that perform feature-specific convolutional operations. A convolutional layer employs a set of filters, which is shifted over the input volume. An element-wise computation corresponding to the outer product of two matrices is computed. If we have an input layer $\tilde{X} = (\tilde{x}_{ij})$ and apply an $m \times m$ filter[1] $\tilde{W} = \tilde{w}_{ab}$, the convolutional layer applies:

$$\tilde{y}_{ij} = \sum_{a=0}^{m-1} \sum_{b=0}^{m-1} \tilde{w}_{ab} \cdot \tilde{x}_{(i+a)(j+b)}. \tag{5}$$

The resulting activation map entries are high if the input (receptive field) is similar to the filter. The outputs of convolutional layers are higher-level features for the following neural layer. They are subject to activation functions like ReLU and usually also to dropout to prevent overfitting. Stacking of convolutional layers yields a feature hierarchy of increasing level. Pooling processes like max pooling decrease the dimensionality by averaging or maximizing over convolutional layer outputs.

---

[1] Also rectangle shapes are possible.

5   SOM with Convolutional Layers

The SOM with convolutional layers (CONVSOM) requires a CNN trained with a labeled data set, e.g., from the same or related domain or in a semi-supervised scenario. Figure 1 illustrates the CONVSOM architecture. After training, the raw input patterns $\mathbf{x}_i \in \mathbb{R}^d$, reshaped to 2-dimensional rectangles, are fed to the convolutional layers resulting in patterns $\hat{\mathbf{x}}_i \in \mathbb{R}^{\hat{d}}$, which are subject to the SOM training process. A multiplexer (MUX) allows the selection of different input modes for the CONVSOM training process, i.e., based on the raw inputs or after one of the convolutional layers.[2] The CONVSOM is exclusively trained on images from an independent data set that has not been used for CNN training.

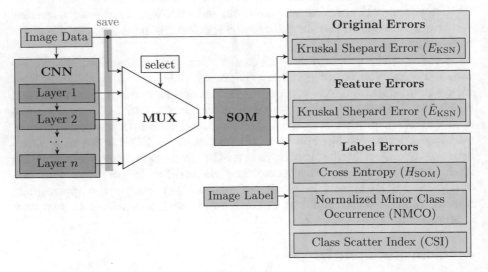

Fig. 1. Illustration of CONVSOM architecture with quality measure setup. Features from different CNN layers serve as inputs to the SOM. The information flow for the quality measures illustrates the origin of information (original, feature, and label errors).

6   Quality Measures

To measure the quality of the SOM learning result, we consider five errors in the following. Figure 1 illustrates, which information is considered for each measure. Standard SOM error measures like topographic and quantization errors are not appropriate for the comparison of feature-transformation-based SOMs as they are based on a comparison with neighborhoods in standard metrics like Euclidean distances. But the argument that maintenance of distances and neighborhoods of the original data space are desirable is not valid in general. Instead, valuable maps should put patterns with similar labels in neighboring positions and

---

[2] MUX and quality measures are currently not used for the training process itself.

patterns showing semantically different content far away on the map. As SOMs are unsupervised, label information is not available during the learning process. However, using labeled benchmark problems, we can train the SOM in an unsupervised fashion without label information, i.e., with patterns $\mathbf{x}_1, \ldots, \mathbf{x}_N$, but evaluate the training result using the labels $y_1, \ldots, y_N$ as a semantic measure.

6.1   Kruskal Shepard Error

The Kruskal Shepard error E_{KS} is an error measure employed in multidimensional scaling [7]. It measures the preservation of distance information in the low-dimensional representation:

$$E_{KS} := \|D_{\text{data}} - D_{\text{SOM}}\|_F^2 \tag{6}$$

where $\|\cdot\|_F^2$ is the squared Frobenius norm, D_{data} is the distance matrix in data space, and D_{SOM} is the distance matrix[3] on the map, both are normalized. The values of the distance matrices are scaled to a range of $[0, 1]$ by dividing by the largest value [11]. Albeit E_{KS} does not take into account semantic information, it allows us to evaluate if the SOM is able to compute low-dimensional representations of the convolutional layer features $\hat{\mathbf{x}}$. The E_{KS} has a theoretical upper limit of $N^2 - N$, when the difference of the distance matrices consists only of ones, apart from the diagonal. The normalized version E_{KSN} of the E_{KS} is scaled taking into account the number of samples N:

$$E_{KSN} := \frac{E_{KS}}{N^2 - N}. \tag{7}$$

An E_{KSN} of zero would mean that all distances were perfectly transferred to the low-dimensional space. The E_{KS} is used twice (see Fig. 1), i.e., E_{KSN} means the outputs of the SOM are compared to the original patterns $\mathbf{x}$, and $\hat{E}_{KSN}$ means they are compared to the transformed features $\hat{\mathbf{x}}$ of the CNN.

6.2   Cross Entropy

The first label-based measure we propose to apply to evaluate the ConvSOM training results is the cross entropy known from classification. It measures the desired distribution of classes for each individual neuron. Due to the variability of the SOM, fixed distributions cannot be applied. Instead, we assume that the most common class of a neuron always corresponds to the target class. If different classes occur the same number of times, one is selected randomly. For each neuron n_i we count the number of patterns $\mathbf{c}_i = (c_{i,1}, \ldots, c_{i,C})$ of each class.

---

[3] The distance matrix contains the pairwise distances between patterns respectively positions on the map.

To obtain the cross entropy of the SOM, the relative frequencies of the expected most common classes of all neurons are summed up:

$$H_{\text{SOM}} := \sum_i -\log \frac{\max\{c_{i,j} \mid c_{i,j} \in \mathbf{c}_i\}}{\sum_j c_{i,j}}. \tag{8}$$

If there are only samples of one class for each neuron, the measure H_{SOM} becomes zero. This may be the case, e.g., if there are fewer patterns than neurons. It is desirable to achieve the lowest possible H_{SOM} value, which applies to all quality measures we employ.

6.3   Minor Class Occurrence

We propose a new measure based on label information. We assume that there should be as few overlaps of patterns of different classes within individual neurons as possible. To measure this objective, we introduce the Minor Class Occurrence (MCO). For each neuron it counts the number of patterns, whose labels do not belong to its majority class:

$$\text{MCO} := \sum_i \left(\sum_j (c_{i,j}) - \max_j(c_{i,j}) \right). \tag{9}$$

For multiple classes occurring the same number of times, one is chosen randomly. This measure is similar to the cross entropy, while not evaluating differently on the basis of the distribution of the errors: To compare the measure for different numbers of patterns, a normalization yields the normalized minor class occurrence NMCO := MCO $/N$.

6.4   Class Scatter Index

Finally, we propose a measure that takes into account the goal that labels should be grouped in as few clusters as possible rather than being scattered on the map. For this sake, we count the number of clusters for each class. A cluster is defined as follows: When a class k is considered, two neurons belong to the same cluster iff they are adjacent and patterns of the same class k are assigned to them, i.e., $\|\mathbf{p}_i - \mathbf{p}_j\|_2 \leq 1 \wedge c_{i,k} > 0 \wedge c_{j,k} > 0$. For each class k, the number of clusters s_k is computed. We define the class scatter index CSI as the mean number of clusters on the map for all classes in the data set:

$$\text{CSI} := \frac{1}{C} \sum_i s_k. \tag{10}$$

7   Experimental Analysis

In this section, we analyze the SOM with convolutional layers experimentally on the well-known data sets MNIST and CIFAR-10.

7.1 Experimental Settings

For the MNIST experiments we use a CNN with two layers with 16 and 32 kernels, kcrnel size 3×3, stride 1, ReLU and non overlapping 2×2 max pooling. The second layer employs dropout with rate of 0.25. Finally, a dense layer with 128 neurons and ReLU is followed by the output layer using SoftMax with 10 neurons. It is trained for 100 epochs using AdaDelta as optimizer on the first 10,000 patterns achieving an accuracy rate of 98.5% on an independent test set.

For CIFAR-10, we use four convolutional layers, two with 32 filters and the subsequent two with 64 filters, each of which have the shape 3×3, use ReLU, employ non-overlapping 2×2 max pooling and a dropout rate of 0.25 for the second and third, and 0.5 for the fourth layer. A dense layer follows the convolutions with 512 neurons and ReLU, while the output layer again uses SoftMax with 10 neurons. This network achieves an accuracy of 80.5% with RMSProp after 1,000 epochs of training with the complete data set.

The 10×10-SOM is trained in batch mode with 100 epochs, batch size 100, learning rate $\alpha = 0.5$ linearly decreased to 0.05, and neighborhood radius $\sigma = 5.0$ linearly decreased to 0.5 in the course of the training process. The CONVSOM is implemented in TENSORFLOW, the CNNs are based on KERAS.

7.2 Quality Measure Results

We compare the CONVSOM approach with an original SOM without convolutional layers based on the quality measures introduced in Sect. 6. Table 1 shows the experimental results of the CONVSOM variants (layer 1–3) and the original SOM (orig) on the MNIST data set on different training set sizes.

Table 1. CONVSOM on MNIST, 20 repetitions, training set sizes (#). Results are presented as mean $\pm$ SD. Standard deviation is omitted if it falls below 10^{-2}.

#	Input	$\hat{E}_{KSN}$	E_{KSN}	H_{SOM}	NMCO	CSI
500	orig	0.094	0.094	26.002 ± 1.935	0.180	4.435 ± 0.365
	layer 1	0.090	**0.093**	22.659 ± 1.658	0.154	4.175 ± 0.340
	layer 2	0.085	0.095	12.137 ± 1.653	0.077	3.110 ± 0.377
	layer 3	**0.031**	0.097	$\mathbf{4.645 \pm 1.116}$	**0.023**	$\mathbf{1.620 \pm 0.194}$
1000	orig	0.230	0.230	27.862 ± 1.650	0.180	4.535 ± 0.292
	layer 1	**0.229**	**0.229**	25.808 ± 2.023	0.158	4.660 ± 0.244
	layer 2	0.234	0.234	15.715 ± 1.383	0.089	3.760 ± 0.235
	layer 3	0.237	0.238	$\mathbf{7.484 \pm 1.198}$	0.029	$\mathbf{1.705 \pm 0.190}$
2000	orig	0.225	0.225	31.106 ± 2.359	0.204	5.380 ± 0.424
	layer 1	**0.222**	**0.222**	26.240 ± 1.717	0.166	6.045 ± 0.606
	layer 2	0.226	0.226	16.534 ± 1.555	0.089	4.885 ± 0.318
	layer 3	0.232	0.232	$\mathbf{8.735 \pm 0.996}$	0.034	$\mathbf{2.360 \pm 0.223}$

Table 2. CONVSOM on CIFAR-10, 20 repetitions, training set sizes (#). Results are presented as mean ± SD. Standard deviation is omitted if it falls below 10^{-2}.

#	Input	$\hat{E}_{\text{KSN}}$	E_{KSN}	H_{SOM}	NMCO	CSI
500	orig	**0.025**	**0.025**	88.079 ± 2.922	0.571	11.250 ± 0.477
	layer 1	0.063	0.039	75.229 ± 2.699	0.514	8.530 ± 0.682
	layer 2	0.030	0.050	88.285 ± 2.524	0.578	11.505 ± 0.672
	layer 3	0.035	0.046	**52.571 ± 3.040**	**0.381**	**4.865 ± 0.545**
1000	orig	0.216	0.216	102.758 ± 2.059	0.625	6.450 ± 0.556
	layer 1	0.214	0.213	84.712 ± 2.798	0.549	7.085 ± 0.488
	layer 2	0.213	0.213	106.634 ± 2.280	0.645	6.735 ± 0.621
	layer 3	**0.201**	**0.201**	**56.452 ± 1.578**	**0.401**	**3.655 ± 0.488**
2000	orig	0.206	0.206	111.509 ± 2.503	0.658	3.540 ± 0.393
	layer 1	0.207	0.207	88.246 ± 1.705	0.563	5.500 ± 0.521
	layer 2	0.209	0.209	118.471 ± 1.960	0.679	3.315 ± 0.575
	layer 3	**0.192**	**0.1925**	**57.969 ± 1.353**	**0.413**	**3.280 ± 0.284**

The $\hat{E}_{\text{KSN}}$ and E_{KSN} results are very similar to each other for # > 500 and show that the CONVSOM approaches achieve reasonable learning results w.r.t. the maintenance of distances. Regarding the label errors H_{SOM}, NMCO, and CSI, the CONVSOM variants achieve significantly better results.

These observations also hold on CIFAR-10, see Table 2. Here, the quality measures are continuously worse than on MNIST due to the higher complexity of the data set.

For a detailed analysis of the CSI, the box plots of Fig. 2 visualize the class-wise CSI values of CONVSOM on MNIST and CIFAR-10 respectively. Each experiment has been repeated 20 times. The comparisons show that the CSIs achieved with CONVSOM (layer 3) are lower than the corresponding CSIs of the original SOM confirming the expectation that CONVSOM achieves better maps w.r.t. higher-level information. Some classes are significantly more scattered on the maps than others, e.g., number 8 of MNIST.

7.3   Visualization Results

Figure 3 presents a visualization of exemplary original SOM and CONVSOM learning results. After training, for each weight **w** the closest pattern (image) from the test set is displayed at the corresponding neuron location.[4] The figures show that the CONVSOM maps show a broader representation of images including different classes and achieving smoother neighborhoods. The CONV-SOM tends to form a large cluster for each class instead of many small clusters. This corresponds to smaller CSI values, which we have observed in the previous section.

---

[4] The weights of the neurons cannot be used directly for visualization in the CONV-SOM as they have a different format.

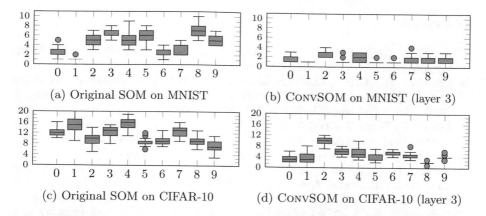

(a) Original SOM on MNIST

(b) ConvSOM on MNIST (layer 3)

(c) Original SOM on CIFAR-10

(d) ConvSOM on CIFAR-10 (layer 3)

Fig. 2. Box-plots of class-wise CSI of original SOM and ConvSOM on MNIST and CIFAR-10 with 500 samples. Outliers are displayed as circles.

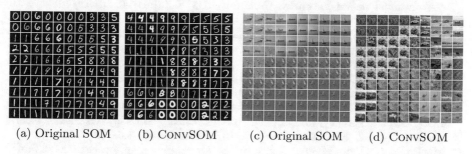

(a) Original SOM   (b) ConvSOM   (c) Original SOM   (d) ConvSOM

Fig. 3. Comparison of exemplary original SOM and ConvSOM training results on MNIST (left) and CIFAR-10 (right) with 500 samples. For each weight **w** the closest pattern (image) from the test set is displayed at the corresponding neuron location.

8   Conclusion

This paper introduces convolutional layers to the basic SOM approach. The experimental analysis has shown that the higher-level features of the convolutional layers add value to the SOM training process yielding better dimensionality reduction quality results, in particular measured by analyzing label-oriented learning results. For example, the CSI is reduced showing that the map training process profits from the higher-level features. The visualizations of the trained map confirm these observations. There are fewer distributed clusters of the same class.

In the future, we plan to consider semantic information between labels for evaluation of the SOM results. An application to various kinds of patterns is also an interesting research direction. Further, we plan to apply the approach to semi-supervised data sets with partially labeled patterns and analyze the learning results based on layers from recent state-of-the art networks.

References

1. András P (2002) Kernel-kohonen networks. Int J Neural Syst 12(02):117–135. https://doi.org/10.1142/S0129065702001084
2. Coelho, D.N., Barreto, G.A., Medeiros, C.M.S.: Detection of short circuit faults in 3-phase converter-fed induction motors using kernel SOMs. In: WSOM, pp 1–7 (2017). https://doi.org/10.1109/WSOM.2017.8020016
3. Dozono H, Niina G, Araki S (2016) Convolutional self organizing map. In: 2016 international conference on computational science and computational intelligence (CSCI), pp 767–771. https://doi.org/10.1109/CSCI.2016.0149
4. Ferles C, Papanikolaou Y, Naidoo KJ (2018) Denoising autoencoder self-organizing map (DASOM). Neural Netw 105:112–131. https://doi.org/10.1016/j.neunet.2018.04.016
5. Kohonen T (1982) Self-organized formation of topologically correct feature maps. Biol Cybern 43(1):59–69. https://doi.org/10.1007/BF00337288
6. Krizhevsky A, Sutskever I, Hinton GE (2017) ImageNet classification with deep convolutional neural networks. Commun ACM 60(6):84–90. https://doi.org/10.1145/3065386
7. Kruskal JB (1964) Nonmetric multidimensional scaling: a numerical method. Psychometrika 29(2):115–129. https://doi.org/10.1007/BF02289694
8. Kutics A, O'Connell C, Nakagawa A (2013) Segment-based image classification using layered-SOM. In: 2013 IEEE international conference on image processing, pp 2430–2434. https://doi.org/10.1109/ICIP.2013.6738501
9. LeCun Y, Boser B, Denker JS, Henderson D, Howard RE, Hubbard W, Jackel LD (1989) Backpropagation applied to handwritten zip code recognition. Neural Comput 1(4):541–551. https://doi.org/10.1162/neco.1989.1.4.541
10. MacDonald D, Fyfe C (2000) The kernel self-organising map. In: KES 2000, fourth international conference on knowledge-based intelligent engineering systems and allied technologies, proceedings (Cat. No. 00TH8516), vol 1, pp 317–320. https://doi.org/10.1109/KES.2000.885820
11. Meier A, Kramer O (2017) An experimental study of dimensionality reduction methods. In: Kern-Isberner G, Fürnkranz J, Thimm M (eds) KI 2017, advances in artificial intelligence. Springer International Publishing, pp 178–192. https://doi.org/10.1007/978-3-319-67190-1_14
12. Miikkulainen R (1990) Script recognition with hierarchical feature maps. Connection Sci 2(1–2):83–101. https://doi.org/10.1080/09540099008915664
13. Platon L, Zehraoui F, Tahi F (2017) Self-organizing maps with supervised layer. In: WSOM, pp 1–8. https://doi.org/10.1109/WSOM.2017.8020022
14. Schleif F (2017) Small sets of random Fourier features by kernelized matrix LVQ. In: WSOM, pp 1–5. https://doi.org/10.1109/WSOM.2017.8020026
15. Villmann T, Biehl M, Villmann A, Saralajew S (2017) Fusion of deep learning architectures, multilayer feedforward networks and learning vector quantizers for deep classification learning. In: 2017 12th international workshop on self-organizing maps and learning vector quantization, clustering and data visualization (WSOM), pp 1–8. https://doi.org/10.1109/WSOM.2017.8020009
16. Wang M, Zhou W, Tian Q, Pu J, Li H (2017) Deep supervised quantization by self-organizing map. In: Proceedings of the 25th ACM international conference on multimedia, MM 2017. ACM, New York, pp 1707–1715. https://doi.org/10.1145/3123266.3123415

Cellular Self-Organising Maps - CSOM

Bernard Girau[1]([✉]) and Andres Upegui[2]

[1] Université de Lorraine, CNRS, LORIA, 54000 Nancy, France
bernard.girau@loria.fr
[2] inIT, HEPIA, University of Applied Sciences of Western Switzerland,
Delémont, Switzerland
andres.upegui@hesge.ch

Abstract. This paper presents CSOM, a Cellular Self-Organising Map which performs weight update in a cellular manner. Instead of updating weights towards new input vectors, it uses a signal propagation originated from the best matching unit to every other neuron in the network. Interactions between neurons are thus local and distributed. In this paper we present performance results showing than CSOM can obtain faster and better quantisation than classical SOM when used on high-dimensional vectors. We also present an application on video compression based on vector quantisation, in which CSOM outperforms SOM.

Keywords: Neural networks · Self-organising maps · Cellular computing

1 Introduction

The work presented in this paper is part of the SOMA project[1] (*Self-organising Machine Architecture* [1]). This project aims at developing an architecture based on brain-inspired self-organisation principles in a digital reconfigurable hardware. Self-organising neural models play a central role in this project. As we wish to deploy the architecture in a manycore substrate, the scalability and decentralisation of the architecture needs to be ensured.

Cellular computing approaches [2] appear as a promising solution for tackling the scalability limitations of current hardware implementations of self-organising maps. Cellular computing refers to biological cells that interact with their neighbour cells, but not with remote ones, at least not directly. While parallel computing deals with a small number (tens up to tens of thousands in supercomputers) of powerful processors able to perform a single complex task in a sequential manner, cellular computing is based on another philosophy: simplicity of basic processing cells, their vast parallelism, and their locality. The two latter properties induce another fundamental property: decentralisation. In the SOMA project,

---

[1] The authors thank the Swiss National Science Foundation (SNSF) and the French National Research Agency (ANR) for funding the SOMA project ANR-17-CE24-0036.

© Springer Nature Switzerland AG 2020
A. Vellido et al. (Eds.): WSOM 2019, AISC 976, pp. 33–43, 2020.
https://doi.org/10.1007/978-3-030-19642-4_4

we propose to combine cellular and neural properties to enable cellular structures to behave like biologically-inspired neural models. In this context, we have defined a cellular version of self-organising maps, that we call CSOM. The aim of this paper is to study the specific properties of this model in terms of vector quantisation performance. We do not focus on scalability and decentralisation aspects, though such requirements are our first motivation to define CSOM.

This paper describes CSOM in Sect. 2 by comparing it to standard SOM. We present the algorithm in detail and we explain its behaviour. Then in Sect. 3, we present two experiments. The first aims at comparing and analysing the behaviour of CSOM and SOM when learning increasingly dimensional and sparse data. The second experiment compares SOM and CSOM algorithms for video compression by replacing a sequence of video frames (images) by a codebook of thumbnails learnt by SOM and CSOM. Section 4 concludes with perspectives on other aspects of the SOMA project.

2   Self-Organising Maps: SOM and Cellular SOM

2.1   SOM: Self-Organising Maps

Self-organising maps (SOM), initially proposed by Kohonen [3], consist of neurons organised on low-dimensional arrays (most often 2D) that project patterns of arbitrary dimensionality onto a lower dimensional array of neurons. Each neuron has an associated weight vector, or prototype vector. Its dimension is defined by the nature of input data, not by the dimensionality of the neural array. All neurons receive the same input pattern and an iterative mechanism updates the neuron's weights so as to learn to quantise the input space in an unsupervised way. This mechanism first selects the neuron whose weights best fit the given input pattern, and then brings the weights of all neurons more or less closer to the current input pattern, depending on the distance to the winner neuron in the neural array (Fig. 1). As with any vector quantisation method, the result is a codebook: it is the set of all neuron weight vectors, or codewords. Self-organising maps are known to define "topological" codebooks: after learning, two vectors that are close in the input space will be represented by codewords of close neurons in the neural map. Thus, neighbouring neurons in the map have similar weight vectors.

2.2   CSOM: Cellular Self-Organising Maps

As Kohonen's SOM, CSOM is a vector quantisation algorithm which aims to represent a probability density function into a codebook, i.e. a set of prototype vectors or codewords. Is is also a neural network composed of an n-dimensional array of neurons. In this paper we will only consider the case of the 2-dimensional architecture. The main difference with SOM is that the neighbourhood notion is replaced by a notion of local connectivity and propagation of influence. Each neuron has a number of associated synapses that define which neuron will have an influence onto which other. Synapses can be seen as interconnection matrices. In this paper, we assume that synapses are simply interconnecting every neuron to its four physical neighbours. With this restriction, the neural structure of CSOM is similar to the 2D grid of the usual SOM. But more complex synaptic interconnections can be used, such as resulting from pruning as in [4].

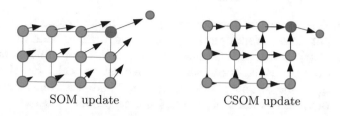

SOM update   CSOM update

Fig. 1. Weight update rule on SOM and CSOM.

Unlike SOM, CSOM performs a weight update rule which moves network weights towards neighbour neurons weights instead of input patterns. Figure 1 illustrates this update: the green node represents a new input pattern, the orange node represent the BMU, which updates its weight towards the new pattern in both cases, and the blue nodes represent the remaining neurons on the SOM. While in SOM, weight update is influenced by every new input pattern in a direct manner, in CSOM the influence is indirect: neuron weights are influenced by the neighbour weights, starting from the BMU.

2.3   Algorithms

Common Notations and Mechanisms. Each neuron in a SOM or CSOM is represented by a d-dimensional weight vector, $\mathbf{m} \in \mathbb{R}^d$, also known as prototype vector or codeword, $\mathbf{m} = [m_1, ..., m_d]$, where d is the dimension of the input vectors, $\mathbf{x}$. Neurons are located in a n-dimensional map (Usually $n = 2$). In this map, $\mathbf{r}_i$ are the coordinates of neuron i and $\mathbf{m}_i$ is its weight vector. For SOM, these coordinates are the basis of distance computations to define neighbourhood relations. For CSOM, we simply consider in this paper that synaptic interconnections are set to connect all adjacent neurons in the map.

In the algorithm of both SOM and CSOM, learning starts with an appropriate (usually random) initialisation of the weight vectors, $\mathbf{m}_i$. Input vectors are presented to the neural map. For each input vector $\mathbf{x}$, the distance from $\mathbf{x}$ to all the weight vectors is calculated using a distance measure noted $\|\cdot\|$, which is typically the Euclidean distance. The neuron whose weight vector gives the smallest distance to the input vector $\mathbf{x}$ is called the best matching unit (BMU), denoted by c, and determined according to: $\|\mathbf{x} - \mathbf{m}_c\| = \min_i \|\mathbf{x} - \mathbf{m}_i\|$.

SOM Algorithm. The basic version of SOM learning is an on-line stochastic process[2] which has been inspired by neurobiological learning paradigms, but other extensions exist [5,6]. For each iteration, an input is randomly drawn, and its BMU c is computed. Then all neurons i update their weights according to:

$$\mathbf{m}_i(t+1) = \mathbf{m}_i(t) + \epsilon(t)(\mathbf{x}(t) - \mathbf{m}_i(t))e^{-\frac{\|\mathbf{r}_c - \mathbf{r}_i\|^2}{2\sigma^2(t)}} \tag{1}$$

where t denotes the time/iteration, $\mathbf{x}(t)$ is an input vector randomly drawn from the input data set at time t, and $\epsilon(t)$ the learning rate at time t. The learning rate $\epsilon(t)$ defines the strength of the adaptation, which is application-dependent. Commonly $\epsilon(t)$ is constant or a decreasing scalar function of t. The term $\|\mathbf{r}_c - \mathbf{r}_i\|$ is the distance between neuron i and the winner neuron c, and $\sigma(t)$ is the standard deviation of the gaussian neighbouring kernel.

CSOM Algorithm. The CSOM learning algorithm is an on-line mode training. For each iteration, an input is randomly drawn, and its BMU c is computed. First the BMU updates its weight vector according to: $\mathbf{m}_c = \mathbf{m}_c + \alpha(t)(\mathbf{x}(t) - \mathbf{m}_c)$ where $\alpha(t)$ is the learning rate at time/iteration t. Then every other neuron $i \neq c$ in the map updates its weight vector as follows:

$$\mathbf{m}_i = \mathbf{m}_i + \alpha(t)\frac{1}{\#Infl(i)}\sum_{j \in Infl(i)}(\mathbf{m}_j - \mathbf{m}_i)e^{(-\frac{\sqrt{d}}{\eta(t)}\frac{hops(c,i)}{\|\mathbf{m}_i - \mathbf{m}_j\|})} \tag{2}$$

where $hops(c,i)$ is the number of propagation hops to reach neuron i from the BMU through synaptic connections, $Infl(i)$ is the set of the influential neurons of neuron i, and $\#Infl(i)$ is the number of influential neurons for neuron i. An influential neuron of neuron i is defined as the neuron from which neuron i received the propagating learning signal. For a neuron with $hops(c,i) = h$, the influential neuron(s) will be every neuron j connected to i with $hops(c,j) = h-1$. The parameter $\alpha(t)$ is the learning rate at time t, and $\eta(t)$ is the elasticity of the network at time t. This latter parameter is modulated by $\sqrt{d}$ so as to take into account the range $[0, \sqrt{d}]$ of euclidean distances in dimension d.

The principle of this algorithm results from a combination of cellular computing principles and dynamic adaptation as found in the Dynamic SOM (DSOM) model [7]. Each neuron weight vector gets influenced by the weight vectors of its influential neurons through which the cellular propagation signal has been received, instead of being influenced by the input vector through its BMU. The learning rate is modulated by the proximity of the weight vector with the weight vectors of its influential vectors, so that learning gets faster when close neurons do not have close weight vectors. The BMU first influences lateral neurons directly connected to it (for them $hops = 1$) through synapses. Equation 2 updates the weight of neighbouring neurons by attracting their weight towards the winning

---

[2] A batch version of SOM learning exists, but the on-line mode training provides us with a more detailed evolution to observe, and we apply our models to applications with huge redundant learning databases for which batch-learning is quite inefficient.

codeword $\mathbf{m}_c$. After being updated, these neighbouring neurons will update their own neighbouring neurons by means of the same Eq. 2. This results in a gradient permitting the overall network weights to get influenced by every new input vector through local influences. The elasticity parameter can be understood as how much the resulting map is expected to maintain its underlying structure in the codeword space. In Fig. 2, for instance, a lower elasticity results in a final network topology closer to its initial random configuration.

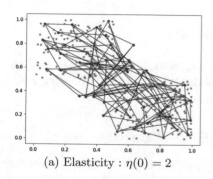
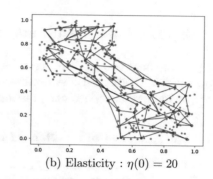

(a) Elasticity : $\eta(0) = 2$ (b) Elasticity : $\eta(0) = 20$

Fig. 2. Learned codebook for CSOM with a 2D distribution with two main clusters, using different elasticity parameters

Complexity. From a sequential[3] complexity point of view, SOM and CSOM have a comparable cost. If we consider that exponential values are pre-computed and stored (neural distances as well as numbers of *hops* can only take a finite number of values), the main computational difference between them is that all influential neurons are taken into account in CSOM, thus inducing an approximately doubled complexity for CSOM with respect to SOM since most neurons have two influential neurons in a 2D structure. Let us mention the fact that randomly choosing only one influential neuron for each neuron leads to similar behaviors and leads to the same computational complexity for the two models[4].

Parameter Tuning. SOM and CSOM algorithms depend on the number of learning iterations I. They also depend on parameters that evolve along training: $\epsilon(t)$ and $\sigma(t)$ for SOM, $\alpha(t)$ and $\eta(t)$ for CSOM. Classically, these parameters linearly decrease during training from an initial value to a final value. Thus, a fair comparison of SOM and CSOM raises the question of choosing ϵ_{init}, ϵ_{final},

---

[3] The question of a parallel complexity comparison between SOM and CSOM is not discussed here, even if the cellular version targets parallel hardware implementations.

[4] It is important to notice that this complexity equivalence requires anyway a careful handling of the iteration order among neurons of CSOM: influential neurons must be updated before the neuron they influence. A simple possible way to ensure this property is to consider that the BMU position defines four sectors (top-left, top-right, bottom-left, bottom-right) in the neural map and to update weights sector per sector, always starting from the BMU corner.

σ_{init}, σ_{final}, α_{init}, α_{final}, η_{init}, η_{final}, I_{SOM} and I_{CSOM}. In this paper, we set these parameters by means of a simple "manual" exploration for our tests on artificial distributions, and by means of an optimisation by genetic algorithm in the case of a video compression application. In these two cases, using the same optimisation technique for SOM and CSOM thus provides performance comparisons on a fair basis even if the choice of each optimisation technique rather than another one can be discussed.

3   Experimental Setup and Results

We propose here two kinds of tests: learning more or less sparse data generated from artificial statistical distributions in variable dimensions, then learning to quantise images in a video compression application.

3.1   Quantisation of Artificial d-dimensional Distributions

We have tested our algorithm by presenting input vectors drawn from statistical distributions in various dimensions. In dimension d, the chosen distribution uses d uniform clusters of data with a δ sparsity coefficient: data are generated as $\{x_1 = U([0, 1 - \delta]), \ldots, x_{j-1} = U([0, 1 - \delta]), x_j = U([\delta, 1]), x_{j+1} = U([0, 1 - \delta]), \ldots, x_d = U([0, 1 - \delta])\}$, where the coordinate number j is randomly chosen for each new generated vector, and $U([a, b])$ is the uniform distribution in $[a, b]$. When $\delta = 0$, it results in d-dimensional uniform distributions. When $\delta = 0.8$ data are distributed in d rather small clusters for which all coordinates are between 0 and 0.2, except one that is between 0.8 and 1. These distributions are illustrated in 3D on Fig. 3.

In the reported experiments, the network topology has been defined as a 2-D array mesh of 8×8 neurons. Each neuron has 4 synapses connected to neighbour neurons. Weights have been uniformly initialised with $U([0, 1])$. Parameters ϵ_{init}, ϵ_{final}, σ_{init}, σ_{final}, α_{init}, α_{final}, η_{init}, η_{final}, I_{SOM} and I_{CSOM} have been optimised by means of a simple "manual" exploration method, considering the small computation time. The performance is measured by the average quantisation error (AQE) computed on independent test sets drawn from the same distributions:

$$AQE = \frac{1}{K} \sum_{k=1}^{K} ||\mathbf{x}_k - \mathbf{m}_{BMU(\mathbf{x}_k)}||^2$$

where K is a number of input test vectors, and $\mathbf{x}_k$ is one of them. In the experiments presented here, K has been set to 500, while each learning epoch uses 100 input vectors, randomly drawn from the distribution at each learning iteration.

Figure 4 shows the ratio of the results $\frac{AQE_{CSOM}}{AQE_{SOM}}$ obtained for 50 runs on several dimensions and for sparsity coefficients $\delta = 0$ and $\delta = 0.9$. An increase in dimensions favours CSOM, especially for sparse distributions. The main property of CSOM is a faster convergence in most cases. We assume that the decorrelation between elasticity and weight attraction in CSOM makes it able to tolerate less organised codewords, at least temporary, so that this model does not require to attract all neuron weights close to each other at the beginning of learning before unfolding in the input space, as it is usually observed with SOM. Nevertheless we have not yet been able to quantify this phenomenon and further studies are still required to validate this assumption.

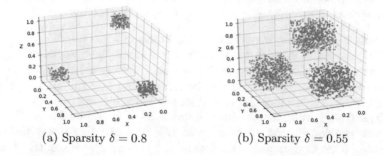

(a) Sparsity $\delta = 0.8$ (b) Sparsity $\delta = 0.55$

Fig. 3. Statistical distribution in 3D for $\delta = 0.8$ and $\delta = 0.55$

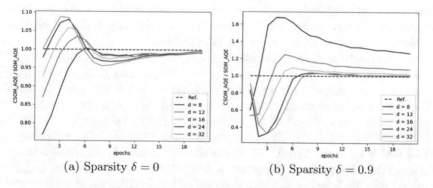

(a) Sparsity $\delta = 0$ (b) Sparsity $\delta = 0.9$

Fig. 4. AQE ratio $\frac{AQE_{CSOM}}{AQE_{SOM}}$ evolution with various dimensions

3.2   Video Compression

In order to evaluate how CSOM behave in real-world applications, we have decided to train them on data extracted from image sequences. Self-organising maps can be interestingly applied to lossy image compression [4,8–10]. The principle is to split the image into non overlapping thumbnails, then learn a good quantisation of these thumbnails. Compressing the image is performed by replacing each thumbnail by the index of the closest codeword (or codeword of its BMU). The list of thumbnail indexes is further compressed by classical methods such as entropy coding. Uncompressing the image can be performed by replacing each decoded index by the corresponding thumbnail (codeword). The result is similar to the original image, but with every thumbnail replaced by the codeword learned by its BMU (see further an example on Fig. 5(c)). Compared to other quantisation techniques, SOM preserve topological properties, so that close parts of an image that are often similar will be coded by neighbouring neurons. It makes it possible to further reduce the size of the compressed file by efficiently coding codeword indexes thanks to a differential coding performed before entropy coding [9].

In order to see how fast CSOM and SOM respectively adapt to changing data, we extend this application to video compression as follows:

1. The first $L \times H$ frame is split into thumbnails of $l \times h$ pixels.
2. A SOM and a CSOM of $n \times n$ neurons learn these thumbnails, using a set of parameters optimised for processing a first frame (see below). Each neuron has a $l \times h$ weight vector, that can be interpreted as a thumbnail codeword.
3. Each consecutive frame is processed in the following way:
   (a) The $L \times H$ frame is split into thumbnails of $l \times h$ pixels.
   (b) The previously learned SOM and CSOM learn these new thumbnails, using a set of parameters optimised for processing consecutive frames.

As an example, we consider sequences of 384×288 images subdivided into 6912 thumbnails of 4×4 pixels. Using a 8×8 SOM or CSOM ($b = 6$), the compression ratio already reaches a very high value of 7 before even further compressing the index list using differential and entropy coding (see [10] for details on computing the exact compression ratio).

For each model and image sequence, we optimise the main parameters of the learning algorithms (ϵ_{init}, ϵ_{final}, σ_{init}, σ_{final}, α_{init}, α_{final}, η_{init}, η_{final}, I_{SOM} and I_{CSOM}) by means of a genetic algorithm with 0.1 mutation probability, 20% of the population considered as elite, 32 individuals in the population and 20 generations, each individual fitness being evaluated by learning 20 randomly initialised SOM or CSOM. We compute the performance of the resulting SOM and CSOM after each "epoch" of 216 randomly chosen thumbnails among the 6912 available. Two sets of parameters are evolved for each model and image sequence: one for learning the quantisation of the first frame, the other one for learning the quantisation of a frame after having learned the quantisation of the previous frame in the sequence. In this second case, the number of learning iterations per "epoch" is reduced to 43, since learning the next frame is quite fast when the SOM or CSOM has already learned the previous frame so that its codewords are already very satisfactory with respect to the new frame.

Based on experiments using various image sequences from the CAVIAR project [11] as well as from the CVD project [12], CSOM significantly improve the mean pixel error (normalised in [0, 255]) with respect to standard SOM. We also observe a significant perceptual improvement of the visual result that is illustrated by the usual peak signal-to-noise ratio (PSNR) to be maximised:

$$\text{PSNR} = 10 \times \log_{10}(\frac{255^2}{\text{MSE}}) \quad \text{where} \quad \text{MSE} = \frac{1}{L \times H} \sum_{x=1}^{L} \sum_{y=1}^{H} (P_{x,y} - P'_{x,y})^2$$

Figure 5 shows results of SOM and CSOM learning to compress a video taken in a city with many buildings and moving pedestrians in addition to a circular

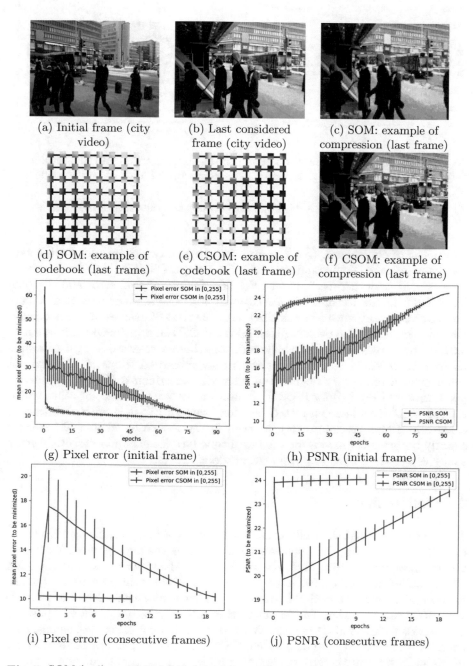

(a) Initial frame (city video)

(b) Last considered frame (city video)

(c) SOM: example of compression (last frame)

(d) SOM: example of codebook (last frame)

(e) CSOM: example of codebook (last frame)

(f) CSOM: example of compression (last frame)

(g) Pixel error (initial frame)

(h) PSNR (initial frame)

(i) Pixel error (consecutive frames)

(j) PSNR (consecutive frames)

Fig. 5. SOM (red) vs CSOM (blue) for lossy video compression (city video example)

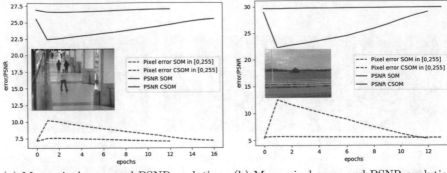

(a) Mean pixel error and PSNR evolution   (b) Mean pixel error and PSNR evolution
(consecutive frames of corridor video)   (consecutive frames of outdoor video)

Fig. 6. SOM (red) vs CSOM (blue) for corridor and outdoor video examples

movement of the camera (see Fig. 5(a) and (b)). Learning results for the first
frame are averaged on 32 randomly initialised maps. CSOM main advantages are
a faster and smoother learning (see Fig. 5(g) for mean pixel error and Fig. 5(h)
for PSNR). CSOM learning of consecutive frames also takes advantage of a more
progressive performance when dealing with the next 30 frames of this sequence,
always using the same evolving 32 SOM and CSOM: while SOM first seem to
unlearn their previous satisfactory codebook before converging again towards
an even more satisfactory one for the newly presented frame, CSOM do not
suffer such a temporary degradation of their performance (Fig. 5(i) for mean
pixel error and Fig. 5(j) for PSNR). Similar results are obtained for other videos
though sometimes less contrasted (see for example Fig. 6(a) with a stationary
camera in a commercial centre), or even in the few cases when SOM reach a
slightly better performance for the first frame (see for example Fig. 6(b) with a
shaky camera in a circular movement outdoor).

4   Conclusion

This paper presented the Cellular SOM algorithm, a Self-Organising Map based
on cellular computing. The behaviour exhibited by the algorithm on the experiments presented in this paper shows that CSOM outperforms classical SOM
when applied to high dimensional sparse data. It can achieve a better quantisation and is faster than SOM. However, on lower dimensional and less sparse
data SOM can be better. The fast algorithm convergence of CSOM is of particular interest for dynamic problems. When data statistical properties change over
time, at it can be the case on a video, CSOM is able to re-adapt faster in order
to match new input vectors, maintaining a low quantisation error over time.

Further work will focus on including pruning and sprouting mechanisms [4]
in order to remove and re-create synapses or create connections with remote
neurons. This feature should lead to a better quantisation of sparse data, and

rebuild the network when the dynamicity of the problem may impose it. CSOM has been initially designed taking into account the constraints imposed by cellular computing: a cellular substratum of processing elements. Future works will implement CSOM on a real cellular hardware architecture in order to drive the self-organisation of a manycore hardware according to new input data, as targeted by the SOMA project.

References

1. Khacef L, Girau B, Rougier NP, Upegui, A, Miramond B (July 2018) Neuromorphic hardware as a self-organizing computing system. In: NHPU: neuromorphic hardware in practice and use. IEEE, pp. 1–4
2. Sipper M (1999) The emergence of cellular computing. Computer 32(7):18–26
3. Kohonen T (1990) The self-organizing map, vol 78, no 9, pp. 1464–1480
4. Upegui A, Vannel F, Girau B, Rougier N, Miramond B (2018) Pruning self-organizing maps for cellular hardware architectures. In: NASA/ESA conference on adaptive hardware and systems
5. Cottrell M, Olteanu M, Rossi F, Villa-Vialaneix N (2016) Theoretical and applied aspects of the self-organizing maps. In: Advances in self-organizing maps and learning vector quantization
6. Kohonen T (2013) Essentials of the self-organizing map. Neural Netw 37:52–65
7. Rougier NP, Boniface Y (2011) Dynamic self-organising map. Neurocomputing 74(11):1840–1847
8. Huang Z, Zhang X, Chen L, Zhu Y, An F, Wang H, Feng S (2017) A hardware-efficient vector quantizer based on self-organizing map for high-speed image compression. Appl Sci 7(11):1106
9. Amerijckx C, Legat J-D, Verleysen M (2003) Image compression using self-organizing maps. Syst Anal Model Simul 43(11):1529–1543
10. Bernard Y, Buoy E, Fois A, Girau B (2018) NP-SOM: network programmable self-organizing maps. In: 2018 IEEE 30th international conference on tools with artificial intelligence (ICTAI), pp. 908–915
11. EC funded CAVIAR project. http://homepages.inf.ed.ac.uk/rbf/CAVIAR/
12. CVD video database. http://www.helsinki.fi/~tiovirta/Resources/CVD2014/

A Probabilistic Method for Pruning CADJ Graphs with Applications to SOM Clustering

Josh Taylor[1(✉)] and Erzsébet Merényi[2]

[1] Department of Statistics, Rice University, Houston, TX 77005, USA
jtay@rice.edu
[2] Department of Statistics and Department of Electrical and Computer Engineering,
Rice University, Houston, TX 77005, USA
erzsebet@rice.edu

Abstract. We introduce a Bayesian Dirichlet-Multinomial model of the edge weights of the Cumulative ADJacency ($CADJ$) graph [1] with the goal of intelligent graph pruning. As a topology representing graph, $CADJ$ is an effective tool for cluster extraction from the learned prototypes of SOMs, but for complex data the graph must typically be pruned to elicit meaningful cluster structure. This work is a first attempt to guide this pruning in a formal modeling framework. Our model, dubbed **DM-Prune**, earmarks edges for removal via comparisons to a *novel null model* and provides an internal assessment of information loss resulting from iterative removal of edges. We show that **DM-Prune**d $CADJ$ graphs lead to clusterings comparable to the best previously achieved on highly structured real data.

Keywords: SOM clustering · Topology representing network · Graph sparsity

1 Introduction: The $CADJ$ Graph

$CADJ$ was introduced in [1] as a weighted version of the Induced Delaunay Triangulation of Martinetz and Schulten [2]. As a subgraph of the Delaunay Triangulation, $CADJ$ represents topological adjacencies of the prototypes $\{w_i\}_{i=1}^{W}$ of a vector quantizer. Given data $X = \{x_s\}_{s=1}^{N}$ drawn from manifold $\mathcal{M} \subseteq \mathbb{R}^d$, the $CADJ$ weight of the edge connecting prototypes w_i and w_j is $CADJ_{ij} = \#\{x \in X : BMU1(x) = i, BMU2(x) = j\}$, where $BMU1(\cdot)$ and $BMU2(\cdot)$ return the index of the Best and second-Best Matching Unit (or prototype, respectively) of the datum x. Positive $CADJ_{ij}$ values reflect the strength of topological connectivity of prototypes w_i and w_j (and, consequently, connectivity of the portion of the manifold $\mathcal{M}$ they represent), whereas values of 0 indicate disconnected portions of the manifold. When used in conjunction with the learned prototypes of a Self-Organizing Map [3], $CADJ$–in particular its

© Springer Nature Switzerland AG 2020
A. Vellido et al. (Eds.): WSOM 2019, AISC 976, pp. 44–54, 2020.
https://doi.org/10.1007/978-3-030-19642-4_5

symmetrized version $CONN = CADJ + CADJ^T$ along with the $CONNvis$ visualization ([1], an example of which is depicted in Fig. 1)–has been a successful tool for cluster discovery ([1,4] and references therein) via identification of its closed, connected communities. However, the $CONN$ graph representing complex, high-dimensional datasets typically does not contain readily identifiable communities. In [1] a method for inducing sparsity in $CONN$ is proposed to remedy this by evaluating *global* and *local* importance of edge weights relative to each other, removing edges by thresholding well-understood graph parameters on a grid, and judging the impact by post-evaluation of the quality of extracted clusters at the various levels of thresholding. This is tedious, and in the current framework the analyst has little formalized feedback regarding the effects of the thresholding on the representation of the data manifold as a whole. This work is a first attempt to provide formal, principled suggestions for thresholding, before extracting clusters. We will earmark connections for removal by comparing their strengths to those which would be expected if the data were generated under uniform noise; a metric afforded to us by DM-Prune will report the impact of their successive removal as a means of locating a stopping point in the pruning process. Our goal is toward intelligent automation of the pruning process.

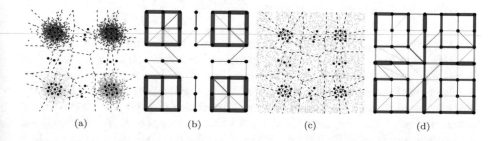

(a)   (b)   (c)   (d)

Fig. 1. (a) Data drawn from four Gaussian clusters (colored points, 1000 draws per cluster) and their learned SOM prototypes (black points). The Voronoi tessellation is outlined in dashed black lines. (b) The $CONNvis$ visualization of the Gaussian data mapped to its SOM lattice, clearly outlining the four clusters (see [1] for an explanation of the line colorings and widths). (c) A sample of 4000 uniformly distributed points depicted against the same Voronoi tessellation as (a). (d) The $CONNvis$ visualization of the uniform data recalled on the SOM learned by the Gaussian data, highlighting the structure expected from the Uniform Null Model $\mathcal{U}$

2   A Probabilistic Model for $CADJ$

By definition, the edge weights $CADJ_{ij}$ are counts of N observations falling into "bins," where the bins are the second-order Voronoi cells V_{ij} generated by the tessellation induced by $\{w_i\}$ (see [5] for a more complete reference for higher-order Voronoi tessellations). To simplify the notation, let $k \in \{1, \ldots, K\}$ re-index the $(i,j) : CADJ_{ij} > 0$, so that K is the number of non-empty cells V_{ij}. Similarly, let $n_k = CADJ_{ij}$, $N = \sum_{k=1}^{K} n_k$; then the collection $\boldsymbol{n} = (n_1, n_2, \ldots, n_K)$

of counts in our K bins (categories) can be modeled via the Multinomial distribution with parameters $\boldsymbol{p} = (p_1, \ldots, p_K)$, which give the probabilities of observation $\boldsymbol{x}$ falling into bins $1, \ldots, K$.

For our data X, the true probability of bin k is $p_k = \int_{V_k} f_{\mathcal{M}}(\boldsymbol{x})$, where $f_{\mathcal{M}}(\boldsymbol{x})$ is the probability density of manifold $\mathcal{M}$ from which X was drawn. Since $f_{\mathcal{M}}(\boldsymbol{x})$ is unknown we must estimate $\boldsymbol{p}$ in order to parameterize the Multinomial distribution. This is easily accomplished in a Bayesian framework by appealing to the conjugacy of the Dirichlet distribution to the Multinomial. The Dirichlet distribution of dimension K, denoted $Dir(\boldsymbol{\alpha})$, is parameterized by vector $\boldsymbol{\alpha} = (\alpha_1, \ldots, \alpha_K)$, where all $\alpha_k > 0$, and describes probability distributions over the $K - 1$ dimensional unit simplex (i.e., $\{\boldsymbol{p} : \sum_{k=1}^{K} p_k = 1\}$). We note for later use that each marginal p_k has a Beta distribution with parameters $\beta_1 = \alpha_k$ and $\beta_2 = \alpha_0 - \alpha_k$ (where $\alpha_0 = \sum_{k=1}^{K} \alpha_k$).

Invoking Dirichlet-Multinomial (DM) conjugacy, the posterior distribution of unknown $\boldsymbol{p}$ after observing multinomial count data $\boldsymbol{n}$ with prior distribution $Dir(\boldsymbol{\alpha})$ is also Dirichlet, but with modified posterior parameter $\bar{\boldsymbol{\alpha}} = \boldsymbol{\alpha} + \boldsymbol{n}$. Under the DM setup, the marginal distribution of the observed counts (i.e., the data likelihood, or Bayesian *evidence*) is given by [6]

$$f_{DM}(\boldsymbol{n}|\boldsymbol{\alpha}) = \int f_M(\boldsymbol{n}|\boldsymbol{p}, N) f_D(\boldsymbol{p}|\boldsymbol{\alpha}) \, d\boldsymbol{p} = \frac{\Gamma(\alpha_0)\Gamma(N+1)}{\Gamma(N+\alpha_0)} \prod_{k=1}^{K} \frac{\Gamma(n_k + \alpha_k)}{\Gamma(\alpha_k)\Gamma(n_k + 1)}. \quad (1)$$

The Dirichlet parameter $\boldsymbol{\alpha}$ controls both the mean and covariance of the resulting distribution, and consequently affects our estimation of $\boldsymbol{p}$. A typical choice for an uninformative prior has $\alpha_k = a = 1 \forall k$, meaning the resulting Dirichlet distribution considers all probability vectors equally likely. As $a \to \infty$ the Dirichlet more strongly favors uniform $\boldsymbol{p}$. Our choice of prior will be influenced by our choice of learning algorithm, where the prototypes $\{\boldsymbol{w}_i\}$ are placed via the *Conscience*-SOM (CSOM) algorithm of DeSieno [7], which aims to produce a maximum entropy mapping $\mathcal{M} \to \{\boldsymbol{w}_i\}$. Because of this, we have reason to expect equi-probable $\boldsymbol{p}$ and set $\alpha_k = a > 1 \forall k$. In the experiments which follow, $\alpha_k = 10$ was used as a default.

3   A Multi-focal View

The model in Sect. 2 was constructed by considering all $CADJ$ connections simultaneously, which we refer to as a *global* view of manifold connectivity (and denote by subscripts G in what follows). However, depending on characteristics of the manifold as represented by the prototypes $\{\boldsymbol{w}_i\}$, *local* connectivity might be more informative at certain points of the manifold than others, and vice-versa. In our framework, we have a separate local model of connectivity emanating from each (non-empty) Voronoi cell $i \in 1, \ldots, W$, with corresponding local versions of our quantities of interest: K_i = the number of bins in each local model, which is given by the number of $CADJ$ neighbors of prototype i; $\boldsymbol{n}_i = (n_{i,1}, \ldots, n_{i,K_i})$ are the observed counts of these K_i bins; $N_i = \sum_{j=1}^{K_i} n_{i,j}$; $\boldsymbol{\alpha}_i$ and $\bar{\boldsymbol{\alpha}}_i$ denote the vectors of prior and posterior parameters associated with these local models, respectively. We stress that only the probability space (the number of bins K) changes between the global and local views, not the counts or parameters. In what follows, an expression $f_{DM}(\boldsymbol{n}|\boldsymbol{\alpha})$ should be interpreted as evaluating the DM density of the global model at observation $\boldsymbol{n}$ with parameter $\boldsymbol{\alpha}$, and $f_{DM}(\boldsymbol{n}_i|\boldsymbol{\alpha}_i)$ denotes evaluation of the DM density of local model i, at counts $\boldsymbol{n}_i$ with parameters $\boldsymbol{\alpha}_i$.

4   The Λ Metric

The evidence provided by the DM model (1) for our $CADJ$ values provides a natural starting point to monitor the degradation caused as we invoke sparsity into the prototype connectivity graph and, as discussed in the previous section, we would like to simultaneously monitor both the global (G) and overall local (L) nature of these impacts. In our setup, removing a $CADJ$ connection k, can be represented by sparsifying the $\boldsymbol{\alpha}$ parameters of our model (i.e., by setting α_k to some small value $\epsilon \approx 0$, we used $1e - 20$). The impact of pruning can then be monitored via the Bayesian odds of the sparse prior model to the full, which forms the basis of our metric Λ.

The pruning process will be iterative over time t. Before beginning, we assign ranks r_k to each connection, which denote its order of removal (by convention, $r_k \in \{1, \ldots, R\}$, where $R \le K$ is the number of distinct ranks and $r_k = 1$ means connection k is slated for removal at time $t = 1$). Let $\rho_t = \{k : r_k \le t\}$, so that ρ_t returns the set of indices slated for removal at all times $\le t$. Since, for us, removing connections corresponds to sparsifying $\boldsymbol{\alpha}$, we also define $\alpha_k(t) = \{\epsilon$ if $k \in \rho_t$, α_k else$\}$ and their vectorized versions $\boldsymbol{\alpha}(t)$ and $\boldsymbol{\alpha}_i(t)$. The impacts on our global and local models from pruning at step t are

$$\Lambda_G(t) = \left(\frac{f_{DM}(\boldsymbol{n}|\boldsymbol{\alpha}(t))}{f_{DM}(\boldsymbol{n}|\boldsymbol{\alpha}(0))} \right)^{1/N} \quad ; \quad \Lambda_L(t) = \prod_{i=1}^{W} \left(\frac{f_{DM}(\boldsymbol{n}_i|\boldsymbol{\alpha}_i(t))}{f_{DM}(\boldsymbol{n}_i|\boldsymbol{\alpha}_i(0))} \right)^{1/N_i} \tag{2}$$

and we aggregate these two views into a combined measure $\Lambda = \Lambda_G \times \Lambda_L$. Λ_G and Λ_L (and, consequently Λ) describe the evidence of the sparse model, relative to the full model. To account for the different natural scales of the data likelihoods and produce a comparable global and local view we have normalized each of the above odds by their effective sample size. We call the curve traced by $\Lambda(t)$ over time the Λ-Path and monitor its trajectory as we successively prune.

An example of a Λ-Path is given in Fig. 2 displaying the global, local and combined relative likelihoods of pruning over time. For viewing purposes, all three curves have been scaled from $[\Lambda(R), 1]$ to $[0, 1]$ (the likelihood of the fully pruned model at $t = R$ is not 0). Red dotted guidelines show selected % levels (e.g, every 5%) of the total likelihood retained, along with their corresponding steps t, at the bottom of the path plot. These guidelines suggest where to prune the graph in order to conserve a particular level of the likelihood. Note that a loss in data *likelihood* is not necessarily proportional to a total loss in data (or connections) from the model: certain connections are more important than others, as will be reflected by their evidence (1). Of particular interest to us is the point of intersection where Λ_G and Λ_L attain equal relative evidence, since removing a connection harms its corresponding local likelihood more than it does Λ_G. At the point $t^* : \Lambda_G(t^*) = \Lambda_L(t^*)$, we have damaged our local model fit to the point where each of the component views contributes equal evidence; we postulate this might serve as an upper bound on t when implementing this pruning procedure.

5   Ranking Connections for Removal

The thresholding procedure described in [1] utilizes relative strengths of the observed $CADJ$ counts themselves as a guide for thresholding, with low-ranking connections (either globally or locally) removed first. This is essentially a rank analysis, where the justification for removing a certain connection is derived by comparing its strength (i.e., weight) to others, at different focal areas (globally or locally). Our basis for comparison will be informed by conditioning our notion of "strength" on a different set, which we call the Uniform Null Model $\mathcal{U}$. Specifically, we take $\mathcal{U}$ to be a fictitious, unobserved manifold that has Uniform measure $f_{\mathcal{U}}(\boldsymbol{x})$ over the support of $\mathcal{M}$. Under $\mathcal{U}$, the probabilities associated with each second-order Voronoi cell V_k are given by $q_k = \int_{V_k} f_{\mathcal{U}}(\boldsymbol{x}) \, d\boldsymbol{x}$. Using the q_k we construct, for each connection k and its estimated probability distribution p_k, the quantity $Q_k = P_B(p_k > q_k)$, which serves as an indicator of strength relative to $\mathcal{U}$. The measure $P_B(\cdot)$ denotes probability with respect to k's marginal Beta distribution (see Sect. 2). Obviously, $0 \leq Q_k \leq 1$, where values ≈ 0 (or 1) indicate posterior bin probabilities much less (or greater) than would be expected under $\mathcal{U}$.

This choice of null model compares our observed $CADJ$ graph, which represents structured $\mathcal{M}$, to a $CADJ$ (with respect to the same set of prototypes $\{\boldsymbol{w}_i\}$) derived from data that has no structure at all. Namely, we are trying to distinguish observed structure in $CADJ$ from that which would result from uniform noise. Due completely to the *size* (volume) of the second-order Voronoi cells V_k, uniform noise (such as is displayed in Fig. 1c) can still produce the appearance of structure in the $CONNvis$ visualization (shown in Fig. 1d, which was generated by recalling the uniform noise using the SOM trained on the Gaussian clusters in Fig. 1a, and recording their $BMU1$ and $BMU2$). In our context, connections k whose Q_k values are the lowest are prime candidates for pruning (and are assigned rank $r_k = 1$, putting them first in line for removal). Under our multi-focal view we actually must consider $Q_{G,k}$ and $Q_{L,k}$ for each connection k, as the global and local models dictate separate marginal distributions. The ranking r_k is defined as the geometric mean of the two so $Q_k = \sqrt{Q_{G,k} \times Q_{L,k}}$ and $r_k = \text{rank}(Q_k)$.

As computing q_k is equivalent to determining the volume of convex polytopes (i.e., the second-order Voronoi cells) we must turn to estimation procedures for practical implementation. In low to medium dimension, rejection sampling can work; we have had some success in creating tight sampling bounds around the second-order Voronoi cells via the solution to the extrema linear program of [8]. Alternatively, Markov-Chain Monte Carlo methods [9, 10] can be employed. We believe even crude estimates may still be useful for our purposes as the q_k for the Ocean City data discussed in Sect. 6.2 were estimated by considering the relative proportions of their bounding hypercubes and the results there are promising.

6 Clustering Applications

To determine the success of our method we compare the cluster structure recovered from the DM-Pruned $CADJ$ graph to known cluster structure in both synthetic and real-world data. Cluster extraction from an SOM has typically been of the highest quality when interactively performed from $CONNvis$. This is time-consuming, requires expertise, and is somewhat subjective. To mitigate these bottlenecks we employ a leading graph segmentation algorithm (Walktrap [11], available in [12]) to segment the pruned $CADJ$, which we have shown produces clusterings of similar quality to those of the human analyst [13]. Results are judged via Unweighted and Weighted Overall Accuracies (UOA and WOA) as compared to reference images. UOA is the average of class-wise %-accuracies whereas WOA is the average pixel-wise accuracy. Utilizing Walktrap as a clustering oracle has its own effects on the quality of the recovered partitioning as it requires further parameterization (which we do not optimize here), but the *relative* accuracies between different pruning levels should still indicate their comparative suitability for cluster discovery.

For each of our data applications we compute the Λ-path and select as candidates for pruning those connections which, when removed, result in 95% and 90% of the "full" model likelihood (denoted Λ-95 and Λ-90, respectively). We also consider the graph resulting from pruning at an additional t-step of interest in each case, as explained below. For reference we compare to a more naive scheme of simply keeping the top locally-ranked connections which contain 100, 95 and 90 percent of the data (denoted $tn - 100, 95, 90$, respectively).

6.1 6d Synthetic Spectral Image

Our synthetic data is a 128×128 pixel synthetic spectral image (6d synthetic, 11-class, [14]) of 6 bands containing 11 clusters of various sizes. The spatial layout of the class labels, distinguished by 11 colors, is shown in Fig. 2a. The 15×15 SOM trained on this data is shown in Fig. 2b, where each lattice cell is painted with the color code of the majority pixel labels which were mapped to the cell's neuron. Overlain on the SOM is a visualization called TopoView [4], which depicts the $CONN$ graph without showing the edge weightings, for clarity. The islands which are readily visible indicate that this $CONN$ graph contains clear closed communities, even without pruning. The one exception is the connection from the single neurons representing clusters R (pink) and Y (fluorescent yellow); our goal with this experiment is to ensure that DM-Pruneing does not destroy the clearly delineated structure visible in Fig. 2b.

The Λ-path resulting from a Dirichlet-Multinomial fit of these data is shown in Fig. 2d. In it, we see that a steep drop in likelihood occurs between $t = 35$ and $t_{\mathrm{drop}} = 36$. Pruning at $t = 25$ (Λ-95), $t = 35$ (Λ-90), and $t_{\mathrm{drop}} = 36$ produces the accuracies reported in Fig. 2c, which shows perfect cluster capture at both the Λ-95 and Λ-90 levels (since each of their resulting pruned graphs removed the connection between R and Y). Pruning at $t_{\mathrm{drop}} = 36$ also removed connection R-Y, but destroyed the local connectivity holding the large white cluster (B) together (not shown here), resulting in its lower accuracies.

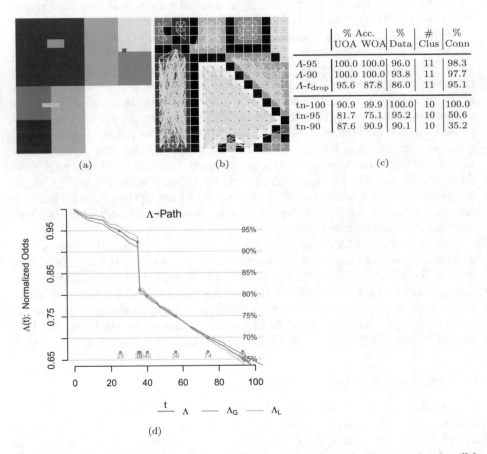

	% Acc.		%	#	%
	UOA	WOA	Data	Clus	Conn
Λ-95	100.0	100.0	96.0	11	98.3
Λ-90	100.0	100.0	93.8	11	97.7
Λ-t_{drop}	95.6	87.8	86.0	11	95.1
tn-100	90.9	99.9	100.0	10	100.0
tn-95	81.7	75.1	95.2	10	50.6
tn-90	87.6	90.9	90.1	10	35.2

(a)   (b)   (c)

(d)

Fig. 2. (a) The spatial layout of the 11 classes of our 6d synthetic spectral cube. (b) The undirected TopoView visualization of $CADJ$ overlain on the SOM lattice which learned the data visualized in (a). Each lattice neuron is connected to the neurons which represent adjacent prototypes in the Induced Delaunay Triangulation of [2]. (c) The accuracies achieved after clustering the SOM depicted in (b) via pruning by both `DM-Prune` and a simpler scheme involving local connection rankings. Along with accuracy, the table also reports the % of data and connections which remain represented in the resulting pruned graph, as well as the number of true clusters which were identified by each method. (d) The Λ-Path (Sect. 4) depicting degradation of the Dirichlet-Multinomial relative likelihood as the $CADJ$ graph is progressively pruned.

6.2   Real Data: Ocean City Spectral Image

For experiments with real data, we use a 512×512 pixel, 8-band spectral image of Ocean City, Maryland, with 1.5 m/pixel resolution. References to data collection, description of pre-processing and mean signatures of verified land-cover classes are given in [15] along with an earlier interactive clustering as in [1]. This image contains many clusters with widely varying statistical properties

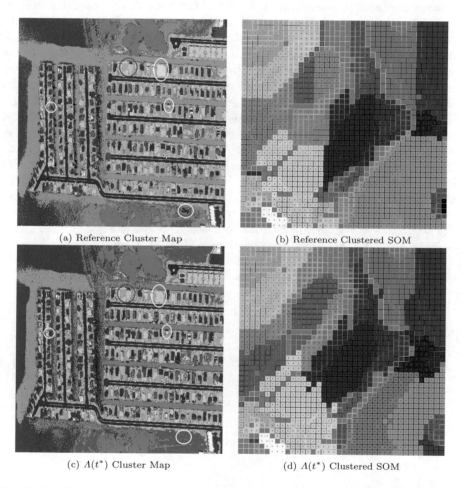

(a) Reference Cluster Map   (b) Reference Clustered SOM

(c) $\Lambda(t^*)$ Cluster Map   (d) $\Lambda(t^*)$ Clustered SOM

Fig. 3. (a) Interactive clustering of the Ocean City spectral image from CONNvis visualization of SOM prototypes, from [15]. The 28 clusters include ocean, bay, canal, pool water, (medium to dark blue colors); roofing materials (red, white, light pink, hot pink, magenta); grass, shrubs around houses (green, yellow), other vegetation (orange, brown), and several rare clusters: roofs a, m (in blue circle), roofs c, V, shrub g, dry grass M (in white ovals). Asphalt (magenta, G) and reflective paint (neon blue, X) occur on roads as well as on roofs. (b) The 40×40 clustered SOM for Ocean City, with the labels and colors of each reference cluster depicted in the lattice cells. (c) The clustered image resulting from pruning the Ocean City $CADJ$ at step $t^* = 398$. (d) The clustered SOM which generated (c).

(see Table 1 of [4]), and it is very noisy. We compare `DM-Prune` segmentations with the segmentation of this image from [15], shown in Figs. 3a, b. The $CONN$ representation of this data can be seen in [1].

The `DM-Prune` clustering from step $t^* = 398$ (where Λ_L dips below Λ_G in Fig. 4a) in Fig. 3c is very similar to that in Fig. 3a by visual inspection. All major landmarks, water components, all houses, roads, and vegetation are very well delineated. Major differences include the absence of cluster T (salmon color); the absence of small clusters c, g, M, V, X; and the presence of a few extra clusters in Fig. 3c. The locations of c, g, M, V are in white ovals; X was merged to E. These diferences are easiest to find in the SOMs in Figs. 3b and d. Cluster T (a particular roof type) has been divided among clusters G, P, Q, and S, which is reasonable given their neighboring locations in the SOM and their similar cluster signatures in the reference clustering (as seen in Fig. 3 of [15]). In the spatial image cluster T pixels are mostly merged to cluster G, a similar type of roof. Given that in [16] cluster T was detected by Walktrap using the unthresholded $CONN$ graph, the absence of T here could be the result of `DM-Prune`ing, as its null model probabilities q_k were very crudely estimated. The missing small clusters can be ascribed to Walktrap performance since in [16] the same are also absent when Walktrap (with the same default number of steps, 4) is applied to the unthresholded $CONN$ graph. The small clusters a and m (circled in blue) are found, probably due to their more unique signatures. Class-wise and pixel-wise accuracies in Fig. 4b also indicate good match between Fig. 3c and a. While they are relatively low (best case just below 70%) in absolute terms, we need to consider that, in an image with 1.5 m/pixel footprint, many pixels contain multiple materials at object boundaries. These mixed pixels may get assigned to

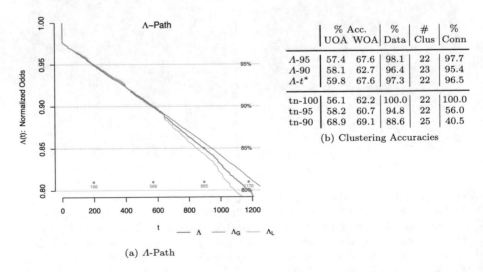

	% Acc.		%	#	%
	UOA	WOA	Data	Clus	Conn
Λ-95	57.4	67.6	98.1	22	97.7
Λ-90	58.1	62.7	96.4	23	95.4
Λ-t^*	59.8	67.6	97.3	22	96.5
tn-100	56.1	62.2	100.0	22	100.0
tn-95	58.2	60.7	94.8	22	56.0
tn-90	68.9	69.1	88.6	25	40.5

(b) Clustering Accuracies

(a) Λ-Path

Fig. 4. (a) 1200 steps of the Λ-Path of pruning the Ocean City $CADJ$ with `DM-Prune`. (b) Accuracies of resulting clustering from `DM-Prune` and simpler thresholding schemes, as compared to Fig. 3a.

different clusters based on different pruning/thresholding of the $CADJ$ graph. The differences can cause fairly large accuracy loss even though they are not distracting to the eye. The accuracy measures ignore the spatial coherence of the errors in areas where mixing is expected. I.e., if the non-matching pixels were scattered throughout the image in a salt-and-pepper fashion, we would consider the match much poorer while the accuracies would remain the same. Overall, we can say the DM-Prune clustering is approaching the quality of the reference clustering.

7   Conclusions and Outlook

We have introduced DM-Prune as a Bayesian Dirichlet-Multinomial model of the edge weights of the $CADJ$ graph for the purpose of pruning its edges to facilitate cluster discovery from SOM prototypes. The modeling framework provides a natural mechanism for both selecting edges ripe for pruning and assessing the overall impact from doing so. Experiments with both synthetic and real-world spectral image data confirm that graph sparsification governed by DM-Prune is capable of retaining vital local connectivities, which if removed would destroy our view of meaningful cluster structure, while simultaneously shedding spurious connections which cloud this view. Future work involves integrating more sophisticated estimation of the null model probabilities q_k and further experiments with the methodology to gain confidence for setting the DM α parameter and, most importantly, to provide more concrete recommendations for identifying optimal pruning levels from the Λ-Path metric.

Acknowledgment. We thank Dr. Beáta Csathó, University of Buffalo, for the Ocean City spectral image and accompanying truth. This project was partially supported by a North American ALMA Development Cycle 5 Study Program, administered by the National Radio Astronomy Observatory, with the consent of the U.S. National Science Foundation.

References

1. Taşdemir K, Merényi E (2009) Exploiting data topology in visualization and clustering of Self-Organizing Maps. IEEE Trans. Neur. Netw. 20(4):549–562. ISSN 1045-9227
2. Martinetz T, Schulten K (1994) Topology representing networks. Neural Netw. 7(3):507–522. ISSN 0893-6080
3. Kohonen T (1997) Self-Organizing Maps, 2nd edn. Springer, Heidelberg
4. Merényi E, Taşdemir K, Zhang L (2009) Similarity-based clustering. chapter Learning Highly Structured Manifolds: Harnessing the Power of SOMs. Springer, Heidelberg, pp 138–168. ISBN 978-3-642-01804-6
5. Okabe A, Boots B, Sugihara K (1992) Spatial tessellations: concepts and applications of voronoi diagrams. John Wiley & Sons Inc., New York
6. Mosimann J (1962) On the compound multinomial distribution, the multivariate β-distribution, and correlations among proportions. Biometrika 49(1/2):65–82. ISSN 00063444

7. DeSieno D (March 1988) Adding a conscience to competitive learning. In: Proceedings of the international conference neural network (ICNN), New York, vol. I, pp I–117–124

8. Agrell E (January 1993) A method for examining vector quantizer structures. In: Proceedings of the IEEE international symposium information theory. IEEE, pp 394–394

9. Dyer M, Frieze A (1991) Computing the volume of convex bodies: a case where randomness probably helps. Probab. Comb. Appl. 44:123–170

10. Lovász L, Vempala S (2006) Simulated annealing in convex bodies and an o(n4) volume algorithm. J. Comput. Syst. Sci. 72(2):392–417

11. Pons P, Latapy M (2005) Computing communities in large networks using random walks. In: Proceedings of the 20th international conference on computer information science, ISCIS 2005, Springer, Heidelberg, pp 284–293

12. Nepusz T, Csardi G (2006) The igraph software package for complex network research. Int J Complex Syst 1695(5):1–9

13. Merényi E, Taylor J (April 2018) Empowering graph segmentation methods with SOMs and CONN similarity for clustering large and complex data. Neural Comput. Appl. (forthcoming)

14. Jain A (2004) Issues Related to Data Mining with Self-Organizing Maps. Rice University, 2004. M.Sc. thesis

15. Merényi E, Csató B, Taşdemir K (2007) Knowledge discovery in urban environments from fused multi-dimensional imagery. In: Gamba P, Crawford M (eds) Proceedings of the IEEE GRSS/ISPRS joint workshop on remote sensing and data fusion over urban areas (URBAN 2007), Paris, France, 11–13 April 2007, pp 1–13

16. Merényi E, Taylor J (June 2017) SOM-empowered graph segmentation for fast automatic clustering of large and complex data. In: 12th international workshop on self-organizing maps and learning vector quantization, clustering and data visualization (WSOM+2017), pp 1–9

Practical Applications of Self-Organizing Maps, Learning Vector Quantization and Clustering

SOM-Based Anomaly Detection
and Localization for Space Subsystems

Maia Rosengarten$^{(\boxtimes)}$ and Sowmya Ramachandran

Stottler Henke Associates, Inc., San Mateo, CA, USA
mrosengarten@stottlerhenke.com

Abstract. The aim of this paper is to contribute to machine-learning technology that expands real-time and offline Integrated System Health Management capabilities for future deep-space exploration efforts. To this end, we have developed Anomaly Detection via Topological feature-Map (ADTM), which leverages a Self-Organizing Map (SOM)-based architecture to produce high-resolution clusters of nominal system behavior. What distinguishes ADTM from more common clustering techniques (e.g. k-means) is that it maps high-dimensional input vectors to a 2D grid while preserving the topology of the original dataset. The result is a 'semantic map' that serves as a powerful tool for uncovering latent relationships between features of the incoming data points. We successfully modeled and analyzed datasets from a NASA Ames Research Center Graywater Recycling System which documents a real hardware system fault. Our results show that ADTM effectively detects both known and unknown anomalies and identifies the correlated measurands from models trained using just nominal data.

Keywords: Self-Organizing Map · Anomaly detection and localization · Integrated System Health Management

1 Introduction

Integrated System Health Management (ISHM) technologies are mission-critical for space exploration. Space habitats are made up of a complex web of subsystems, and the rising demand for rapid fault detection and response in deep-space habitats calls for autonomous monitoring software that can run on board. In particular, communication delays between onboard crews and Earthbound experts (lasting up to 44 min) could make the difference between a successful and failed mission, risking the loss of both equipment and crew [1]. Expansion of both machine learning and data mining techniques in this field is therefore of the utmost importance to ensuring mission safety.

In this paper we discuss the application of a semi-supervised approach to anomaly detection and localization called Anomaly Detection via Topological feature-Map (ADTM), which combines a Self-Organizing Map (SOM) for anomaly detection with a Random Forest of Decision Trees to identify the most salient measurands contributing to data flagged as anomalous. Our research has largely been inspired by a successful body of work leveraging SOMs for anomaly detection within the aeronautics domain [2]. Our contribution is the application of this approach to the space domain. To the

© Springer Nature Switzerland AG 2020
A. Vellido et al. (Eds.): WSOM 2019, AISC 976, pp. 57–69, 2020.
https://doi.org/10.1007/978-3-030-19642-4_6

best of our knowledge, the use of SOMs in conjunction with decision trees for system health monitoring has never been applied to the space domain before.

The remainder of this paper is organized as follows. Section 2 reviews related research employing machine learning and statistical techniques for anomaly detection. Section 3 provides the background for Self-Organizing Maps. Section 4 provides the technical details and methods of our ADTM model. Section 5 describes the experiment we ran on NASA ARC subsystem data to test the feasibility of ADTM within the space ISHM domain. Section 6 concludes our work with a summary of key research findings and plans for future work.

2 Related Work

The focus of this work was on unsupervised anomaly detection for discrete sequences of subsystem data using SOM-based models trained on nominal subsystem behavior. Similar approaches to anomaly detection have been applied in existing research. Principal Component Analysis has been a widely used algorithm for anomaly detection across a wide breadth of applications, including diagnosing offshore wind turbines [3], cyber networks [4], and space telemetry [5]. Furthermore, Gaddam used a supervised approach to anomaly detection by combining K-Means clustering with ID3 decision tree classification [6]. The classification decisions across the clusters and decision trees were combined for a final decision on class membership. The main drawback of such an approach in the space domain is the limited availability of labeled fault data needed for training and validation.

NASA Ames Research Center (ARC) uses k-means and density-based clustering techniques for system monitoring in its IMS and ODVEC software systems [7]. Similarly, Gao, Yang, and Xing used a K-Nearest-Neighbor (kNN) approach for anomaly detection of an in-orbit satellite using telemetry data [8]. SOMs have been used for fault detection and diagnosis in several industries. Datta, Mabroidis and Hosek combine SOMs with Quality Thresholding (QT) to refine the resolution of clusters learned by SOMs within the semi-conductor industry [9]. Similarly, Tian, Azarian, and Pecht train a SOM on nominal cooling fan bearing data but use a kNN approach in place of the Minimum Quantization Error (MQE) to assign test data anomaly scores based on their distance to centroids learned by the kNN model [10]. Cottrell and Gaubert apply anomaly scores to aircraft engine test data using the MQE approach that we have used in this paper (see Sect. 4) and leverage the visualization capabilities of SOMs to visualize the transition states of engines from run-to-failure datasets [2].

ADTM contributes to this existing bed of clustering research by combining a Self-Organizing Map in combination with Extra Tree Classifier for both detecting and localizing faults, which has rarely (if at all) been used in the ISHM space domain.

3 Self-Organizing Map Background

Also known as a Kohonen map, a Self-Organizing Map (SOM) is a two-layer artificial neural network (ANN) that uses unsupervised learning to produce a low-dimensional representation of the training samples [11]. The goal is to transform incoming inputs to a 1- or 2-dimensional map in a topically ordered fashion such that points that are close together in the higher-dimensional input space are also close together in the lower-dimensional output space. This mapping allows us to detect patterns of normal or anomalous behavior in a system, as different types of behavior map to different output units, called "neurons."

Specifically, the N-dimensional input data is fed into the SOM in the first layer and fully connected to a lattice of $(l \times p)$ output neurons O_i in the second layer [10]. Each neuron O_i is associated with a N-dimensional weight vector w_i. We represent O_i by a two-dimensional coordinate of its position in the $(l \times p)$ grid, e.g., $O_i = (x_i, y_i)$. The values of l and p are parameters that are tuned during model validation. Based on the literature [10], we chose $l = p = \sqrt{(5\sqrt{N})}$, though we intend to further tune these parameters in future work. Unlike k-means, the clusters learned during SOM training are topologically ordered through a competitive learning rule.

The topological ordering happens with the following training process: each input vector $m \in M$ is compared with the weight vector w_i associated with each neuron O_i. The closest O_c is chosen as the winner, or 'Best Matching Unit' (BMU), where 'close' is defined by a distance function between the input vector m and the closest w_c associated with O_c. The smallest distance is called the *Minimum Quantization Error* (MQE). Each BMU in the output layer is related to an entire *neighborhood* of neurons through a 'neighborhood function' $h(c, k)$ that computes the relation between the BMU O_c and neuron O_k. The weight vectors within a neighborhood are updated in proportion to their distance to the BMU in the 2D output lattice. Because entire neighborhoods of related neurons get updated in the direction of the input data that is closest to them, the topology of the N-dimensional input space is preserved in the 2-dimensional output space.

Our research used open-source Python libraries for data processing and building a baseline SOM model. Though there are several SOM-based open source libraries available, we chose Somoclu [12] because it leverages a highly parallel implementation in the C programming language. Without performing cross-validation for hyperparameter tuning, our SOM-based anomaly detectors still showed promising results in the experiment detailed in Sect. 5. This suggests a significant opportunity for additional performance and efficiency gains through fine-tuning our baseline algorithms in future work.

4   Methods

At a high level, our methods use a SOM trained on nominal subsystem behavior to identify anomalous data, followed by a Random Forrest to identify the salient measurands implicated in the flagged anomaly. ADTM is implemented with the following procedure (each step is detailed in the sections that follow):

1. Given nominal data and fault data sets for testing, divide the nominal data into a training and a test set.
2. Normalize all the data with decimal scaling, using the training data as the scaling reference.
3. Train one SOM per subsystem using nominal training data.
4. For all data sets (training sets, nominal test sets, fault test sets), calculate the MQE for each sample point.
5. Set the kth and $(1 - k^{th})$ percentiles of the MQE scores of the training data as the nominal MQE thresholds for flagging anomalies, where $0 < k < 1$. In our experiments $k = 0.99$.
6. For each data point in each test set, flag the point as anomalous if its MQE is < lower nominal MQE threshold or > the upper nominal MQE threshold.
7. Find the most salient measurands contributing to data flagged as anomalous with a supervised feature extractor, ordered by importance.

Steps 1–2 are detailed in Sect. 4.1. Step 3 uses the SOM training process described in Sect. 3. Steps 4–6 are detailed in Sect. 4.2, while Step 7 is detailed in Sect. 4.3.

4.1   Data Processing

Because the SOMs require numerical data, we converted categorical variables (e.g., "ON/OFF") to quantitative variables (e.g., 1/0). We also dropped the columns related to the timestamp of data collection, as we were not concerned with multi-scale time-series analysis. Such analysis is a research focus of future work, however, specifically for the purpose of conducting cross-subsystem analysis given datasets measured in different timescales. Additionally, we normalized the data to prevent measurands with larger ranges from out-weighing measurands with smaller ranges. We used decimal scaling to scale the values so that all values fell within the range −1 to 1 [13]. Normalization was done on the training sets for each subsystem, and the test sets were scaled relative to this normalization.

4.2   Anomaly Detection via MQE

Once trained on nominal data, the SOM maps new data seen during testing to the most similar weight vector w_c of the output neurons O_i, using Euclidean Distance as the similarity metric. A low MQE implies that the new sample closely aligns with a previously seen sample from the training data and is therefore nominal, whereas a higher MQE connotes that the point is anomalous, either because it contains a true fault or because it captures novel nominal behavior unseen during training. We defined a range of nominal MQE scores and classified all samples as anomalous during testing if they fell outside that range. The range was chosen by re-running the training data through an already-trained SOM and setting the 1-percentile value and the 99-percentile value of the resulting MQEs as the lower and upper bounds respectively. Admittedly, these thresholds were chosen rather arbitrarily from our observations of the available data. We intend to include a more principled approach to threshold tuning in future work and anticipate doing so will improve the generality of our results.

4.3 Anomaly Localization via Supervised Feature Extraction

In addition to identifying regions of anomaly, it would be helpful to localize the anomalies to a small subset of the measurands that explain observed behavior deviation and best distinguish between the two regions (anomalous vs. non-anomalous). For this, we rely on the insight that the two regions can be treated as two classes and supervised classification methods can be used to identify the features that distinguish them. For this analysis, we are not concerned about the accuracy of anomaly identification, i.e., the external consistency of anomaly detection with respect to ground truth. All we are concerned with is determining the features that accurately separate two given segments of data (i.e., internal consistency). The data points labeled as anomalies are grouped into one class, and the weight-vectors learned by the SOMs during training form the data for the second class. This highlights one of the benefits of the SOM-based approach: the weight vectors of the SOM are effectively a reduced representation of the training data and lead to efficient storage for future analysis.

A number of supervised feature extraction approaches are available, though we chose Extra Tree Classifier [14], which is a variant of the Random Forest approach [15]. Our experiment in Sect. 5 demonstrates the utility of this approach to anomaly localization, though we intend to experiment with other techniques such as Recursive Feature Elimination (RFE) [16] in future work. Our anomaly localization goal was to identify subsets of measurands that contributed the most to deviation in anomalous data. For this we employed a Random Forest (RF) to classify nominal and anomalous data points and output the measurands that resulted in the greatest reduction in Gini Impurity scores across all decision trees employed [15]. We used the default parameters that came from a Python machine-learning package, in which the number of trees was set to 10. We intend to tune this parameter (e.g., employing 100s of trees) to further improve our anomaly localization capabilities in future work.

The weights associated with the SOM output neurons are known as "codebook vectors," as they represent prototypical nominal activity learned from the training data. Thus, we labeled these codebook vectors as "nominal" for our Extra Tree Classifier model. Similarly, we labeled the data samples from our test sets that were flagged as anomalous as the class "anomaly." Finally, we trained a RF on the labeled data and output the list of measurands that resulted in the best splits between the two classes, ranked by their (normalized) reductions in GI scores across all trees. For this paper, we arbitrarily chose the subset of measurands with a feature importance score of at least 10 to characterize the anomalies from each test set, though this threshold is an additional hyperparameter that we will tune in future work.

5   Experiments and Discussion

We divide the results of our experiment into the following three subsections: *Data Collection, Anomaly Detection Analysis, Anomaly Localization Analysis.*

5.1   Data Collection

We acquired data of a Graywater Recycling System from NASA Ames Research Center installed at Stanford University. This data documented a real system failure that propagated across two interconnected subsystems. The Forward Osmosis (FO) membrane became fouled with a bacterial sludge, and the system shut down due to a low OA tank float alarm. The data we received decomposed the Graywater Recycling System into two subsystems, "Subsystem 1" and "Subsystem 2." For each subsystem, we received two days' worth of nominal data and four days of faulty data. We divided the nominal data into a training set and a nominal test set for each subsystem, the latter of which was used to compare against the fault test sets. The shapes of the datasets used for training and testing are described in Table 1. Although both Subsystem 1 and Subsystem 2 were running for the same length of time during the October experiments, Subsystem 1 has significantly more data points than Subsystem 2 due to differences in the sampling rate for each subsystem.

Table 1. Datasets used for training and testing SOMs for Subsystems 1 & 2

Subsystem	Train data (#rows, #features)	Test data (#rows, #features)
Subsys1 SOM	(47031, 32)	fault: (274100, 32) nominal: (23643, 32)
Subsys2 SOM	(789, 7)	fault: (4595, 7) nominal: (394, 7)

Beyond detecting the fault, our algorithm was also able to output the specific sensors that contributed most to the anomaly, as shown in Subsect. 5.3.

5.2   Anomaly Detection Analysis

For each test set, our SOMs flagged a sample point as anomalous based on its MQE score, using the 99th percentile of the training MQEs as a threshold. The results are displayed in Table 2. We see that the SOMs flagged >99% and 84% in the fault test sets of Subsystem 1 and Subsystem 2 respectively. By comparison, the Subsystem 1 SOM flagged ∼12% of the nominal test set as anomalous, while the Subsystem 2 SOM detected no anomalies in the nominal Subsystem 2 test set.

Table 2. Percentage anomalies detected in Graywater Recycling System Data with 99% confidence interval

Subsystem	Test dataset	Percentage anomalies detected
Subsystem 1	A. nominal test set	A. 12.2%
	B. fault test set	B. 99.86%
Subsystem 2	A. nominal test set	A. 0%
	B. fault test set	B. 84%

Subsystem 1 SOM MQE Results

Nominal Test Set

The Subsystem 1 SOM detected a relatively high percentage of anomalies in the nominal test set ($\sim 12\%$). This was due to anomalous behavior in the tail-end of the test set. Observe the plot comparing the MQE scores on the Subsystem 1 nominal test set (blue line) with that of the Subsystem 1 training set (orange line) in Fig. 1. The MQE scores of the nominal test set spike around ~ 21000 data points at the end of the run. We observed that this was due to many sensors in Subsystem 1 simultaneously exhibiting low or stopped activity, likely due to a "shut down" procedure. Though this behavior is not necessarily faulty, it was not captured by the training data so represents an anomaly that we would expect our SOM to flag, as it did.

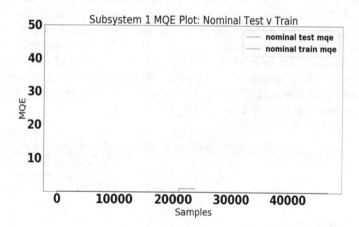

Fig. 1. Graywater Recycling Subsystem 1 Nominal Test MQE (blue) vs Train MQE (orange) Plot. Spike in nominal test MQE occurs at end of run due to shut down procedure.

Fault Test Set

Observe from Fig. 2 that the MQE scores across the Subsystem 1 fault test set (blue) are significantly greater than the maximum MQE score for the Subsystem 1 training set (orange), indicating that the SOM trained on Subsystem 1 nominal data detects significant deviation in the faulty test set.

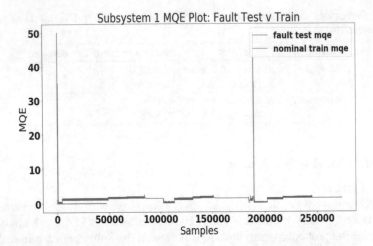

Fig. 2. Graywater Recycling Subsystem 1 Fault Test MQE (blue) vs Train MQE (orange) Plot. Fault MQE is substantially larger than nominal train MQE for entirety of run.

Subsystem 2 SOM MQE Results

The SOM trained on Subsystem 2 data did not detect any anomalies in the Subsystem 2 nominal test set. Observe in Fig. 3A that this is because the MQE scores of the nominal test set closely align with the MQE scores of the training set (~ 0)—that is, they fall within the 99th-percentile of nominal MQE scores. By comparison, observe the significant deviation between the MQE scores of the Subsystem 2 fault set (blue) and those of the Subsystem 2 training set (orange) in Fig. 3B. We see behavior similar to a stair-step function in the interval marked by [A]. In between [A] and [B], the MQE briefly drops to within nominal range before spiking again in [B]. It then oscillates between *nominal* in intervals [C] and [E] (compare with the nominal training set MQE scores, in orange) and *high* in intervals [B] and [D].

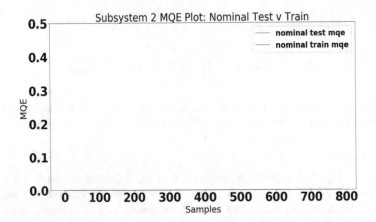

Fig. 3A. Graywater Recycling Subsystem 2 Nominal Test MQE (blue) vs Train MQE (orange) Plot. Both nominal test and train MQE's are ~ 0, as expected.

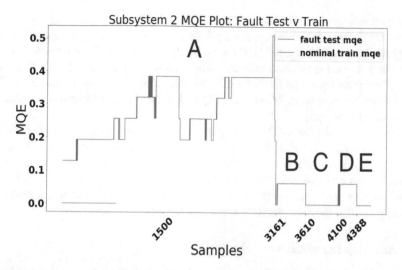

Fig. 3B. Graywater Recycling Subsystem 2 Fault Test MQE (blue) vs Train MQE (orange) Plot. Test MQE deviates significantly in intervals [A], [B], and [D].

We observed that the Subsystem 2 MQE plot closely aligns with the behavior of the CONDUCTIVITY SCALED OA measurand plotted in Fig. 4, comparing the sensor values from the Subsystem 2 fault test set (in red) with those from the Subsystem 2 training set (in black). Compare Figs. 3B and 4 and notice how the changes in the MQE scores in Fig. 3B across intervals [A – E] correlate with the changes in behavior of the CONDUCTIVITY SCALED OA measurand in Fig. 4 across the same intervals.

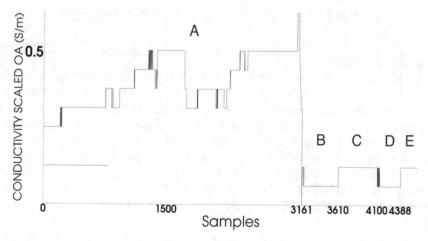

Fig. 4. Graywater Recycling Subsystem 2 CONDUCTIVITY SCALED OA: red (test), black (train). Test data deviates significantly from training in intervals [A], [B], and [D].

In particular, the CONDUCITIVITY SCALED OA measurand in Fig. 4 experiences dramatic deviation from the nominal training data (in black) in a stepwise fashion throughout interval [A], while MQE scores in Fig. 3B deviate from nominal range in a similar way. It then oscillates between a low reading during the intervals [B] and [D], and nominal readings in the intervals [C] and [E]. Similarly, the MQE scores in Fig. 3B deviate from nominal range in intervals [B] and [D] and return to within nominal range in intervals [C] and [E].

This demonstrates the SOMs ability to not only detect deviations in the fault data, but also to capture the relative degree of deviation exhibited in a fault. That is, the greater the deviation from nominal behavior seen in training, the greater the MQE score will be. This is a significant innovation for an effective ISHM tool, as it distinguishes between severe (and potentially fatal) faults and more mild anomalies, allowing end-users to prioritize their overhauling and response activities accordingly.

5.3   Anomaly Localization Analysis

Our ExtraTreeClassifier (ETC) algorithm identified the sensors in Table 3 as contributing the most to the anomalies detected in each subsystem's fault test set. We only listed the measurands with a feature importance score of at least 10. We validated our results with a subject matter expert (SME) who worked at Kennedy Space Center for 32 years, including on the actual International Space Station (and other spacecraft) systems and consumables while they were on the ground.

Table 3. Salient features of flagged anomalies from ETC for Graywater Recycling Subsystems

Subsystem	Test data	Salient measurands identified
Subsystem 1	fault	RO PUMP SPEED: 68.6
Subsystem 2	fault	CONDUCTIVITY SCALED OA: 94.0

Subsystem 1 Validation

Our SME confirmed that the RO PUMP SPEED is one of the most important measurands to detect for a clogged FO Membrane fault, as the pump is what moves fluid from the FO Membrane through the OA Tank to the RO Membrane. When the FO Membrane is clogged by sludge, the excess build-up prevents fluid from circulating properly through the pump, so deviation from nominal pump behavior is expected and should be flagged. Such underactivity is clearly displayed in Fig. 5, in which the RO PUMP SPEED from the fault test set (green) is plotted against that from the train set (black). The TURN SYSTEM ON/OFF INDICATOR is also plotted for the test set (red). Notice that the RO PUMP SPEED deviates from nominal each time the system is turned on.

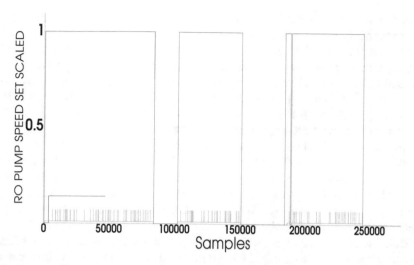

Fig. 5. Graywater Recycling Subsystem 1 RO PUMP SPEED SET SCALED: green (test), purple (train). TURN SYSTEM ON/OFF INDICTOR: red (test).

Subsystem 2 Validation

The CONDUCTIVITY SCALED OA measurand measures the electrical conductivity of the fluid moving through the recycling system. During a clog, the water contains greater mineral and salt deposits, which affects the conductivity of the fluid. Our SME confirmed that we would expect to see deviation in the CONDUCTIVITY SCALED OA measurand for the OA system, since it receives fluid directly from the clogged FO Membrane implicated in the fault. Furthermore, refer back to Fig. 4 which displays the significant deviation in CONDUCTIVITY SCALED OA in the test set from training, and to Fig. 3B which shows the corresponding MQE scores for the Subsystem 2 fault test set. The correlation between these two plots indicates that the CONDUCTIVITY SCALED OA was a large contributor to the deviation in the fault data. The fact that our ETC algorithm correctly identified it as the top salient feature proves the effectiveness of our fault localization approach.

6 Conclusions and Future Work

Our research demonstrates the feasibility of applying an unsupervised, SOM-based anomaly detection approach to the Integrated System Health Management (ISHM) domain for space subsystems and lays the foundation for behavior diagnosis through our anomaly localization techniques that isolate measurands most correlated with flagged anomalies. We were able to demonstrate these results on a NASA ARC Graywater Recycling System dataset implicated by a known fault. Moreover, our research makes use of Python packages that use highly parallel processing techniques to produce computationally efficient results.

As this work was a relatively small feasibility study to investigate the utility of SOM-based analysis for space subsystem health monitoring, we have relegated several important research questions to future work, which we mention here briefly. In addition to taking a more principled approach to hyperparameter tuning, e.g. through cross-validation or Bayesian optimization, we will build upon our existing methods through incorporating SOM-based prognostics capabilities based on the work of [17] and implementing multi-timescale analysis in order to cross-correlate anomalies across subsystems from data collected across different timescales.

Furthermore, we intend to investigate methods for exploiting the visualization capabilities of SOMs as in [2] for the purpose of fault localization and characterization. For instance, displaying the component planes of sensors highly correlated with data implicated in a fault may assist human operators in more quickly diagnosing and responding to flagged anomalies. We will continue discussions with NASA engineers to better understand the desirability and effectiveness of such visualizations.

While the NASA data we received did not contain confounding anomalies, nor severely unbalanced classes (e.g. 99% nominal samples versus 1% anomalous samples), effectively handling such cases is important for generalizing ADTM to new subsystems, and we have included such analysis as part of ongoing work. Finally, we intend to compare the results of our techniques to other unsupervised approaches, such as k-means and PCA, in order to further establish the utility of our approach to real applications.

References

1. Crusan J (2016) Habitation Module, NASA Advisory Council, Human Exploration and Operations Committee. 3
2. Cottrell M, Gaubert P, Eloy C et al. (2009) Fault prediction in aircraft engines using self-organizing maps. Adv Self-Organizing Maps, 37–44. https://doi.org/10.1007/978-3-642-02397-2_5
3. Bennouna O, Heraud N, Leonowicz Z (2012) Condition monitoring & fault diagnosis system for Offshore Wind Turbines. In: 2012 11th international conference on environment and electrical engineering. https://doi.org/10.1109/eeeic.2012.6221389
4. Pascoal C, de Oliveira M, Valadas R, et al. (2012) Robust feature selection and robust PCA for internet traffic anomaly detection. In: 2012 proceedings IEEE INFOCOM. https://doi.org/10.1109/infcom.2012.6195548
5. Nassar B, Hussein W, Mokhtar M (2019) Space telemetry anomaly detection based on statistical PCA algorithm. In: Zenodo. http://doi.org/10.5281/zenodo.1109667
6. Gaddam S, Phoha V, Balagani K (2007) K-Means+ID3: a novel method for supervised anomaly detection by cascading K-Means clustering and ID3 decision tree learning methods. IEEE Trans Knowl Data Eng 19:345–354. https://doi.org/10.1109/tkde.2007.44
7. Iverson D, Martin R, Schwabacher M, et al (2009) General purpose data-driven system monitoring for space operations. In: AIAA Infotech@Aerospace conference. https://doi.org/10.2514/6.2009-1909
8. Gao Y, Yang T, Xu M, Xing N (2012) An unsupervised anomaly detection approach for spacecraft based on normal behavior clustering. In: 2012 fifth international conference on intelligent computation technology and automation. https://doi.org/10.1109/icicta.2012.126

9. Datta A, Mavroidis C, Hosek M (2007) A role of unsupervised clustering for intelligent fault diagnosis, vol. 9. Mechanical Systems and Control, Parts A, B, and C. https://doi.org/10. 1115/imece2007-43492
10. Tian J, Azarian M, Pecht M (2014) Anomaly detection using self-organizing maps-based K-Nearest neighbour algorithm. In: European conference of the prognostics and health management society 5
11. Kohonen T (1982) Self-organized formation of topologically correct feature maps. Biol Cybern 43:59–69. https://doi.org/10.1007/bf00337288
12. Wittek P, Gao S, Lim I, Zhao L (2017) Somoclu: an efficient parallel library for self-organizing maps. J. Stat. Softw. https://doi.org/10.18637/jss.v078.i09
13. Saranya C, Manikandan G (2013) A study on normalization techniques for privacy preserving data mining. Int. J. Eng. Technol. 5:2701–2704
14. Geurts P, Ernst D, Wehenkel L (2006) Extremely randomized trees. Mach Learn 63:3–42. https://doi.org/10.1007/s10994-006-6226-1
15. Breiman L (2001) Mach Learn 45:5–32. https://doi.org/10.1023/a:1010933404324
16. Guyon I, Weston J, Barnhill S, Vapnik V (2002) Gene selection for cancer classification using support vector machines. Mach Learn 46:389–422. https://doi.org/10.1023/a: 1012487302797
17. Rai A, Upadhyay S (2017) Intelligent bearing performance degradation assessment and remaining useful life prediction based on self-organising map and support vector regression. Proc Inst Mech Eng Part C: J Mech Eng Sci 232:1118–1132. https://doi.org/10.1177/ 0954406217700180

Self-Organizing Maps in Earth Observation Data Cubes Analysis

Lorena Santos[✉], Karine Reis Ferreira, Michelle Picoli, and Gilberto Camara

National Institute for Space Research (INPE), São Jose dos Campos, Brazil
{lorena.santos,karine.ferreira,gilberto.camara}@inpe.br,
mipicoli@gmail.com

Abstract. Earth Observation (EO) Data Cubes infrastructures model analysis-ready data generated from remote sensing images as multidimensional cubes (space, time and properties), especially for satellite image time series analysis. These infrastructures take advantage of big data technologies and methods to store, process and analyze the big amount of Earth observation satellite images freely available nowadays. Recently, EO Data Cubes infrastructures and satellite image time series analysis have brought new opportunities and challenges for the Land Use and Cover Change (LUCC) monitoring over large areas. LUCC have caused a great impact on tropical ecosystems, increasing global greenhouse gases emissions and reducing the planet's biodiversity. This paper presents the utility of Self-Organizing Maps (SOM) neural network method in the process to extract LUCC information from EO Data Cubes infrastructures, using image time series analysis. Most classification techniques to create LUCC maps from satellite image time series are based on supervised learning methods. In this context, SOM is used as a method to assess land use and cover samples and to evaluate which spectral bands and vegetation indexes are best suitable for the separability of land use and cover classes. A case study is described in this work and shows the potential of SOM in this application.

Keywords: Self-Organizing Maps ·
Earth Observation Data Cubes Analysis · Satellite image time series ·
Land Use and Cover Changes

1 Introduction

Earth Observation Data Cubes (EO Data Cubes) are emergent infrastructures that model analysis-ready data generated from remote sensing images as multidimensional cubes, especially for satellite image time series analysis [1]. Such data cubes have three or more dimensions that include space, time and properties. EO Data Cubes can be defined as a set of time series associated to spatially aligned pixels ready for analysis.

EO Data Cubes infrastructure is an innovative way to organize the big amount of Earth observation satellite images freely available nowadays and to

© Springer Nature Switzerland AG 2020
A. Vellido et al. (Eds.): WSOM 2019, AISC 976, pp. 70–79, 2020.
https://doi.org/10.1007/978-3-030-19642-4_7

take advantage of big data technologies and methods to store, process and analyze time series extracted from these images. Examples of computational platforms for EO Data Cubes are the Open Data Cube (ODC) [2], the Joint Research Centre (JRC) Earth Observation Data and Processing Platform (JEODPP) [3] and the System for Earth Observation Data Access, Processing and Analysis for Land Monitoring (SEPAL) [4].

A typical application that benefits from EO Data Cubes infrastructures and satellite image time series analysis is LUCC monitoring. Characterizing and mapping changes in land surface is essential for planing and managing natural resources. The growing pressures for food and energy production promoted by increasing population make humans modify the Earth's environment in a rapid pace. LUCC can affect hydrological and biological process causing great impacts on tropical ecosystems [6].

Recently, EO Data Cubes infrastructures and satellite image time series analysis have brought new opportunities and challenges for LUCC mapping over large areas. Time series derived from Earth observation satellite images allow us to detect complex underlying processes that would be difficult to identify using bitemporal or other traditional change detection approaches [6]. The use of remote sensing image time series analysis to produce LUCC information has increased greatly in the recent years [7].

Most classification techniques to create LUCC maps from satellite image time series are based on supervised learning methods. Such methods require a training phase using land use and cover samples labeled *apriori*. These training samples must properly represent the land use and cover classes to be identified by the classifier. The quality of these samples is crucial in the classification process. Representative samples lead to good LUCC maps.

This paper presents the utility of Self-Organizing Maps (SOM) neural network method in the process to extract LUCC maps from EO Data Cubes infrastructures. SOM is a clustering method suitable for time series data sets. This work describes the use of SOM in the training phase to produce metrics that indicate the quality of the land use and cover samples and to evaluate which spectral bands and vegetation indexes are best suitable for the separability of land use and cover classes. A case study is described in this work and shows the potential of SOM in this context.

In the LUCC domain, SOM has not being widely exploited for image time series analysis. The good review provided by [7] cites [8] as the main reference in this context. However, [8] proposed an approach to classify land cover from MODIS EVI time series using SOM. Besides that, [9] proposes the use of supervised SOM for pure and mixed pixels, called soft supervised self-organizing map to improve the classification of MODIS-EVI time series. Both references, [8] and [9], propose the use of SOM to classify one agricultural year using only EVI attribute. Differently from them, our proposal use SOM to explore the separability time series using several attributes in order to improve the classification.

2   Land Use and Cover Change Information from Earth Observation Data Cubes

This section describes the process, illustrated in Fig. 1, to extract LUCC information from Earth Observation Data Cubes using image time series analysis and the utility of SOM method in this process.

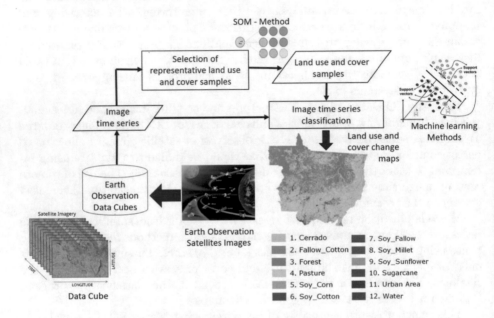

Fig. 1. LUCC information from EO Data Cubes

To perform LUCC classification using Earth observation image time series, machine learning methods such as Support Vector Machine (SVM) and Random Forest (RF) have been used quite frequently [12]. Most of these methods are based on supervised learning methods which require a training phase using land use and cover samples labeled *apriori*.

The selection of representative samples is crucial to obtain good classification accuracy. The exploratory analysis using time series clustering techniques, such as SOM method, assist users to improve the quality of land cover change samples.

2.1   Earth Observation Satellite Image Time Series

Remote sensing satellite revisit the same place on Earth during their life cycle. The measure of same place can be obtained in different times. These measures are mapped to three-dimensional array in space-time [10], as shown in Fig. 2(a). The time series is made from values obtained of each pixel location $I(x, y)$ over time, as presented in Fig. 2(b). From these time series, LUCC can be extracted through vegetation phenology. Figure 2(b) shows an example of an area that was covered by forest during 2000 to 2001 the area, then it was deforested and during three years it was maintained as pasture. From 2006 to 2008 it was used for crop production.

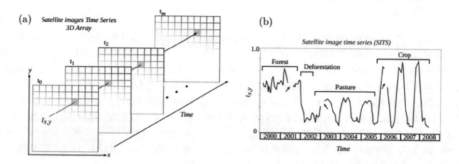

Fig. 2. (a) A dimensional array of satellite images, (b) vegetation index time series at pixel location (x, y). Source: [10]

2.2 Vegetation Indexes

Vegetation phenology is a biological event that indicate the stages of growth and development of plants during the life cycles. Remote sensing satellites are becoming essential for remotely capturing phenological variations in large scale and extract phenological metrics from time series data of vegetation parameters. The most commonly parameters used are the vegetation indexes.

Vegetation Indexes (VI) derived from spectral information of Earth observation satellite images are widely used to generate LUCC information. Vegetation indices are spectral transformations of two or more bands designed to enhance vegetation properties. Two examples of most used vegetation indices are NDVI (Normalized Difference Vegetation Index) and EVI (Enhanced Vegetation Index). Some spatial agencies provide these vegetation indexes as products derived from their satellite images. An example is the product MOD13Q1 of MODIS (Moderate Resolution Imaging Spectroradiometer) sensor provided by NASA with temporal resolution of 16 days and spatial resolution of 250 m [11].

During plant growth periods, different vegetation styles can be distinguished by time-series vegetation indexes [13]. Along with the VI, MODIS provides surface reflectance bands as RED, BLUE Near Infrared (NIR) and Mid Infrared (MIR). The VI are derived from these reflectance bands.

Limitations of the NDVI include sensitivity to atmospheric conditions, soil background and saturation tendency in closed vegetation canopies with large leaf area index values [13]. The EVI signal has improved sensitivity in high biomass regions and improved vegetation monitoring. The blue band is used to remove residual atmosphere contamination caused by smoke and sub-pixel thin cloud [14]. While NDVI is chlorophyll sensitive, EVI is more responsive to canopy structural variations, including leaf area index (LAI), canopy type, plant physiognomy, and canopy architecture [11]. The two vegetation indices complement each other in global vegetation studies.

2.3   Using SOM to Improve the Quality of Land Use and Cover Samples

In the process to extract LUCC information from EO Data Cubes, SOM is used to improve the training step of the land cover change classification. It is used to assess the quality of the land use and cover samples and to evaluate which spectral bands and vegetation indexes are best suitable for the separability of land use and cover classes. This approach explores two main feature of SOM: (1) the topological preservation of neighborhood, which generates spatial clusters of similar patterns in the output space; and (2) the property of adaptation, where the winner neuron and its neighbors are updated to make the weight vectors more similar to the input.

SOM can deal with the variability of vegetation phenology better than other methods that do not have these two features. Due to climatic phenomena, the vegetation phenology can suffer variations over time. Phenological patterns can vary spatially across a region and are strongly correlated with climate variations over time [5]. For example, rainy years may have a pattern for pasture different from a non-rainy year. Therefore, it is necessary methods that can take into account these small variations. The SOM method is able to learn with new patterns during the training process. Besides that, the use of multiples attributes such as combined vegetation indices and multiples spectral bands can improve that patterns generated by SOM.

Instead of use one attribute as input for SOM, the super-organized maps were implemented by [15] in order to use several attributes in a separate layer during the network training. Considering a sample x with two attributes, x_{a1} and x_{a2}, for example the vegetation indices NDVI and EVI. For each attribute a output layer consisting of a 2-D grid of neurons is created. The layers A_1 and A_2 are associated with attributes a_1 and a_2 respectively. The weight vectors, ω_{a1} and ω_{a2}, for each attribute are initialized with random values as shown in Fig. 3. To find the Best Matching Unit (BMU) the distances between input vectors and weight vectors are computed separately for each layer and then they are summed in order to define an overall distance of an sample to a neuron. The Eq. 1 shows how to calculate the distance for multiple layers.

$$D_l = \sum_{l=1}^{n_l} D_l(i,j). \tag{1}$$

where l is the layer and n_l is the number of layer.

After the training step of SOM, to evaluate the separability of samples is necessary to label the neurons in order to create clusters. A cluster can be one neuron or a set of neurons that belongs to the same class. In this step, each neuron is labeled using the majority vote technique. Each neuron receives the

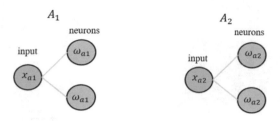

Fig. 3. Structure of SOM with two attributes

label of the majority of the samples associated to it. In some cases, no samples is associated to a neuron. Then, this empty neuron receives the label 'Noclass'. To verify the quality of clusters generated by SOM, the confusion matrix can be accessed. From the confusion matrix, percentage of mixture within a cluster is calculated.

3   Case Study

To show the potential of SOM method in the selection of good quality land use and cover samples from satellite image time series, this section describes a case study using VI time series of the product MOD13Q1 of the MODIS sensor from 2001 to 2016. The area of study is the Mato Grosso State in Brazil, as shown in Fig. 4. Each sample has a spatial location (latitude and longitude), start and end date that corresponds to agricultural year (from August to September), the label of the class that corresponds to the sample, and the set of time series with multiple attributes. In this case study, we used the attributes EVI, NDVI, NIR, MIR, BLUE and RED. The ground samples include natural vegetation and agricultural classes for the Mato Grosso state of Brazil. The data set includes 2215 ground samples divided in nine land use and cover classes: (1) Cerrado, (2) Pasture, (3) Forest, (4) Soy-Corn, (5) Soy-Cotton, (6) Soy-Fallow, (7) Soy-Millet, (8) Fallow-Cotton and (9) Soy-Sunflower. The ground samples were collected by [12].

To evaluate the separability of these classes using SOM, clusters combining spectral bands and vegetation indices were generated in three cases: (1) Case I: NVDI and EVI; (2) Case II: NDVI, EVI, NIR and MIR; (3) Case III: NDVI, EVI, NIR, MIR, RED and BLUE. The SOM parameters that we used were: grid size = 25 × 25, learning rate = 1, and number of iteration = 100.

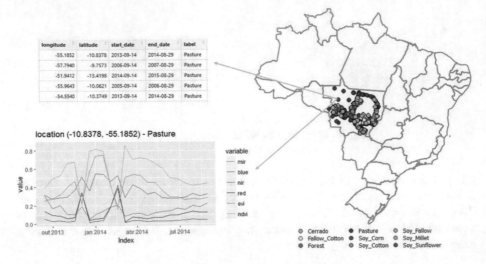

Fig. 4. Samples dataset

Figure 5 shows the maps created for each case. As we have the label of samples, the confusion matrix for each case was generated. Although the large variability within the land use and cover classes and the phenological patterns similarity among the classes, SOM was able to separate these land use and cover classes with good accuracy. For the first case the accuracy was 88%, the second case was 93% and the third was 90%. Besides that, we can note that the most of neurons that belongs to a neighborhood are the same category, but the time series samples contains small variations.

Table 1. Quality of clusters

Cluster	Case I	Case II	Case III
Cerrado	84	97.3	93.3
Fallow-Cotton	72.2	85.7	78.9
Forest	100	99.3	89.9
Pasture	92.7	97.3	93.7
Soy-Corn	82.0	84.0	85.4
Soy-Cotton	94.6	95.5	93.5
Soy-Fallow	97.8	100	98.9
Soy-Millet	85.5	90.3	88.2
Soy-Sunflower	77.1	76.9	72.9

From the confusion matrix, we can evaluate the quality of each land use and cover cluster generated by SOM. For each case, Table 1 shows the percentage of samples that were assigned to the right cluster, that is, the class associated to

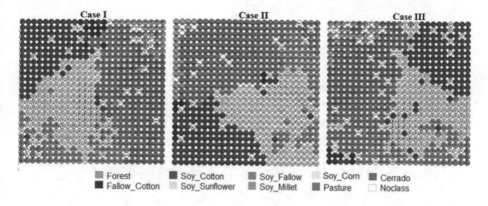

Fig. 5. Grids generated for each case

the sample is the same class associated to the cluster. For example, the Cerrado cluster has 97.3% of samples labelled as Cerrado in the Case II; 84% in the Case I and 93.3% in the Case III.

In general, we can notice that in the Case II, where the attributes MIR and NIR were considered, the quality of clusters were improved. The separability had a significant increase in the Cerrado and Fallow-Cotton clusters. For Forest and Soy-Sunflower clusters there were a loss of quality of separability, but is not so significant. In the same way, in the Case III, the attributes BLUE and RED improved the separability of some clusters when compared with the Case I but not so significant.

For the Case II, the confusion of each cluster is shown in Fig. 6. The clusters Fallow-Cotton, Soy-Corn and Soy-Sunflower are the most confusing, that are crop classes. Crop classes have similar phenological patterns. This confusion can be noted in the maps of Fig. 5 where there are neurons labelled as Fallow-Cotton and Soy-Sunflower within the neighborhood of Soy-Corn. Some samples of Cerrado and Pasture have similar spectral curves but the attributes MIR and NIR reduced the confusion between these samples as shown in Fig. 6.

4   Final Remarks

This paper presents the utility of SOM method to improve LUCC classification from satellite image time series using EO Data Cubes infrastructures. The proposed approach uses SOM to evaluate which spectral bands and vegetation indexes are best suitable for the separability of land use and cover classes and to improve the quality of the land use and cover samples.

We present a case study that evaluates the combination of six attributes, EVI, NDVI, NIR, MIR, RED and BLUE, using MODIS time series of land use and cover samples in the Mato Grosso State in Brazil. The results show the potential of SOM to identify the separability of land use and cover types.

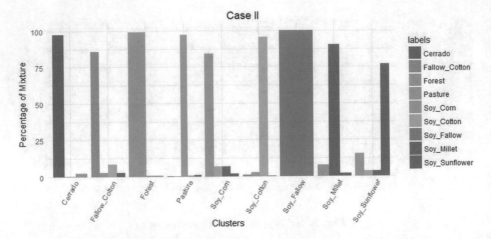

Fig. 6. Confusion among the classes

Despite the general accuracy of separability using only NDVI and EVI attributes was 88%, the classification in big scale areas can generate a great amount of errors. Including the attributes MIR and NIR, we noticed a great improvement in the accuracy of separability that was 93%. Considering that a neuron is a cluster, when the MIR and NIR attributes were added, the percentage of samples that were assigned to right clusters increased. A sample assigned to a right cluster means that the class associated to the sample is the same class associated to the cluster.

In the third case, when we used all attributes including BLUE and RED, the accuracy of separability was worse (90%) than the second case. This analysis is important because we can conclude that adding more attributes does not mean increasing the results. Besides that, the computational cost is proportional to the number of attributes used in the LUCC classification. It is crucial to identify the minimal number of attributes that leads to the best results.

Finally, we have implemented our approach in **R** using the SOM method available in the Kohonen R package [15]. This package has the online and batch approaches of SOM, however we use the online method for this work. The Kohonen package was integrated in the Satellite Image Time Series (sits) package [16]. The package sits package was developed in the e-sensing project developed by the Brazilian Institute for Space Research (INPE) in order to provide a set of tools for working with analyses, clustering and classification of satellite image time series. Its source code is available at https://github.com/e-sensing/sits.

References

1. Nativi S, Mazzetti P, Craglia M (2017) A view-based model of data-cube to support big Earth data systems interoperability. Big Earth Data 1:75–99
2. Lewis A, Oliver S, Lymburner L, Evans B, Wyborn L, Mueller N, Wu W (2017) The Australian geoscience data cube — foundations and lessons learned. Remote Sens Environ 202:276–292
3. Soille P, Burger A, De Marchi D, Kempeneers P, Rodriguez D, Syrris V, Vasilev V (2018) A versatile data-intensive computing platform for information retrieval from big geospatial data. Future Gener Comput Syst 81:30–40
4. FAO: Sepal repository (2018). https://github.com/openforis/sepal. Accessed 14 Dec 2018
5. Suepa T, Qi J, Lawawirojwong S, Messina P (2016) Understanding spatio-temporal variation of vegetation phenology and rainfall seasonality in the monsoon Southeast Asia. Environ Res 147:621–629
6. Pasquarella J, Holden E, Kaufman L, Woodcock E (2016) From imagery to ecology: leveraging time series of all available landsat observations to map and monitor ecosystem state and dynamics. Remote Sens Ecol Conserv 2(3):152–170
7. Gomez C, White C, Wulder A (2016) Optical remotely sensed time series data for land cover classification: a review. J Photogram Remote Sens 116:55–72
8. Bagan H, Wang Q, Watanabe M, Yang Y, Ma J (2005) Land cover classification from Modis EVI time-series data using SOM neural network. Int J Remote Sens 26:4999–5012
9. Siam L (2013) Soft supervised self-organizing mapping (3SOM) for improving land cover classification with MODIS time-series. PhD thesis, Michigan State University, Michigan
10. Maus V, Camara G, Cartaxo R, Sanchez A, Ramos M, Queiroz G (2016) A time-weighted dynamic time warping method for land-use and land-cover mapping. IEEE J Sel Top Appl Earth Observ Remote Sens 9(8):3729–3739
11. Huete A, Didan K, Miura T, Rodriguez E, Gao X, Fereira L (2002) Overview of the radiometric and biophysical performance of the MODIS vegetation indices. Remote Sens Environ 86:195–213
12. Picoli M, Camara G, Sanches I, Simoes R, Carvalho A, Maciel A, Coutinho A, Esquerdo J, Antunes J, Begotti R, Arvor D, Almeida C (2018) Big Earth observation time series analysis for monitoring Brazilian agriculture. ISPRS J Photogram Remote Sens 145:328–339
13. Boles H, Xiao X, Liu J, Zhang Q, Munktuya S, Chen S, Ojima D (2004) Land cover characterization of temperate East Asia using multi-temporal vegetation sensor data. Remote Sens Environ 90(4):477–489
14. Udelhoven T, Stellmes M, Rodes A (2015) Assessing rainfall-EVI relationships in the Okavango catchment employing MODIS time series data and distributed lag models. In: Revealing land surface dynamics. Remote sensing time series. Springer, Cham, pp 225–245
15. Wehrens R, Buydens L (2007) Self and super-organizing maps in R: the Kohonen package. J Stat Softw 21:1–19
16. Camara G, Simoes R, Andrade P, Maus V, Sanchez A, Assis L, Santos L, Ywata A, Maciel A, Vinhas L, Ferreira K, Queiroz G (2018) Sits e-sensing/sits: Version 1.12.5, December 2018. https://doi.org/10.5281/zenodo.1974065

Competencies in Higher Education: A Feature Analysis with Self-Organizing Maps

Alberto Nogales[1(✉)], Álvaro José García-Tejedor[1],
Noemy Martín Sanz[2], and Teresa de Dios Alija[2]

[1] CEIEC Research Institute, Universidad Francisco de Vitoria, Ctra. M-515
Pozuelo-Majadahonda km. 1,800, 28223 Pozuelo de Alarcón, Spain
{alberto.nogales,a.gtejedor}@ceiec.es
[2] Psychology School, Universidad Francisco de Vitoria, Ctra. M-515
Pozuelo-Majadahonda km. 1,800, 28223 Pozuelo de Alarcón, Spain
{n.martin.prof,t.dedios.prof}@ufv.es

Abstract. Students are supposed to accomplish with a set of generic competencies when they finish their studies. One of the major challenges in Universities is to detect shortcomings in students in order to strengthen them, so they could accomplish with the competencies required for a professional career. In this paper, unsupervised machine learning techniques as Self-Organizing Maps are used to analyze features of students from the bachelor's degree in Psychology. The approach is clusterization students' profiles in their first course of college to identify potential improvement areas. The dataset contains 16 features from 54 individuals. Results show that clusters differentiate mostly on the organizational and social competencies on one side, and neuroticism and agreeableness on the other.

Keywords: Self-Organizing Maps · Generic competencies · Higher education

1 Introduction

European Higher Education Area (EHEA) was launched in 2010 to train better professionals within the European Union. Its aim was to develop a flexible system by recognizing higher education obtained in different member countries. Because of EHEA, organizations started to demand Universities an evaluation model based on competencies common to the different educational systems. Competencies are defined nowadays as sets of behaviors that are instrumental in the delivery of desired results or outcomes [1], are related to work performance and distinguish excellent from average performers [2]. In the frame of EHEA, a project called Tuning Education Structures was developed in order to define generic competences that students must accomplish as a complement to their technical skills [3].

To objectively evaluate the degree of accomplishment with a set of given competencies, it is necessary to analyze big datasets with characteristics that normally can not be seen at a glance. It is at this point where Artificial Neural Networks (ANN), a subset of Machine Learning techniques, come into play. These techniques are defined in [4] as "the

© Springer Nature Switzerland AG 2020
A. Vellido et al. (Eds.): WSOM 2019, AISC 976, pp. 80–89, 2020.
https://doi.org/10.1007/978-3-030-19642-4_8

capability of the computer program to acquire or develop new knowledge or skills from existing or non-existing examples for the sake of optimizing performance criterion".

ANN extract relationships (both linear and non-linear) from datasets presented during training process and are able to deal with complex changing information due to generalization and self-organization capabilities. Considering their learning paradigms, they can be divided into two broad categories: supervised and unsupervised. The second ones work with datasets that have not been previously labelled. Self-Organizing Maps (SOM), a kind of ANN with unsupervised training algorithm, reveal as an efficient and widely used categorization method due to its capability to build non-linear mappings between high-dimensional characteristics space and low-dimensional class spaces without external guidelines. SOM were first introduced by Kohonen in 1982, [5]. The model is based on the idea that input data are mapped into a 2D layer of Kohonen neurons taking into account data features. Weight vectors distribution tends to approximate the density function of the input vectors, generating clusters of similar individuals.

This paper presents an unsupervised neural system based on SOM that analyses interrelations of emotional intelligence, sense of life and employability features. The system is applied to a dataset of psychological profiles from students in their first year of the bachelor's degree in Psychology. Clusters of similar students will be later used to properly design academic itineraries so they can acquire their graduate generic competencies when finishing their studies at college.

The rest of the paper is divided as follows: Sect. 2 reviews the state of the art taking into account SOM applied to students and competences measurement, Sect. 3 describes data and methods applied during the experiment, Sect. 4 explains the results obtained and finally, Sect. 5 discusses the conclusions on these experiments and proposes future works.

2   State of the Art

As it has been said before, this paper consists of applying Machine Learning techniques and in particular Kohonen Maps to a dataset of student features in order to improve their skills based on accomplishing generic competencies. In the next paragraphs a review of some papers related with the different fields will be presented.

Kohonen Maps have been applied in several fields. In [6], SOM have been used to detect level of corruption by analyzing a set of macro-economic variables. It has also been used in the field of Natural Language Processing (NLP), by clustering Chinese morphological families in [7]. Another application has been the analysis of demographic changes as can be seen in [8]. In [9], Kohonen Maps are applied to education by analyzing the behavior of students and their academic results. The dataset contains logs from an e-learning platform. In [10], researchers cluster students by cognitive structural models by answering a conceptual understanding test. SOM are also used in [11] where an adaptive and intelligent tutoring system is developed. The tool proposes learning materials depending on the student. An analysis of the students' behavior is made in [12]. In the paper SOM are obtained based on the interaction of students in Moodle. Finally, in [13], a visualization tool using SOM is developed for inducing awareness to students based on self and peer-assessment data.

Related with the use of competencies, [14] makes a study of three models of competencies, the benefits of their implementations and how they are applied in New Zealand. In [15], it is made a quantitative and exploratory study of Information and Communications Technology (ICT) based on students of two Universities in Slovakia and Serbia. Another paper working with competencies is [16], where a set of students and graduates in different business programs ranked 24 competencies extracted from the literature. Finally, [17] makes a systematic review in order to develop a framework based on teams and team competencies in information systems education.

This paper differs from the others as it makes a SOM analysis based on a set of psychological tests that includes features as sense of life or emotional intelligence. The final aim of the work is to decide which reinforcement will be needed by each student to accomplish the competencies of the Psychology bachelor's degree.

3 Materials and Methods

This section presents an in-depth description of the materials and methods. The first subsection explains how the data has been obtained and the meaning of the different features. Also, information about the individuals of the dataset is provided. The second subsection describes all the methods used to obtain the results. Information about the used libraries can also be found.

3.1 Training Dataset

The experiment described in this paper uses information compiled in a set of questionnaires delivered through Internet using a safe connection. Questionnaires were filled by students in the first year of a bachelor's degree in Psychology in a University room, controlled by members of the research group. Participation was voluntary and none of them received financial compensation. They signed their informed consent, where they were notified about the aim of the study, their willfulness and data confidentiality.

The dataset used for the study includes 54 students and 16 integer variables described below:

- 8 variables correspond to the perception of performance in the competencies of the Bartram model. This model is used to validate several potential predictors of workplace performance [1]. The variables used are: Leadership, Cooperation, Communication, Information_Analysis, Learning, Planning, Adaptation and Achievement. These variables have been evaluated with the PRISM 4D test, [18], their values ranging from 1 to 4.
- 1 variable, Sense_of_Life, evaluated with the PIL test [19, 20]. This variable may range from 1 to 7
- 5 variables from Big Five personality traits or OCEAN model. Evaluated features were Openness_to_experience (O), Conscientiousness (C), Extraversion (E), Agreeableness (A) and Neuroticism (N). These variables have been evaluated with a Spanish adaptation of the Five-Factor model presented in [21], with values ranging from 1 to 5.

- 2 variables related with emotional intelligence (WLEIS and MSCEIT), evaluated with the WLEIS test [22] and MSCEIT test [23]. In this case the values range from 1 to 5 in the first case and 1 to 7 in the second.

3.2 Clustering Students and Obtaining Main Features

The experiment aims to obtain clusters of students that share common competencies characteristics based on 16 students' features. These clusters will let researchers understand commonalities among the students and elaborate a possible itinerary to reinforce or modify competencies performance. SOM reveals as a good technique to classify and represent multidimensional datasets into a 2D feature map. Clusters are created and later on analyzed in order to obtain which are the most representative features of each one. A Python library called GEMA has been developed to build and train SOM [24]. GEMA allows building SOM attending to several key parameters that are going to be explained below.

Preprocessing stage is the first step of building a SOM classifier and several decisions have to be made. As the different features are distributed in different ranges, there is a need to standardize the values. In this case, it has been decided to use the Min Max scaler from scikit-learn, a Python library for data mining and data analysis [25]. This standardization transforms the values to a range from 0 to 1 taking into account the maximum and the minimum value that can be found.

SOM hyperparameters configuration is the second step. Initial learning rate and total number of iterations are 0.01 and 1000 respectively. Size of the map can be estimated accordingly to [26]. In this paper, the size of the squared map depends on the number of neurons M that can be calculated with the number of individuals N, as it is noted in Eq. 1

$$M \approx 5\sqrt{N} \qquad (1)$$

SOM quality has been measured following quantization and topography error criteria [27]. Quantization error is defined as the average distance between each data vector and its Best Matching Unit (the neuron with the closest weight vector to a given data vector) and measures map resolution. Topographic error is the proportion of all data vectors for which first and second BMUs are not adjacent units and measures topology preservation.

Combining both quality criteria and Eq. 1 the best map size for our problem can be determined. As the number of students is 54, the number of neurons M is 36 yielding a map side of 6. In Table 1, quantization and topological errors obtained for experiments with increasing SOM sides corroborate this decision.

Table 1. Map side experiments with quantification and topological error.

Side	Quantification error	Topological error
3	0.094	0.129
4	0.089	0.037
5	0.086	0.092
6	0.084	0.018
7	0.080	0.018
8	0.079	0.018

4   Results

Once GEMA has been properly trained, it is used to classify the training dataset itself. A map size of 6 × 6 is obtained, having in total 23 activated neurons with different numbers of students in each one. Each activated neuron represents a set of students as can be seen in the activation map depicted in Fig. 1. This figure shows the number of activations per neuron also coded as a heat-map.

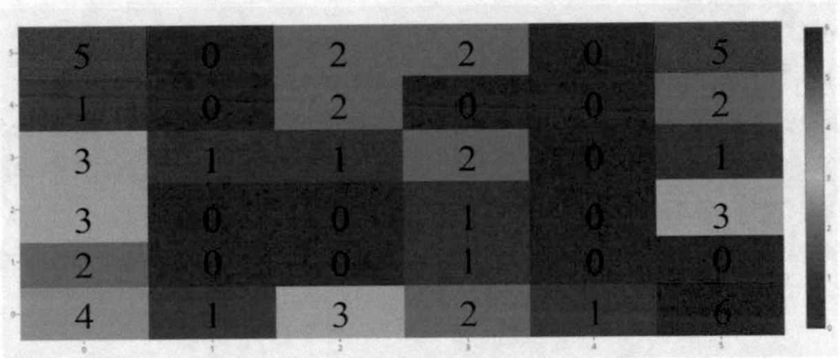

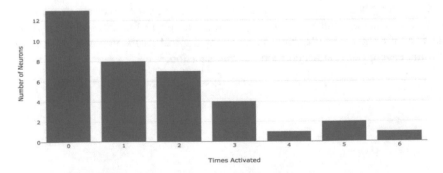

Fig. 1. Activation Map for the best experimental result (Map side of 6).

Figure 2 shows the frequency histogram of how many neurons have recognized zero input, one input and so on, each input being a student profile. In this case, there are more than 12 neurons that have recognized no students. That is, they have not been activated at all.

Fig. 2. Frequency histogram of neurons activation.

Codebook vectors in a SOM are organized accordingly to their topology relationship. Close patterns in our 16-dimension space are recognized by nearby neurons. A codebook vector is a neuron weight vector that represents the consensus for all input data (students' profile) grouped by this neuron. So, if all non-empty codebook vectors are drawn as a line-graph map, the whole students' dataset can be interpreted. This representation is depicted in Fig. 3. The resulting map has been analyzed and interpreted by the same psychology research team that obtained the data.

Neuron (5,5) represents one of the most interesting activated neurons. It can be interpreted as a group of students characterized for having a high competencies performance in analysis and usage of knowledge, planification and organization, and achievement and entrepreneurship. Moreover, they are students having high scores in neuroticism and low scores in agreeableness.

These results are coherent with the theoretical interpretation of the evaluated variables. Neuroticism is a variable of personality that is characterized among other things for a continuous state of concern. This state can make students having more need to plan, and related with that, they have to analyze the information in a most exhaustive way. Due to that, scores in the competencies of analysis and usage of knowledge, and planification and organization are high. Additionally, neuroticism is also usually characterized by the tendency to blame the results, which could lead to having students more focused in achieving the expected results. This will be related with the high scores obtained in the competencies of achievement and entrepreneurship. Otherwise, low scores in agreeableness are also coherent with the results in the competencies' variables. This is related with the high scores that these students obtain in the competencies that are more organizational and less social. Figure 3 shows that the lower scores can be found in the competencies of leadership and decision, cooperation and respect and communication and relationship.

Starting in position (5,5), it can be seen the evolution to position (5,0) in the variables related with the performance of competencies. It goes from having high scores in the organizational competencies and low scores in the more social ones, to having high scores in the social competencies and low scores in the organizational competencies. In particular, the high scores can be found in the leadership and decision competencies, cooperation and respect, communication and relationship and learning and innovation.

Otherwise, the evolution can also be seen in neuroticism. In this case the differences are related with the personality variables. In position (5,5) students show high scores in neuroticism and low scores in agreeableness. While in position (0,5), students are characterized for having low scores in neuroticism and high scores in agreeableness. Again, as it has been explained before, results are coherent because as students have higher scores in agreeableness and lower in neuroticism, they start to score higher in competencies that require interpersonal behaviors.

Related with the necessary training to reach the level of performance required by the graduate profile, it will be necessary to strengthen organizational competencies for cluster 4. That means, it will be necessary to propose activities related with the development of planification and organization competencies, tasks that develop their capacity to analyze and which is the process that they should follow to accomplish with them.

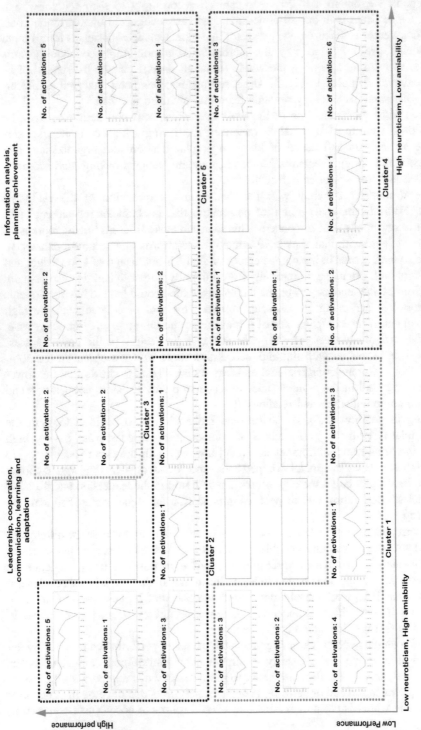

Fig. 3. Topological map with codebook vectors plotted for each activated neuron.

Regarding cluster 5, it is necessary, on the contrary, work in the social competencies. In this case it is necessary to develop their social skills, their communication and leadership capacity. Activities as role-playing, expositions, among others will help.

Clusters 1 and 2, will need a combination of the previous approaches, due to their low score on all competencies. Finally, cluster 3 has not been yet properly defined.

5 Conclusions and Future Works

In this paper, an analysis of a set of psychological features of students has been performed. The training dataset corresponds to students of first year of the bachelor's degree of Psychology. It has been obtained by filling in a set of questionnaires that measures some parameters. The features that have been obtained are related with perception of performances in competencies, sense of life, personality and emotional intelligence. This dataset has been used to train a Kohonen map, after being preprocessed.

Results show that students are mainly organized based on the organizational and social competencies on one side, and neuroticism and agreeableness on the other. Under these two dimensions, five clusters can be defined. This conclusion is an outcome of the self-organization of Kohonen map since clusters have been not explicitly included in the data fed to the network.

Clusters represent students with similar features that requires a different approach to acquire their graduate generic competencies as graduated students. One of them is formed by students who need reinforcement related with planification and organization. Another group needs activities where the students have to expose themselves to others. The other clusters are a mix between the ones shown above, so their reinforcements also need to be a mix of the ones described above.

In future works, new features as responsibility and social compromise of the students will be also used. Then, the analysis will be done in a deeper way taking into account students from the other three years at college. This will let researchers know how students can evolve during the four courses that comprise the bachelor's degree. Another study will use the same variables for the same students since they start in college till they finish, so it can be analyzed the particular evolution of each student. The final goal is to identify relevant features in admission profiles for Psychology students. This will let professors to adopt learning itineraries that will ease the development of the graduate profile promised by the curriculum of the bachelor's degree.

References

1. Bartram D (2005) The great eight competencies: a criterion-centric approach to validation. J Appl Psychol 90(6):1185–1203
2. Sparrow PR (1997) Organisational competencies: creating a strategic behavioural framework for selection and assessment. In: Anderson N, Herriot P (eds) International handbook of selection and assessment. John Wiley, Chichester, pp 343–368

3. Palese A, Zabalegui A, Sigurdardottir AK, Bergin MJ, Dobrowolska BB, Gasser C, Pajnkihar M, Jackson C (2014) Bologna process, more or less: nursing education in the European economic area: a discussion paper. Int J Nurs Educ Sch 11:63–73
4. Alpaydin E (2004) Introduction to machine learning. The MIT Press, Cambridge
5. Kohonen T (1982) Self-organized formation of topologically correct feature maps. Biol Cybern 43:59–69
6. Huysmans J, Martens D, Baesens B, Vanthienen J, Gestel TV (2006) Country corruption analysis with self organizing maps and support vector machines. In: Chen H, Wang F-Y, Yang CC, Zeng DD, Chau M, Chang K (eds) WISI, pp 103–114
7. Galmar B (2011) Using Kohonen maps of Chinese morphological families to visualize the interplay of morphology and semantics in Chinese. ROCLING (Posters). Association for Computational Linguistics and Chinese Language Processing (ACLCLP), Taiwan, pp 240–251
8. Skupin A, Hagelman R (2005) Visualizing demographic trajectories with self-organizing maps. GeoInformatica 9:159–179
9. Bara MW, Ahmad NB, Modu MM, Ali HA (2018) Self-organizing map clustering method for the analysis of e-learning activities. In: 2018 Majan international conference (MIC), pp 1–5
10. Yorek N, Ugulu I, Aydin H (2016) Using self-organizing neural network map combined with ward's clustering algorithm for visualization of students' cognitive structural models about aliveness concept. Comp Int Neurosc 2016:14. Article ID 2476256
11. Zataraín-Cabada R, Barrón-Estrada ML, García CA (2011) EDUCA: a web 2.0 authoring tool for developing adaptive and intelligent tutoring systems using a Kohonen network. Expert Syst Appl 38:9522–9529
12. Alias U, Ahmad N, Hasan S (2015) Student behavior analysis using self-organizing map clustering technique. ARPN J Eng Appl Sci 10:17987–17995
13. Ueki U, Ohnishi K (2016) Visualizing self- and peer-assessment data by a self-organizing map for inducing awareness in learners. Int J Comput Inf Syst Ind Manag Appl (IJCISIM) 8:23–32
14. Markus LH, Cooper-Thomas HD, Allpress KN (2005) Confounded by competencies? An evaluation of the evolution and use of competency models. N Z J Psychol 34(2):117–126
15. Kiss G (2017) Measuring the ICT competencies in Slovakia and in Serbia in the higher education. SHS Web Conf 37:01075. https://doi.org/10.1051/shsconf/20173701075
16. Rainsbury E, Hodges DL, Burchell N (2002) Ranking workplace competencies: student and graduate perceptions. Asia Pac J Coop Educ 3(2):8–18
17. Kathrin F (2010) A systematic review of developing team competencies in information systems education. J Inf Syst Educ 21(3):323–338
18. Aguado D, González A, Antúnez M, De Dios Alija T (2017) Evaluación de competencias transversales en universitarios. Propiedades psicométricas iniciales del cuestionario de competencias transversales. REICE Rev Iberoam Sobre Calid Eficac Cambio Educ 15(2):129–152
19. Noblejas MA (1994) Logoterapia: fundamentos, principios y aplicación. Una experiencia de evaluación del "logro interior de sentido". Disertación doctoral no publicada, Universidad Complutense, Madrid, España
20. Noblejas MA (2000) Fiabilidad de los tests PIL y Logotest. NOUS. Bol Logoterapia Anál Exist 4:81–90
21. Caprara GV, Barbaranelli C, Borgogni L, Perugini M (1993) The "Big Five Questionnaire". A new questionnaire to assess the five-factor model. Pers Individ Differ 15:281–288

22. Wong CS, Law KS (2002) The effects of leader and follower emotional intelligence on performance and attitude: an exploratory study. Leadersh Q 13:243–274. https://doi.org/10.1016/S1048-9843(02)00099-1
23. Extremera N, Fernández-Berrocal P, Salovey P (2006) Spanish version of the Mayer-Salovey-Caruso emotional intelligence test (MSCEIT). Version 2.0: reliabilities, age and gender differences. Psicothema 18(Suplemento):42–48
24. GEMA: a SOM library. https://github.com/ufvceiec/GEMA
25. Scikit-learn: machine learning in Python. https://scikit-learn.org/stable/. Accessed 21 Jan 2019
26. Tian J, Azarian MH, Pecht M (2014) Anomaly detection using self-organizing maps-based k-nearest neighbor algorithm
27. Polzlbauer G (2004) Survey and comparison of quality measures for self-organizing maps. In: Proceedings of the fifth workshop on data analysis (WDA 2004), pp 67–82

Using SOM-Based Visualization to Analyze the Financial Performance of Consumer Discretionary Firms

Zefeng Bai[1], Nitin Jain[1], Ying Wang[1],
and Dominique Haughton[1,2,3($\boxtimes$)]

[1] Bentley University, Waltham, MA, USA
dhaughton@bentley.edu
[2] Université Paris 1 (SAMM), Paris, France
[3] Université Toulouse 1 (TSE-R), Toulouse, France

Abstract. This paper analyzes financial ratios of 27 consumer discretionary firms listed on the S&P 500 over an eleven-year period from 2006–2016. It adopts a two-step approach wherein first a confirmatory factor analysis (CFA) on the financial time-series is conducted and the resulting constructs' scores are then used to perform a cluster analysis using self-organizing maps (SOMs). The consumer discretionary sector is considered an economic and stock market predictor. It consists of non-essential goods and services which in an economic slump are more likely to be foregone. The suggested approach is expected to be a useful reference guide to help understand the past performance of inter- and intra-sector companies. It also enriches the body of literature on the application of machine learning techniques to the analysis of firm- and sectoral-level performance.

Keywords: Consumer discretionary sector · Clustering · Financial ratios · Self-Organizing Maps · Time series

1 Introduction

The advent of machine learning has lent a new dimension to the analysis of financial and accounting ratios. A growing body of research employs machine learning techniques – both supervised and unsupervised – to bring out useful insights from these ratios, beyond the traditional approach to analyzing financial ratios. Our paper contributes to this evolving research by proposing a two-step dynamic process to facilitate understanding firms from the consumer discretionary sector in the US.

We consider the consumer discretionary firms sector due to its inherent nature. It consists of goods and services that are not essential but are desirable if income is enough to purchase them. Lower stock values in this sector, which includes durable goods, apparel, entertainment, and leisure, etc., can be considered as signaling an economic slump. Such stocks tend to outperform other sectors' stocks during strong economic times and underperform them during an economic slump. It is therefore particularly interesting to explore the dynamics of firms from this sector.

© Springer Nature Switzerland AG 2020
A. Vellido et al. (Eds.): WSOM 2019, AISC 976, pp. 90–99, 2020.
https://doi.org/10.1007/978-3-030-19642-4_9

Most of the literature focuses on financial services or the information technology sectors. Hence, this paper also adds value to sectoral analyses. Further, the paper considers a longer time-window of 11 years from 2006–2016. This serves two purposes: it encompasses the sub-prime period and at least one complete economic cycle.

The findings from the analyses bring out interesting perspectives. For example, they show how Amazon differs from its peers in the sector, how Macy's can be differentiated from others on certain financial aspects, etc. Such insights can help investors understand these firms better and make their investment decisions accordingly. The contributions from this paper stem from these findings, the sector of application and the two-step approach that can be easily replicated across other sectors and time periods.

In the next section, we give an overview of the extant literature, which is followed by our methodology. We then analyze the results from CFA and SOM and discuss the findings. Finally, we conclude with a summary of our contributions, limitations and next steps.

2 Literature Review

Multiple streams of literature are referred to in this paper. These include general discussions about partial least squares (PLS) models, time-series of financial ratios, analyses of financial ratios and clustering of time series using SOMs.

The PLS structural equation modeling (SEM) approach is widely used by researchers across diverse applications ranging from consumer behavior [27], HRM [21], market research [10], natural sciences [16] and accounting [19]. We extend its utility to a financial ratio analysis as appropriate for our objectives. The PLS methodology employed here is motivated by [8, 9].

Areas of application of SOM within the area of finance include credit rating [24], credit card fraud detection [29], failure prediction [11], market structure analysis using stock price indices [7], pattern discoveries [6], selection of mutual fund investments, mapping of investment opportunities in emerging markets [4], portfolio selection [25], high frequency data, etc.

[2] applies SOMs to cluster high-frequency financial trade data for a period of 103 trading days. The authors choose the number of nodes for the SOM to be proportional to the square root of the number of trades. A growing body of research applies SOMs to financial statements' analysis. For example, [14] analyzes financial statements using hierarchical SOMs. They argue that concatenating these ratios from different years into a single output vector is challenging. Hence, they employ a two-step approach wherein in the first step, a SOM is trained with the first year's data, and in the second step, a SOM is trained on the firms' coordinates for the two of three next consecutive years on the SOM resulting from the first step.

[17] forecasts credit classes of customers using SOMs on credit agency data from Australia and Germany. [24] proposes an analytical tool to visualize the relationships between 26 financial ratios and credit ratings for the Chinese Taipei market. [25], on the other hand, uses the component panel from a SOM to visualize correlated features in stock prices data and help investors select different portfolios. The authors show that the SOM is able to identify nonlinear relationships. [26] analyzes firms using their

financial ratios by performing a SOM-based clustering analysis. [18] implements a two-level approach consisting of SOM and k-means clustering to group Spanish mutual funds. [23] uses a SOM to visualize transition probabilities of various firms according to different financial ratios. Possible implications include financial performance comparisons and currency crisis predictions. According to [22], topology-preserving data–dimension reduction combinations with predefined, regular grid shapes, such as the self-organizing map, are ideal tools to analyze financial performance visually.

Extant work uses SOMs to cluster firms on financial ratios or derives factors from financial time-series. However, there is a scope for combining these two approaches. Likewise, some articles consider performances over just a few years which may not be very helpful or they chose a handful of ratios only (see [1]). Also, it is a non-trivial task to identify key financial ratios for performance analyses [3, 28]. Research has also been carried out by selecting a handful of financial ratios without a statistical rationale, as done in the analysis by [5].

To bridge these gaps, we propose a two-step approach which contains a CFA and a SOM-based clustering analysis for a period spanning 11 years. The initial set of nineteen financial ratios is reduced to thirteen based on statistical screening. These are then mapped with five-dimensional factors (latent constructs) that capture a firm's financial performance. Further, based on these five financial factors, the SOM clusters firms into different segments, deriving a signature temporal behavior for each cluster for each of the five dimensions.

3 Methodology

The two-step approach adopted in this paper includes a dimension reduction of the indicators, i.e., the financial ratios, and using the dimension scores as input variables for non-linear clustering analyses using SOM.

In the first step, the dimensional reduction is achieved by performing a CFA using the Smart PLS package 2.0. We choose a CFA for multiple reasons. The financial ratios are already known to be aligned to a particular construct based on their definition and construction. Secondly, this procedure reduces the number of financial ratios statistically, while capturing the relevant ones for the analysis. In general, CFA allows us to analytically test a theoretically grounded construct [10]. CFA is pertinent here as we know apriori which financial ratio belongs to which latent construct. For example, current ratio, quick ratio, and cash ratio belong to the liquidity ratio construct. An iterative process based on the multicollinearity, reliability, and variance extracted was undertaken to finalize the reduced set of financial ratios.

We follow the instruction in *PLS-SEM* [8] to construct our latent factors formulated by the financial ratios we selected. For evaluating the CFA results, we consider the average variance extracted (AVE) and the composite reliability as key parameters recommended by [8, 9].

A SOM network, also referred to as Kohonen map, is a type of competitive neural network that projects a high-dimensional input space on prototypes of a low-dimensional regular grid [15, 16]. It accomplishes two goals: to reduce dimensions and to display similarities among prototypes. The algorithm aims at clustering together similar observations while preserving the original topology of the data (i.e., similar observations in the input space are clustered together into the same unit or into neighboring units on the map) [20].

Specifically, in contrast with other artificial neural networks which apply error correction learning, SOM is based on competitive learning, where the output nodes compete among themselves to produce the winning node (or neuron). Only the winning node and its neighborhoods are activated by a particular input observation. This architecture allows SOM to preserve the topological properties of the input space. Therefore, SOM can be effectively utilized to visualize and explore the properties of the data.

A SOM network is a two-layer feedforward and completely connected network as shown in Fig. 1. The data from the input layer are passed along directly to the output layer. The output layer is represented in the form of a grid, usually in one or two dimensions, and typically in the shape of a rectangle or hexagon.

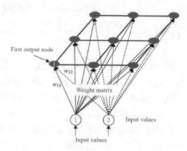

Fig. 1. SOM with 2-D Input, 3 × 3 Output, obtained from [12]

4 Results

The data are sourced from Bloomberg for the 67 consumer discretionary firms listed on the S&P 500 as of 2016. During the data pre-processing stage, it was decided to consider 27 of these 67 firms with 2006–2016 as the time-window and 19 financial ratios since it is prudent to not impute missing values in a financial time-series, which can display unusual jumps. The 11-year period covers at least one full economic cycle and encompasses the sub-prime period, which is an event key to analyzing firms' performance.

However, we do realize that having adequate data, for say, 15+ years (if possible on a quarterly basis) and a wider set of firms would have been very helpful. But, we do think we have covered to a large extent the impact of the sub-prime crisis with our time horizon.

The first-step involves building a confirmatory factor analysis (CFA) model. Nineteen indicators (variables) are used initially and pruned to the thirteen that are listed in Table 1 after removing indicators via an iterative procedure to ensure that multicollinearity, redundancy, reliability, and communality were within acceptable norms. The number of variables associated with each latent construct varies from two to three due to the iterative screening of variables and availability of non-missing data for the time period being analyzed.

Table 1. List of variables for the CFA analysis.

S. no.	Latent construct	Variable name
1	Profitability	Return on assets
2	Profitability	Return on common equity
3	Profitability	Profit margin
4	Liquidity	Current ratio
5	Liquidity	Quick ratio
6	Liquidity	Cash ratio
7	Debt	Total debt to total asset
8	Debt	Long-term debt to common equity
9	Activity	Inventory to sales
10	Activity	Inventory to total assets
11	Activity	Inventory to current assets
12	Market	Basic earnings per share
13	Market	Enterprise value to 12M sales

The final CFA model with outer weights and path coefficients is shown in Fig. 2, and pertinent model statistics are given in Table 2. The model assumes that the three constructs – activity, liquidity, and debt – impact profitability, which in turn affects the market perception of firms. As shown in Fig. 2, we observe a positive association between liquidity and profitability, and we see that profitability decreases as the activity or debt increases for the firms being analyzed. Eventually, profitability contributes positively to the market factor. As mentioned earlier, a bigger sample size, both on the number of firms and number of years, would have been very helpful to generalize such findings.

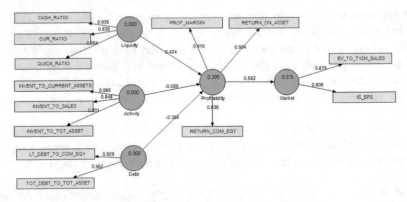

Fig. 2. CFA model output from Smart PLS 2.0

The second step involves constructing a SOM on the 5 indicators extracted from the CFA analysis. In the application of self-organizing maps, we rely on the R based SOMbrero package (see [20]). The map size is decided after testing several sizes (3 by

Table 2. PLS model overview for the finalized CFA model

	Avg. var. extracted	Composite reliability	R square	Cronbach's alpha	Communality	Redundancy
Activity	0.8523	0.9452		0.9298	0.8523	
Debt	0.8852	0.9390		0.8714	0.8852	
Liquidity	0.8334	0.9373		0.8994	0.8334	
Market	0.7102	0.8303	0.3153	0.5960	0.7102	0.2221
Profitability	0.6442	0.8380	0.3954	0.7554	0.6442	0.0081

3, 4 by 4, 5 by 5) of the SOM to check that the cluster structures are shown with sufficient resolution and an acceptable number of empty nodes.

We adopt a 4 by 4 grid in which less than 50% of the nodes are empty. The algorithm used is described in the previous sub-section, with the Euclidean distance and Gaussian neighborhood function. Before implementing, each of the five indicators is centered and rescaled. To visualize the dynamics over the years, we treat each year as one input dimension. Therefore, we have five maps corresponding to the five indicators. For each indicator, we have 11 dimensions representing the data from the year 2006 to 2016. For each dimension, there are 27 observations from the 27 companies.

5 Discussion

For the sake of brevity, we present results related to the debt ratio construct; similar analyses have been conducted for the other constructs. The debt ratio construct provides a quick measure of the amount of debt that the company has on its balance sheets compared to its assets. It shows how much the company relies on debt to finance assets. Usually, the higher the ratio, the greater the risk associated with the firm's operation. A low debt ratio indicates conservative financing with an opportunity to borrow in the future at no significant risk.

From Fig. 3, we see that cluster 1 and cluster 16, at the two opposite corners of the map, possess the largest number of observations; and 6 out of 14 clusters in between are empty. This indicates that the debt ratio in the consumer discretionary sector presents a polarized situation. Both conservative and bold financing strategies are adopted. It is interesting to note that the companies within a cluster present a heterogeneous mix of business content and industry. This means that the temporal dynamics of companies' financial performance seems not determined by traditional classification indicators, such as by industry or line of business, and that these traditional classification indicators are insufficient to underpin an efficient comparison among companies.

In terms of temporal dynamics, it seems that most firms seek stable financial leverage in the long-run (see Fig. 4); the fluctuation in the short run may reflect the inefficiency in the process of debt issuance. However, we identify two clusters (cluster 2 and cluster 13) that act differently. Cluster 2 exhibits a strong upward momentum at the beginning and end stages while being flat in the middle stage. Cluster 13 shows a downward trend during the first half and gradually becomes flat during the second half.

Cluster 4
LKQ Corp
Leggett & Platt Inc
PVH Corp
Cluster 8
Mattel Inc
Cluster 12
BorgWarner Inc
Tiffany & Co
O'Reilly Automotive Inc
VF Corp
Cluster 16
Starbucks Corp
Garmin Ltd
Genuine Parts Co
TJX Cos Inc
Ralph Lauren Corp
Tapestry Inc
Cluster 3
Hasbro Inc
Cluster 7
Cluster 11
Whirlpool Corp
Cluster 15
Advance Auto Parts Inc
Cluster 2
Home Depot Inc
Cluster 6
Cluster 10
Cluster 14
Cluster 1
Newell Brands Inc
Macy's Inc
MGM Resorts International
Royal Caribbean Cruises Ltd
Harley-Davidson Inc
Nordstrom Inc
Cluster 5
Cluster 9
Cluster 13
Booking Holdings Inc
Mohawk Industries Inc
Amazon.com Inc

Fig. 3. Debt ratio SOM clusters

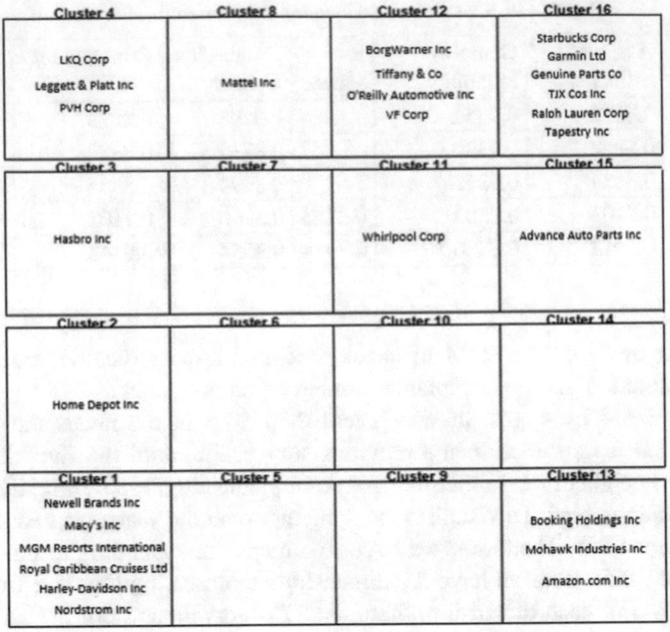

Cluster 4
Cluster 8
Cluster 12
Cluster 16
Cluster 3
Cluster 11
Cluster 15
Cluster 2
Home Depot Inc
Cluster 1
Cluster 13
Mohawk Industries Inc
150
100
50
0
2006 2008 2010 2012 2014 2016

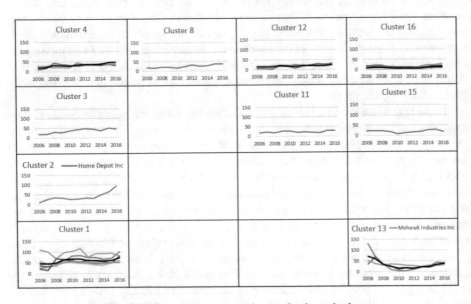

Fig. 4. Debt ratio constructs time series in each cluster

Further research regarding the particular firms in these two clusters indicates that these unique behaviors of financial leverage signal important changes within the firms. For example, the excessive leverage showed in cluster 2 is likely associated with Home Depot's aggressive engagement in buybacks in the stock market; the debt behavior in cluster 13 may be highly associated with Mohawk Industries (MHK)' acquisition in 2005 which was financed with debt.

Under the framework of SOM, the comparison can be done by treating prototypes (Fig. 5) as benchmarks. The prototype is the weight vectors of neurons. It can be understood as the local "average" of the data calculated by SOM. For each prototype line chart in Fig. 5, the x-axis represents the input dimensions (years); the y-axis represents the value of prototypes.

Finally, the distance relationship among prototypes is revealed in Fig. 6. The polygon graph depicts the distances between prototypes with polygons plotted for each neuron. The smaller the distance between a polygon's vertex and a cell border, the closer the pair of prototypes. The color used for filling the polygon shows the number of observations in each neuron. A white polygon means that there is no observation. With the default colors, a red polygon means a high number of observations. If we consider a long distance with neighborhood clusters as a signal of uniqueness, we find that cluster 1 is the most unique cluster.

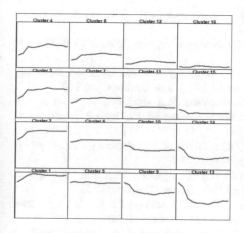

Fig. 5. Debt ratio SOM prototypes

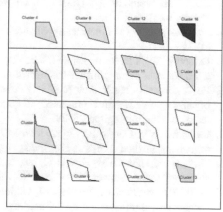

Fig. 6. Debt ratio SOM distance polygons

6 Conclusion

In this paper, we have adopted an innovated machine learning technique, SOMs, to capture and visualize the temporal dynamics of the financial performance of companies in the consumer discretionary sector in the United States. Strong temporal differences among clusters are revealed by SOM. This analysis allows for identifying the signature behavior in clusters of companies, recalling the unique debt behavior (excessive leverage) of Home Depot and the strong profitability secured by TJMaxx during the financial crisis.

The contributions of our paper lie in considering a sector that has not received so much attention and in combining CFA and SOM analyses in an approach which can be replicated to a larger sample and a wide range of situations. However, we do acknowledge the small sample size as a limitation for generalizing the results. The next step would be to follow this approach across multiple sectors, geographies and quarterly financial ratios to obtain a substantial sample and time-window. This is expected to bring out insights on the temporal movement of firms across clusters and thus facilitate a much better and easier understanding of the firms' performance for investors, market analysts while contributing to academic research in this field.

References

1. Back B, Sere K, Vanharanta H (1998) Analyzing financial performance with self-organizing maps. Paper presented at the 1998 IEEE international joint conference on neural networks proceedings. IEEE world congress on computational intelligence (Cat. no. 98CH36227), vol 1, pp 266–270
2. Blazejewski A, Coggins R (2004) Application of self-organizing maps to clustering of high-frequency financial data. Paper presented at the proceedings of the second workshop on Australasian information security, data mining and web intelligence, and software internationalisation, vol 32, pp 85–90
3. Chen KH, Shimerda TA (1981) An empirical analysis of useful financial ratios. Financ Manag 10:51–60
4. Deboeck GJ (1998) Financial applications of self-organizing maps. Neural Netw World 8 (2):213–241
5. Eklund T, Back B, Vanharanta H, Visa A (2002) Assessing the feasibility of self organizing maps for data mining financial information. In: ECIS 2002 proceedings, vol 140
6. Fu T, Chung F, Ng V, Luk R (2001) Pattern discovery from stock time series using self-organizing maps. Paper presented at the workshop notes of KDD 2001 workshop on temporal data mining, pp 26–29
7. Gafiychuk VV, Datsko BY, Izmaylova J (2004) Analysis of data clusters obtained by self-organizing methods. Phys A Stat Mech Appl 341:547–555
8. Hair JF Jr, Hult GTM, Ringle C, Sarstedt M (2016) A primer on partial least squares structural equation modeling (PLS-SEM). Sage Publications, Thousand Oaks
9. Hair JF Jr, Sarstedt M, Ringle CM, Gudergan SP (2017) Advanced issues in partial least squares structural equation modeling. Sage Publications, Thousand Oaks
10. Hair JF, Sarstedt M, Ringle CM, Mena JA (2012) An assessment of the use of partial least squares structural equation modeling in marketing research. J Acad Mark Sci 40(3):414–433
11. Huysmans J, Baesens B, Vanthienen J, Van Gestel T (2006) Failure prediction with self organizing maps. Expert Syst Appl 30(3):479–487
12. Kantardzic M (2011) Data mining: concepts, models, methods, and algorithms. John Wiley & Sons, Hoboken
13. Kiviluoto K, Bergius P (1997) Analyzing financial statements with the self-organizing map. Paper presented at the proceedings of WSOM, vol 97, pp 4–6
14. Kiviluoto K, Bergius P (1998) Two-level self-organizing maps for analysis of financial statements. Paper presented at the 1998 IEEE international joint conference on neural networks proceedings. IEEE world congress on computational intelligence (Cat. no. 98CH36227), vol 1, pp 189–192

15. Kohonen T (1982) Self-organized formation of topologically correct feature maps. Biol Cybern 43(1):59–69
16. Kohonen T (2013) Essentials of the self-organizing map. Neural Netw 37:52–65
17. Merkevičius E, Garšva G, Simutis R (2004) Forecasting of credit classes with the self-organizing maps. Inf Technol Control 33(4)
18. Moreno D, Marco P, Olmeda I (2006) Self-organizing maps could improve the classification of Spanish mutual funds. Eur J Oper Res 174(2):1039–1054
19. Nitzl C (2016) The use of partial least squares structural equation modelling (PLS-SEM) in management accounting research: directions for future theory development. J Account Lit 37:19–35
20. Olteanu M, Villa-Vialaneix N (2015) Using SOMbrero for clustering and visualizing graphs. J Soc Fr Stat 156(3):95–119
21. Ringle CM, Sarstedt M, Mitchell R, Gudergan SP (2018) Partial least squares structural equation modeling in HRM research. Int J Hum Resour Manag 5192:1–27
22. Sarlin P (2015) Data and dimension reduction for visual financial performance analysis. Inf Vis 14(2):148–167
23. Sarlin P, Yao Z, Eklund T (2012) A framework for state transitions on the self-organizing map: some temporal financial applications. Intell Syst Account Finance Manag 19(3):189–203
24. Shih J (2011) Using self-organizing maps for analyzing credit rating and financial ratio data. Paper presented at the 2011 IEEE international summer conference of Asia pacific business innovation and technology management, pp 109–112
25. Silva B, Marques NC (2010) Feature clustering with self-organizing maps and an application to financial time-series for portfolio selection. Paper presented at the IJCCI (ICFC-ICNC), pp 301–309
26. Stankevičius G (2001) Forming of the investment portfolio using the self-organizing maps (SOM). Informatica 12(4):573–584
27. Teo A, Tan GW, Ooi K, Lin B (2015) Why consumers adopt mobile payment? A partial least squares structural equation modelling (PLS-SEM) approach. Int J Mob Commun 13 (5):478–497
28. Wang Y, Lee H (2008) A clustering method to identify representative financial ratios. Inf Sci 178(4):1087–1097
29. Zaslavsky V, Strizhak A (2006) Credit card fraud detection using self-organizing maps. Inf Secur 18:48

Novelty Detection with Self-Organizing Maps for Autonomous Extraction of Salient Tracking Features

Yann Bernard[1,2], Nicolas Hueber[2], and Bernard Girau[1(✉)]

[1] Université de Lorraine, CNRS, LORIA, 54000 Nancy, France
{Yann.Bernard,Bernard.Girau}@loria.fr
[2] French-German Research Institute of Saint Louis, 68300 Saint-Louis, France
Nicolas.Hueber@isl.eu

Abstract. In the image processing field, many tracking algorithms rely on prior knowledge like color, shape or even need a database of the objects to be tracked. This may be a problem for some real world applications that cannot fill those prerequisite. Based on image compression techniques, we propose to use Self-Organizing Maps to robustly detect novelty in the input video stream and to produce a saliency map which will outline unusual objects in the visual environment. This saliency map is then processed by a Dynamic Neural Field to extract a robust and continuous tracking of the position of the object. Our approach is solely based on unsupervised neural networks and does not need any prior knowledge, therefore it has a high adaptability to different inputs and a strong robustness to noisy environments.

Keywords: Self-Organizing Maps · Saliency map · Dynamic Neural Fields

1 Introduction

Visual tracking is currently an important research topic in computer vision. It is a complex problem that requires a strong robustness and adaptation to environmental variability when used in a real world context that current methods do not offer convincingly [6]. The field of computer vision has been historically dominated by models without or with limited learning capabilities, so that the algorithm performances were dependent on prior knowledge of the object to track and a fixed architecture that only took into account a selected number of arbitrarily chosen features. Recent works highlighted the efficiency of deep neural networks to detect and classify objects in a video stream [9] in a supervised way. But this approach relies on a considerable amount of labelled data and considerable computation. Our idea is to use unsupervised neural network properties to efficiently and robustly detect and track objects in a video input stream. Contrary to supervised learning, unsupervised methods need much less

© Springer Nature Switzerland AG 2020
A. Vellido et al. (Eds.): WSOM 2019, AISC 976, pp. 100–109, 2020.
https://doi.org/10.1007/978-3-030-19642-4_10

data and no labels to learn features, which in turn results in much less computation required and opens the way to embedded tracking in video surveillance for instance.

Self-Organizing Maps (SOM) have already been used as a novelty detection tool or rather as a fault or anomaly detection tool as in [15], [7] or [8]. SOMs are well known for their vector quantization and clustering properties, and for preserving neighborhood relations of the input space when projecting data onto the neural map. Novelty detection relies on these properties by detecting elements that are too far from the neural clusters and that do not fit the topology learned. These properties can be interestingly applied to the image processing field, as in [2] or [16]. Our aim is to use these models to perform novelty detection within images without any prior knowledge, so as to be able to extract unexpected targets from image sequences and track them. Current change detection algorithms struggle with problems like a moving camera, intermittent motions and turbulence [14]. With our method, the change detection will be robust to camera movements and turbulences, as it does not rely on precise previous pixel values in the image. It also has the advantage of not relying on local motion information (optical flow) to detect novelty and therefore it is able to track static objects or objects that stopped moving.

Following the seminal work of [3], we choose to couple our autonomous novelty detection tool to a robust bio-inspired tracking technique based on Dynamic Neural Fields (DNF). DNF are populations of partial differential equations first mathematically analyzed by [1] in a continuous framework. We use a discrete DNF built from populations of excitatory and inhibitory neurons that interact continuously, with a on-center off-surround approach modeled as a synaptic kernel computed as a difference of gaussians applied to the distance between neurons in the neural map. These DNF have been successfully applied to sequential visual exploration of an environment [3] or in [13], with great robustness properties that can even improve with some adaptation like the use of simple spiking neurons [12].

The paper is organized as follows. After a short description of the standard SOM model and of the notations used throughout the paper, Sect. 2 explains how SOM can be applied to image compression by means of a quantization of the thumbnails extracted from the image. Section 3 briefly describes the DNF model and its main properties. The proposed coupling between SOM and DNF for tracking novelty in video sequences is detailed in Sect. 4 and preliminary results obtained with real-world images are given in Sect. 5.

2   Image Representation with SOM

2.1   Self-Organizing Maps

In this paper, we use a standard Self-organizing map (SOM) with a 2D grid neural structure as can be found in [5]. Self-organizing maps (SOMs), initially proposed by Kohonen [4], consist of neighbouring neurons commonly organized on one- or two- dimensional arrays that project patterns of arbitrary dimension-ality onto a lower dimensional array of neurons. More precisely, each neuron in a SOM is represented by a d-dimensional weight vector, $\mathbf{m} \in \mathbb{R}^d$, also known as prototype vector, where d is equal to the dimension of the input vectors $\mathbf{x}$. The neurons are connected to adjacent neurons by a neighbourhood relation-ship, which defines the structure of the map. The mechanism for selecting the winning neuron requires a centralized entity, so that the usual Kohonen SOM is not a fully distributed model as in the cortex organization [10]. After learning, or self-organization, two vectors that are close in the input space will be repre-sented by prototypes of the same or of neighbouring neurons on the neural map. Thus the learned prototypes become ordered by the structure of the map, since neighbouring neurons have similar weight vectors.

It starts with an appropriate (usually random) initialization of the weight vectors, $\mathbf{m}_i$. The input vectors are presented to the neural map in multiple iterations. For each iteration, i.e., for each input vector $\mathbf{x}$, the distance from $\mathbf{x}$ to all the weight vectors is calculated using some distance measure. The neuron whose weight vector gives the smallest distance to the input vector $\mathbf{x}$ is usually called the best matching unit (BMU), denoted by c, and determined according to:

$$\|\mathbf{x} - \mathbf{m}_c\| = \min_i \|\mathbf{x} - \mathbf{m}_i\| \tag{1}$$

where $\|\cdot\|$ is the distance measure, typically the Euclidean distance, $\mathbf{x}$ is the input vector and $\mathbf{m}_i$ is the weight vector of neuron i. The winner c and its neighbouring neurons $i \in N_w$ update their weights according to the SOM rule:

$$\mathbf{m}_i(t+1) = \mathbf{m}_i(t) + \alpha(t)h_{ci}(t)[\mathbf{x}(t) - \mathbf{m}_i(t)] \tag{2}$$

where t denotes the time, $\mathbf{x}(t)$ is an input vector randomly drawn from the input data set at time t, $\alpha(t)$ the learning rate at time t, and $h_{ci}(t)$ is the neighbourhood kernel around c. The learning rate $\alpha(t)$ defines the strength of the adaptation, which is application-dependent. Commonly $\alpha(t) < 1$ is a monotonically (e.g. linearly) decreasing scalar function of t.

The neighbouring kernel $h_{ci}(t)$, which is a function of the distance between the winner neuron c and neuron i, can be computed using a Gaussian function:

$$h_{ci}(t) = \exp\left[-\frac{\|\mathbf{r}_c - \mathbf{r}_i\|^2}{2\sigma^2(t)} \right] \tag{3}$$

The term $\|\mathbf{r}_c - \mathbf{r}_i\|$ is the distance between neuron i and the winner neuron c. The precise value of $\sigma(t)$ does not really matter, as long as it is fairly large in the beginning of the process, for instance in the order of 20% of the longer side of the SOM array, after which it is gradually reduced to a small fraction of it, e.g. 5% of the shorter side of the array [5].

We chose to compute the distance between neurons weights and the input vectors as an euclidean distance. We parameterized it with a linearly decreasing σ starting from 0.5 down to 0.001. The learning parameter α starts at 0.6 and linearly decreases to a final value of 0.05. We ran the SOM for 40 epochs for the training and with 10×10 neurons. Increasing the number of epochs or neurons improves the quality of the image representation at the cost of more computation. But as our experimental result could be assimilated to a binary result (the new object is tracked or is not tracked) it is not sensitive to a small performance change. So we chose to limit ourselves to a small but sufficient number of neurons and epochs.

2.2   Image Representation

In order to train the SOM to learn an image, we inspired ourselves from the common application of lossy image compression [2]. A picture or series of pictures to be compressed is split into smaller $k \times k$ pixels wide thumbnails. When the image height or width is not divisible by k, we crop it on the right and bottom. We then use these thumbnails as training samples of a Vector Quantization model. Once the training is finished, the compressed image is composed of the whole codebook, and the index of the Best Matching Unit for each thumbnail extracted from the image or sequence of images. The result is similar to the original image, but with every thumbnail replaced by the codeword learned by its Best Matching Unit. Fig. 1 illustrates this compression process.

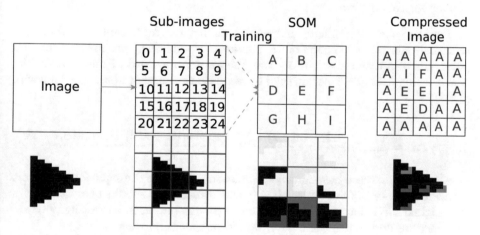

Fig. 1. Simplified scheme of the image compression process (with only 25 sub-images and 9 neurons) with a simple test example underneath.

3   Dynamic Neural Fields

Continuous Neural Fields Theory (CNFT) has lead to the development of two dimensional Dynamic Neural Fields (DNF) [11]. Neural fields are models that represent the evolution of a population of neurons. In our case, we use a two dimensional DNF. The number of neurons is dependent and equal to the size of the input map, because neurons are connected in a retinotopic way to afferent inputs, and are connected in an all-to-all connection scheme between them. All neurons also have a real value attached to them that we call potential. This potential $u(x, t)$, with x being the neuron position in the field and t the time of the simulation, is ruled by the following differential equation:

$$\tau \frac{\partial u(x, t)}{\partial t} = -u(x, t) + \int u(x', t)\omega(\|x - x'\|)\delta y + \text{Input}(x, t)$$

With:

- τ is the time constant.
- $-u(x, t)$ is the decay term. It is meant to suppress already activated neurons when there is no input or lateral excitation.
- $\omega(\|x - x'\|)$ is the lateral interaction. It represents the effect of the other neurons onto this neuron's potential. We are using a difference of gaussian with the excitatory gaussian part being narrow with high intensity and the inhibitory one being wide with low intensity. This leads to close neurons having an excitatory effect onto each other and far away neurons inhibiting themselves.
- $\text{Input}(x, t)$ is the current value of the afferent input extracted from the input map for this neuron.

For the sake of simplicity and computability, we implement a spatially and temporaly discretized version of the previous formula. It is obtained by handling potentials of a discrete set of neurons (neural map instead of neural manifold) and by using a simple Euler method to estimate the state of $u(x, t + \Delta t)$ knowing $u(x, t)$:

$$u(x, t + \Delta t) = u(x, t) + \frac{\Delta t \left(-u(x, t) + \sum u(x', t)\omega(\|x - x'\|) + \text{Input}(x, t + \Delta t)\right)}{\tau}$$

Δt is the time step between two estimations, it can be the same for all neurons (synchronous) or different each time (asynchronous). It should be noted that in the original DNF formula, there are more parameters such as resting potential but since we do not use them here, we did not mention them.

It is often difficult to understand how a DNF will behave just from the formula. We have set it up with optimized parameters in order to have a winner-takes-all behaviour where the most prominent and spatially coherent features in the input map create a local bubble of activation in the neural map and suppress the ability of other such bubbles to appear elsewhere in the map (Fig. 2).

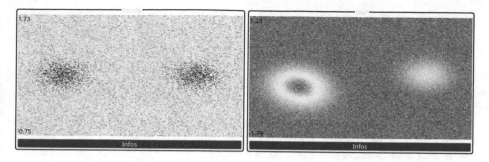

Fig. 2. Example of a DNF. The noisy input with two attractors is on the left, and on the right there is the DNF potentials with a winner takes all behavior.

4   Our Tracking Application

DNF have already been used for tracking applications [3]. It has shown strong resistance to noise and distractions but it needs an a priori knowledge of the features that are to be tracked. SOM on the other hand does not need any a priori knowledge of the input, it can learn and organize itself to represent the input as a concept, meaning that the neurons representing a certain feature will activate when the general pattern of this feature appears. But when there is a completely new input, the distance between it and the BMU will be high. We use this property in order to create a saliency map that robustly outlines novelty in the inputs.

Fig. 3. The learning part of the process. On the left, our "background" image from which we will compare the following received images, composed of white, green and blue stripes. In the center, the codebook learned by the SOM with 16 neurons (4 × 4) and displaying their learned weights or codewords as 10 × 10 pixel thumbnails. On the right, the reconstructed image from the learned SOM weights and the BMU indexes (see Fig. 1 for more details).

The first part of the algorithm relies on the SOM learning the features of the background. This learning will make this SOM able to construct its perception of the main features of the "usual" visual environment. The background can be composed of a single image or a series of similar images in order to have a

Fig. 4. Example of the tracking process. From left to right; the perceived image with a new object (a red star); the reconstructed image from the learned SOM, the star not having been learned, the SOM cannot reconstruct it; the saliency map obtained by making the difference between the first and second images; lastly the DNF output focusing on the new object and eliminating noise.

learning that is more tolerant to small changes in the input. The learning is the slowest part in SOMs, and one advantage of our method is that we only learn once so it is not so penalizing. An example of learning can be found on Fig. 3.

The second part of the algorithm is the tracking. For each new frame captured by our camera, we reconstruct it with the SOM as if we were to compress and decompress it. If no new object has appeared in the image, then the compression will be pretty satisfactory and the result will look similar to the whole image captured. If a new object is present, at the compression will be much less satisfactory in the precise location of this "unexpected" object. We thus compute the salient map as the difference between the current captured image and its reconstruction by the SOM. New objects will stand out on this salient map because of this locally unsatisfactory reconstruction, along with noise on the whole map due to inherently lossy compression (particularly around the edges). Finally to extract and track the "interesting" new object and remove the noise from the saliency map, we use a DNF that will focus on the most prominent and spatially coherent activation on the salient map, and that will be able to track it despite significant variations of saliency between consecutive frames, taking advantage of the natural ability of DNFs to self-maintain bubbles of activity. This is illustrated in Fig. 4.

Our inputs are colored images. Two completely different ways can be considered to handle color by the SOM. The first way is to make each pixel represented by 3 color values (so 10×10 thumbnails will become input vectors of size 300 for instance instead of 100 for grey-scale images). The other way is to use one SOM by color channel. We are going to explore these two possibilities in the following section.

5   Results

In this section we present and discuss results that have been obtained on real camera footage. We have selected a few video clips where the camera is static, that have moving elements in them (like water ripples, wind or snow) and where a new object appears during the clip. This is for now only a proof of concept, so

Background (1st image from sequence)   Camera feed (15th image from sequence)

Fig. 5. We can observe on the saliency map the noise from the lossy image reconstitution by the SOM combined with the new input (the ducks). The DNF manages here to correctly focus on the new target.

the experiments presented here are only showing the potential of this approach. But nonetheless these results are already interesting as they confirm some of our hypotheses and give us hints at what future research on this topic should focus on in order to improve the performance and robustness of this kind of coupled unsupervised novelty detection and neural tracking.

Figures 5 illustrate how the proposed approach performs on some real-world examples. Several observations result from considering the saliency map. The first one is that our assumption that new objects unknown to the SOM appear in the saliency map seems correct. We can also note that the edges of the treeline is badly learned by the SOM because it usually struggles with the sharp edges of the stem of the trees and the chaotic nature of the foliage. We have observed this phenomenon in multiple images where there is no smooth separation of colors. A more expected source of noise is the ripple of the water but we can see in the saliency map that it is nearly invisible. This seems to indicate that there is some sort of generalization of the concept of "water flow" learned in the SOM that makes it robust to small changes there.

The DNF part manages to focus correctly on the object to be tracked when it is there, but in some cases the background noise is too strong and the signal too weak so that the bubble of activity locks itself on badly compressed parts of the image instead of new elements. Let us also note that when there is no new object the DNF focuses on some part of the background. Thus the DNF is not useful to directly detect if something new has appeared, it can only follow

Background   Saliency map   DNF output

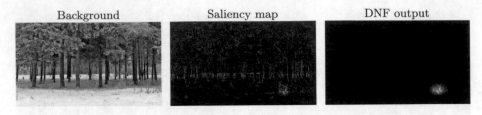

Fig. 6. A second example with a snowy landscape and a dog appearing from the bottom right corner.

Fusioned Colors   Separated Colors

Fig. 7. Learning color channels separately on different SOMs degrades the result with visual artefacts.

the stimulus after it has been detected that something significantly new has appeared. Another example of tracking can be found in Fig. 6.

Another interesting result is that separating colors into channels and learning all of them separately seems to slightly degrade the performance. The visual artefacts observed on Fig. 7 are due to the lack of consistency between the BMUs of different color channels. It strongly suggests that learning colors together should be preferred, even if tracking results are not significantly different in our first experiments. There is also a theoretical argument in favor of not separating colors, considering that diminishing the dimensionality of the codewords also reduces the outlierness of new elements of the image because the distance between them and the closest neuron would be smaller, thus degrading the performance.

6   Conclusion

In this paper, we have presented a new approach for autonomous tracking using Self-Organizing Maps and Dynamic Neural Fields without any pre-requisite information about the target that we want to track. We have shown that novelty detection can be used for tracking, and that some inherent robustness features of SOM and DNF are a good fit for this application. Furthermore, the unsupervised learning base makes us hope that a low computational cost, real time implementation is possible. The current obstacle to a direct application is the unequal compression of the SOM when it comes to edges and chaotic landscapes that deteriorates the quality of the saliency map. For future works, we aim to improve in this area in order to be able to compare our method with current state of the art tracking models.

Acknowledgements. The authors thank the French AID agency (Agence de l'Innovation pour la Défense) for funding the DGA-2018 60 0017 contract.

References

1. Amari S (1977) Dynamics of pattern formation in lateral-inhibition type neural fields. Biol Cybern 27:77
2. Amerijckx C, Legat J-D, Verleysen M (2003) Image compression using self-organizing maps. Syst Anal Model Simulat 43(11):1529–1543
3. Fix J, Rougier NP, Alexandre F (2011) A dynamic neural field approach to the covert and overt deployment of spatial attention. Cognit Comput 3(1):279–293
4. Kohonen T (1990) The self-organizing map. Proc IEEE 78(9):1464–1480
5. Kohonen T (2013) Essentials of the self-organizing map. Neural Netw 37:52–65
6. Kulchandani JS, Dangarwala KJ (2015) Moving object detection: review of recent research trends. In: International conference on pervasive computing (ICPC), Pune, pp 1–5
7. Lee H, Cho S (2005) SOM-based novelty detection using novel data. In: Gallagher M, Hogan JP, Maire F (eds) Intelligent data engineering and automated learning. Lecture notes in computer science, vol 3578. Springer, Heidelberg
8. Lotfi Shahreza M, Moazzami D, Moshiri B, Delavar MR (2011) Anomaly detection using a self-organizing map and particle swarm optimization. Sci Iran 18(6):1460–1468
9. Redmon J, Divvala SK, Girshick RB, Farhadi A (2015) You only look once: unified, real-time object detection. CoRR, abs/1506.02640
10. Rougier NP, Noelle DC, Braver TS, Cohen JD, O'Reilly RC (2005) Prefrontal cortex and flexible cognitive control: rules without symbols. PNAS 102(20):7338–7343
11. Taylor J (1999) Neural bubble dynamics in two dimensions: foundations. Biol Cybern 80:393–409
12. Vazquez R, Girau B, Quinton J-C (2011) Visual attention using spiking neural maps. In: International joint conference on neural networks IJCNN, San José
13. Vitay J, Rougier NP, Alexandre F (2005) A distributed model of spatial visual attention. In: Wermter S, Palm G, Elshaw M (eds) Biomimetic neural learning for intelligent robots. Lecture notes in computer science, vol 3575. Springer, Heidelberg
14. Wang Y, Jodoin P, Porikli F, Konrad J, Benezeth Y, Ishwar P (2014) CDnet 2014: an expanded change detection benchmark dataset In: IEEE conference on computer vision and pattern recognition workshops, Columbus, OH, pp 393–400
15. Wong MLD, Jack LB, Nandi AK (2006) Modified self-organising map for automated novelty detection applied to vibration signal monitoring. Mech Syst Signal Process 20(3):593–610
16. Xiao Y, Leung CS, Lam PM et al (2012) Self-organizing map-based color palette for high-dynamic range texture compression. Neural Comput Appl 21:639

Robust Adaptive SOMs Challenges in a Varied Datasets Analytics

Alaa Ali Hameed[1]($\boxtimes$), Naim Ajlouni[2], and Bekir Karlik[3]

[1] Department of Mathematics and Computer Science,
Istanbul Aydin University, Istanbul, Turkey
aalihameed@aydin.edu.tr
[2] Department of Computer Engineering, Istanbul Aydin University,
Istanbul, Turkey
naimajlouni@aydin.edu.tr
[3] Neurosurgical Simulation Research and Training Centre, McGill University,
Montreal, QC, Canada
bkarlik@hotmail.com

Abstract. The advancement of available technology in use cause the production of huge amounts of data which need to be categorised within an acceptable time for end users and decision makers to be able to make use of the data contents. Present unsupervised algorithms are not capable to process huge amounts of generated data in a short time. This increases the challenges posed by storing, analyzing, recognizing patterns, reducing the dimensionality and processing Data. Self-Organizing Map (SOM) is a specialized clustering technique that has been used in a wide range of applications to solve different problems. Unfortunately, it suffers from slow convergence and high steady-state error. The work presented in this paper is based on the recently proposed modified SOM technique introducing a Robust Adaptive learning approach to the SOM (RA-SOM). RA-SOM helps to overcome many of the current drawbacks of the conventional SOM and is able to efficiently outperform the SOM in obtaining the winner neuron in a lower learning process time. To verify the improved performance of the RA-SOM, it was compared against the performance of other versions of the SOM algorithm, namely GF-SOM, PLSOM, and PLSOM2. The test results proved that the RA-SOM algorithm outperformed the conventional SOM and the other algorithms in terms of the convergence rate, Quantization Error (QE), Topology Error (TE) preserving map using datasets of different sizes. The results also showed that RA-SOM maintained an efficient performance on all the different types of datasets used, while the other algorithms a more inconsistent performance, which means that their performance could be data type-related.

Keywords: Robust · Adaptive-SOM · Clustering · Topology

© Springer Nature Switzerland AG 2020
A. Vellido et al. (Eds.): WSOM 2019, AISC 976, pp. 110–119, 2020.
https://doi.org/10.1007/978-3-030-19642-4_11

1 Introduction

The Self-Organizing Map (SOM) is an unsupervised learning algorithm introduced by Kohonen [1]. In the area of artificial neural networks, the SOM is an excellent data-exploring tool as well [2, 3]. It can project high-dimensional patterns onto a low-dimensional topology map. The SOM map consists of a one or two dimensional (2-D) grid of nodes. These nodes are also called neurons. Each neuron's weight vector has the same dimension as the input vector. The SOM obtains a statistical feature of the input data and is applied to a wide field of data classification [4–6]. SOM is based on competitive learning.

In SOM prior knowledge of the target output is not required for the recognition of process. This algorithm works by finding input features similarities within data objects to define their relation by calculating the distance between them [15]. The nodes output must map to the same weighed vector have been proposed by [7–9]. The winner output is defined as the node with the shortest distance between that node and the input vector. The weighted model continues to be updated to obtain the optimal cluster's topology [10]. The training time depends on dataset-size and the ability to find optimal weights within an acceptable time.

A number of modified SOM versions are developed and proposed for the improvement of vector quantization and the topology preservation performances [11–18]. Brugger et al., and Bogdan et al. proposed a method for detecting clusters by applying the different clustering algorithm to SOM [12, 19].

Berglund and Sitte [20, 21] proposed Parameter-Less SOM (PLSOM) and Parameter-Less SOM2 (PLSOM2) to overcome limitations with Kohonen SOM. PLSOM uses a Quadratic function for error fitting in place of the well-known neighbourhood size and learning parameters, this method suffers from initial weight distribution overreliance and oversensitivity to outliers. The PLSOM2 extended the work of PLSOM by updating the weights by scaling them according to input range observed instead of updating them based on the size of error relative to training maximum error.

In this paper, the performance of the RA-SOM algorithm, which employs a decreasing adaptive learning rate function, is to be tested using a number of different data types. The performance of the RA-SOM will then be compared against well-known algorithms which will be tested using the same datasets. It is expected that the RA-SOM will perform more efficiently than the other algorithms as it will require lower implementation run times to achieve the desired convergence, provide a lower Quantization Error QE, and maintain the topology of the clusters [22]. The test will be carried out on a number of datasets obtained from UCI and KEEL repository.

The remainder of the paper is organized as follows: Sect. 2 reviews the conventional SOM algorithm, Sect. 3 Reviews the RA-SOM algorithm, and Sect. 4 presents the simulation results and a performance comparison between the RA-SOM and other known algorithms including Kohonen SOM, PLSM, PLSOM2, and GF-SOM. The conclusions and future work are presented in Sect. 5.

2 SOM Algorithm

The SOM architecture is composed of input and output layers, connected by link-associated weights. The SOM map uses neuron connections topologies of the hexagonal and rectangular form [15, 16]. SOM output layers contain $n \times m$ neurons arranged as a two-dimensional grid. The original n-dimensional data are transferred to a two dimensional map in SOM. In this case the input vector $x_i = \{x_1, x_2, \ldots, x_n\}$, $i = 1, 2, \ldots, n$, where i is the number of input and n is the input units of the vector. Each i is associated to the map through a weight vector $w = \{w_{n1}, w_{n2}, \ldots, w_{nm}\}$.

SOM adapt a number of processes: First step, the $n \times m$ neuron weight vector is initialized randomly, the second step, an input vector x from the dataset is fed into the SOM network. Input vector x is fed to all neurons, at the same time. Third, the distance between the input and output neurons are calculated, then the closest neuron to the input identified (closest-distance) in this case using Euclidean Distance; this will be called the Best Matching Unit (BMU). The wining neuron is denoted by c.

$$c = arg\min_i(\|w_i(t) - x(t)\|). \qquad (1)$$

This process is iterated for entire input vectors in the dataset. In each iteration, the weight vector is updated by the winning neuron by:

$$w_i(t+1) = w_i(t) + \alpha(t).[x(t) - w_i(t)], \qquad (2)$$

where $\alpha(t)$ is the learning rate. The GF-SOM algorithm utilizes a Gaussian-function which is given by:

$$w_i(t+1) = w_i(t) + h_{c,i}(t).[x(t) - w_i(t)], \qquad (3)$$

where $h_{c,i}$ is the Gaussian neighborhood function given as

$$h_{c,i}(t) = \alpha(t).exp\left(-\frac{\|r_c - r_i\|}{2\sigma^2(t)}\right), \qquad (4)$$

where $\|r_c - r_i\|$ is the Euclidean distance between the positions of the winning neuron c and the neuron i on the grid in each updated weight, and $\sigma(t)$ is the width of Gaussian. $\alpha(t) = \delta_\alpha.\alpha(t)$ and $\sigma(t) = \delta_\sigma.\sigma(t)$ are decreasing gradually during the learning process by constants factors δ_α and δ_σ, respectively.

3 RA-SOM Algorithm

The conventional Kohonen SOM algorithm uses a fixed learning α which is usually between 0–1. The choice of the learning rate affects the speed of the conversion and accuracy of the optimum model. It is known that the higher the learning rate, the faster the convergence. However, this will not guarantee the accuracy of the data topology (clustering), as data accuracy will require a lower leaning rate. Therefore, choosing a high learning rate will provide a high initial convergence, but once this is achieved the algorithm will be forced to diverge to a higher QE due to the inaccuracy in the data topology. On the other hand, choosing a small value for the learning rate will cause a slow divergence which will require many more iterations to achieve the required low QE. This cannot be acceptable in the case of big data.

For this reason, the RA-SOM introduces an adaptive learning rate $\alpha(t)$ [22, 23] which is of a decreasing form, it start by introducing a high learning rate $\alpha(t)$, which is decreased adaptively in subsequent iterations. The adaptively decreasing learning rate achieves high convergence during the first few iterations, this will be followed by a lower learning rate $\alpha(t)$, which will guarantee the continuous high convergence, and will help in defining the accuracy of the data clusters.

The new adaptive learning rate adapted in the RA-SOM is applied in this paper this will be of the form given by

$$w_i(t+1) = w_i(t) + \alpha(t).[x(t) - w_i(t)], \quad t = 0, 1, \ldots \tag{5}$$

where $w_i(t+1)$ is defined as the updating weights, and $\alpha(t)$ is a variable adaptive learning convergence rate is defined as

$$\alpha(t) = \frac{\lambda}{1 - \beta^t} \tag{6}$$

As a result, substitute (6) in (5) deriving a new format of RA-SOM as

$$w_i(t+1) = w_i(t) + \left(\frac{\lambda}{1 - \beta^t}\right).[x(t) - w_i(t)]. \tag{7}$$

In the RA-SOM, the weight vector w is randomly initialized as a grid of $n \times m$ neurons similar to the conventional SOM algorithm. Then, updating the weights is controlled adaptively through the proposed learning algorithm. The optimal weights are obtained in a shorter time compared to the conventional SOM, PLSOM and PLSOM2 algorithms. Moreover, the optimum weight vectors are also improved and provide lower quantization error. The logic behind the adaptive learning function (6) is quite simple: at the start of the function the value of β is large enough while the value of t is small, hence the term $(1 - \beta^t)$ will be relatively small, therefore $\alpha(t)$ will be relatively large, which will result in faster convergence of the updated weights in (7). As time t increases, the term $(1 - \beta^t)$ increases to a value close to unity, and hence $\alpha(t)$ will then be close to or equal to λ, which will result in low error performance in the updated weights of (7).

4 Simulation Results

The proposed algorithm has been tested in four different applications to assess its performance. In this paper, the methods were coded using MATLAB R2010b, and the tests were performed using a Core (TM) i7-3612QM CPU (2.10 GHz) PC equipped with 8,00 GB of RAM with Windows 7 Ultimate operating system. This involved a number of tests which were carried out on different datasets collected from UCI and KEEL repository. Data were divided into 70% training and 30% testing sets. Datasets used in this test were normalized using Min-Max normalization between 0 and 1. A comprehensive comparison between (QE), topology error (TE), and run time was carried out for all the test using all the different datasets. The algorithms tested during the test are conventional Kohonen SOM, GF-SOM, PLSOM, and PLSOM2. The results of these algorithms were then compared against the performance of the RA-SOM under the same tests conditions.

4.1 Balance Dataset

This test was carried out using the Balance dataset. The Balance dataset consists of 625 instances with 4 attributes and 3 classes. The dataset was collected from UCI and KEEL repository. This data set was generated to model psychological experimental results. Each example is classified as having the balance scale tip to the right, tip to the left, or be balanced. The attributes are the left weight, the left distance, the right weight, and the right distance. The Kohonen map, in this case, consists of 4 neurons for the input layer with a 2D grid of 4×3 neurons in the competitive layer. The experimental results are reported in Tables 1, 2, 3, 4 and 5. They show that the RA-SOM outperforms all rest of algorithms, with the PLSOM being the second best.

It must be noted that the conventional SOM had the worst performance in this test. The RA-SOM obtained the lowest QE using $\lambda = 0.5 \times 10^{-2}$ and $\beta = 0.992, 0.991$, however in the following parameters the QE increased, but still was much lower than the QE obtained by other algorithms as shown in Fig. 1. The test also shows that even through RA-SOM outperformed all other algorithms during all subsequent algorithm runs. From the result it can also be seen that the performance of the algorithms is parameter-dependent, this was very clear when considering the performance of SOM, GF-SOM and PLSOM2 algorithms, for example at parameter 1, PLSOM2 with parameters ($\beta = 1.3$, QE $= 0.222$) outperformed both SOM with parameters ($\delta_\alpha = 0.17$, and QE $= 0.24$) and GF-SOM with parameters ($\delta_\alpha = 1$, and $\delta_\sigma = 0.85 \times 10^{-2}, 0.87 \times 10^{-2}$, and QE $= 0.242$), this later changed at parameter 3, PLSOM2 at ($\beta = 1.3$, QE $= 0.243$), SOM at parameters ($\delta_\alpha = 0.15$, and QE $= 0.24$), and GF-SOM at parameters ($\delta_\alpha = 0.8$, and $\delta_\sigma = 0.85 \times 10^{-2}, 0.87 \times 10^{-2}$, and QE $= 0.24$).

4.2 Dermatology Dataset

The test was carried out on the Dermatology dataset. The dataset was collected from UCI and KEEL repository. The differential diagnosis of erythemato-squamous diseases is a real problem in dermatology. The dataset consists of 366 instances, 33 attributes and 6 classes namely (psoriasis, seborrheic dermatitis, lichen planus, pityriasis rosea, chronic dermatitis and pityriasis rubrapilaris). The structure of the Kohonen map used in this case consists of 33 neurons for the input layer and 2D grid size of 33×6 neurons in the competitive layer.

The results of the test are provided in Tables 6, 7, 8, 9 and 10 for the conventional SOM, GF-SOM, PLSOM, PLSOM2, and RA-SOM, respectively. Figure 2 shows the relevant QE against the run time. From this result, it can be concluded that the RA-SOM outperform all algorithms by obtaining lowest initial QE $= 0.174$ with the following parameters ($\lambda = 0.7 \times 10^{-2}$, and $\beta = 0.992$), this was further improved to obtain an optimal QE $= 0.173$ at ($\lambda = 0.5 \times 10^{-2}$, and $\beta = 0.99$); these values remained consistent in all subsequent runs. The result shows that the RA-SOM defined a dataset cluster topology in early run times and managed to maintain this topology throughout.

Table 1. QE results of the conventional SOM algorithm for balance dataset

δ_α	0.17	0.16	0.15
QE	0.24	0.25	0.24

Table 2. QE results of the GF-SOM algorithm for balance dataset

δ_σ	δ_α		
	1	0.9	0.8
0.0087	0.242	0.239	0.24
0.0086	0.242	0.24	0.24
0.0085	0.242	0.238	0.24

Table 3. QE results of the PLSOM algorithm for balance dataset

B	4	3	2
QE	0.22	0.22	0.23

Table 4. QE results of the PLSOM2 algorithm for balance dataset

B	1.5	1.4	1.3
QE	0.222	0.224	0.243

Table 5. QE results of the RA-SOM algorithm for balance dataset

β	λ		
	0.005	0.004	0.003
0.992	0.2	0.21	0.2
0.991	0.2	0.21	0.216
0.99	0.22	0.22	0.2

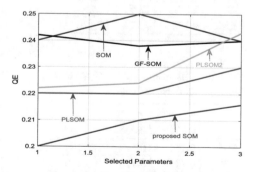

Fig. 1. Comparison of QE measures with various test parameters for balance dataset.

However, the rest of the algorithms did not manage to maintain the topology, as shown in Fig. 2. The PLSOM started at a high QE = 0.188 in the first run at $\beta = 13.7$; this was reduced to QE = 0.177 at the second run at $\beta = 13.6$, and slightly improved by the third run to QE = 0.176 at $\beta = 13.5$. No more improvement was obtained for any further runs and variations of β. The PLSOM2 also started with a QE = 0.182 at $\beta = 7.5$; this was further reduced to QE = 0.175 at $\beta = 8$; however, QE was increased at the third run to QE = 0.179 at $\beta = 8.5$, which indicates that the algorithm had difficulty to maintain the dataset cluster topology. The conventional SOM in this test provided a lower initial QE = 0.181 at $\delta_\alpha = 0.6$, which is better than the performance of both PLSOM and PLSOM2. However, this was not maintained as both of the algorithms performed much better at subsequent runs. The optimal QE for the conventional SOM is QE = 0.18 at $\delta_\alpha = 0.4$, which was the worse between all algorithms. The GF-SOM started at QE = 0.179 at $\delta_\alpha = 0.6$ and $\delta_\sigma = 0.4$. This improved to an optimum of QE = 0.177 at $\delta_\alpha = 0.4$ and $\delta_\sigma = 0.02$. The algorithm best QE was much

Table 6. QE results of the conventional SOM algorithm for dermatology dataset

δ_α	0.6	0.5	0.4
QE	0.181	0.181	0.18

Table 7. QE results of the GF-SOM algorithm for dermatology dataset

δ_σ	δ_α		
	0.6	0.5	0.4
0.04	0.179	0.177	0.177
0.03	0.181	0.181	0.182
0.02	0. 18	0.18	0.177

Table 8. QE results of the PLSOM algorithm for dermatology dataset

B	13.7	13.6	13.5
QE	0.188	0.177	0.176

Table 9. QE results of the PLSOM2 algorithm for dermatology dataset

B	8.5	8	7.5
QE	0.179	0.175	0.182

Table 10. QE results of the RA-SOM algorithm for dermatology dataset

β	λ		
	0.007	0.006	0.005
0.992	0.174	0.179	0.184
0.991	0.176	0.176	0.179
0.99	0.18	0.1732	0.173

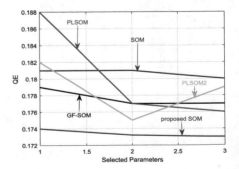

Fig. 2. Comparison of QE measures with various test parameters for dermatology dataset

higher than the global best which was obtained by the RA-SOM which is QE $= 0.173$. However, from Fig. 2, it can be seen that the GF-SOM did not undergo too many topology changes compared to the PLSOM and PLSOM2 algorithms.

4.3   Arcene Dataset

The dataset was collected from UCI and KEEL repository. ARCENE was obtained by merging three mass-spectrometry datasets to obtain enough training and test data for a benchmark. Another dataset we used to examine the efficiency and investigate the performance of the RA-SOM algorithm against other algorithms was the Arcene dataset. It consists of 100 instances, 10000 attributes and 2 classes. The structure of the used Kohonen maps consists of 100 neurons for the input layer with a 2D grid of 100×2 neurons in the competitive layer.

The basic parameters used in this experiment are: for the conventional SOM algorithm: $\delta_\alpha = 0.5 \times 10^{-5}$, for the PLSOM2 algorithm, $\beta = 0.5 \times 10^{-5}$ and for the proposed RA-SOM: $\lambda = 0.5 \times 10^{-10}$, and $\beta = 0.8$. The test results are provided in Table 11. The number of iterations used in this test the same for all three algorithms (SOM, PLSOM2, and RA-SOM). The test results show that RA-SOM outperformed the other two algorithms by obtaining the lowest QE $= 0.067$, with an accuracy of 66.67%. The CPU time shows that the PLSOM2 was the worst, which is expected as the algorithms require many more iterations to complete a cycle compared to both Conventional SOM and RA-SOM, RA-SOM needed extra CPU time as more iterations are needed to calculate the Adaptive learning rate compared to conventional SOM.

Table 11. Performance comparison of the conventional SOM, GF-SOM, PLSOM, PLSOM2 and RA-SOM for Arcene dataset

Appendicitis dataset	Accuracy (%)	# iteration	QE	CPU time
Conventional SOM	60.00	100	0.0645	2.88
PLSOM2	60.00	100	0.0634	3.21
RA-SOM	66.67	100	0.0607	2.90

4.4   Gisette Dataset

The dataset was collected from UCI and KEEL repository. The digits have been size-normalized and centered in a fixed-size image of dimension 28×28. The original data were modified for the purpose of the feature selection challenge. The final dataset used to examine the efficiency and investigate the performance of the proposed SOM algorithm against other algorithms is the Gisette dataset. The Gisette dataset consists of 6000 instances, 5000 attributes and 2 classes. The structure of the used Kohonen maps consists of 100 neurons for the input layer with a 2D grid of 100×2 neurons in the competitive layer.

The basic parameters used in this experiment are: for the conventional SOM algorithm: $\delta_\alpha = 0.5 \times 10^{-7}$, for the PLSOM2 algorithm, $\beta = 0.5 \times 10^{-5}$ and for the proposed SOM: $\lambda = 0.5 \times 10^{-8}$, and $\beta = 0.9$. The test results are provided in Table 12 below. The number of iterations used in this test the same for all three algorithms (SOM, PLSOM2, and RA-SOM). The test results show that RA-SOM has outperformed the other two algorithms by obtaining the lowest QE = 0.0419, with an accuracy of 60.44%. The CPU time shows that the PLSOM2 was again the worst. RA-SOM needed extra CPU time as more iterations are needed to calculate the Adaptive learning rate compared to conventional SOM.

Table 12. Performance comparison of the conventional SOM, GF-SOM, PLSOM, PLSOM2 and RA-SOM for Gisette dataset

Appendicitis dataset	Accuracy (%)	# iteration	QE	CPU time
Conventional SOM	50.83	100	0.0443	5.56
PLSOM2	55.89	100	0.0431	9.29
RA-SOM	60.44	100	0.0419	5.69

5   Conclusion and Future Work

In this work, several alternative algorithms we tested to the proposed RA-SOM under the same conditions. Results showed that the RA-SOM performed more efficiently than the other algorithms in all the datasets tested. It was noticed that the RA-SOM not just outperformed the other algorithms, but it also maintained the dataset variations. The increase or reduction of the number of classes, instances and attributes had no effects on the abilities of the RA-SOM to efficiently converge the QE end the algorithms ability to maintain the dataset topology. It is well known that selecting suitable learning parameters is key to obtain an optimum model with lower clustering topology error. This is one of the main drawbacks in model estimation and bound to be even a bigger issue in big data contexts, as selecting the optimum parameters one needs to run the program many times and each run may be extremely time-consuming. RA-SOM offers more flexibility to obtain the different selection of parameters and thus obtain relevant optimum model quickly and more efficiently.

References

1. Kohonen T (1990) The self-organizing maps. Proc IEEE 78(9):1464–1480
2. Kohonen T, Oja E, Simula O, Visa A, Kangas J (1996) Engineering applications of the self-organizing map. Proc IEEE 84(9):1358–1384
3. Shieh SL, Liao IE (2012) A new approach for data clustering and visualization using self-organizing maps. Expert Syst Appl 39:11924–11933
4. Kohonen T (2001) Self-organizing maps. Springer, Berlin
5. Vesanto J, Alhoniemi E (2002) Clustering of the self-organizing map. IEEE Trans Neural Netw 11(3):586–600
6. Lapidot I, Guterman H, Cohen A (2002) Unsupervised speaker recognition based on competition between self-organizing maps. IEEE Trans Neural Netw 13(4):877–887
7. Kohonen T (1989) Self-organization and associative memory process. Springer, Berlin

8. Hoffmann M (2005) Numerical control of Kohonen neural network for scattered data approximation. Numer Algorithms 39(1):175–186
9. Marini F, Zupan J, Magri AL (2005) Class-modeling using Kohonen artificial neural networks. Anal Chim Acta 544(1):306–314
10. Astel A, Tsakovski S, Barbieri P, Simeonov V (2007) Comparison of self-organizing maps classification approach with cluster and principal components analysis for large environmental data sets. Water Res 41(19):4566–4578
11. Appiah K, Hunter A, Dickinson P, Meng H (2012) Implementation and applications of tri-state self-organizing maps on FPGA. IEEE Trans Circuits Syst Video Technol 22(8):1150–1160
12. Brugger D, Bogdan M, Rosenstiel W (2008) Automatic cluster detection in Kohonen's SOM. IEEE Trans Neural Netw 19(3):442–459
13. Chi SC, Yang CC (2008) A two-stage clustering method combining ant colony SOM and k-means. J Inf Sci Eng 24(5):1445–1460
14. Cottrell M, Gaubert P, Eloy C, Francois D, Hallaux G, Lacaille J, Verleysen M (2009) Fault prediction in aircraft engines using self-organizing maps. In: International workshop on self-organizing maps, vol 5629. Springer, Heidelberg, pp 37–44
15. Tasdemir K, Milenov P, Tapsall B (2011) Topology-based hierarchical clustering of self-organizing maps. IEEE Trans Neural Netw 22(3):474–485
16. Wong MLD, Jack LB, Nandi AK (2006) Modified self-organising map for automated novelty detection applied to vibration signal monitoring. Mech Syst Signal Process 20(3):593–610
17. Yang L, Ouyang Z, Shi Y (2012) A modified clustering method based on self-organizing maps and its applications. Procedia Comput Sci 9:1371–1379
18. Yen GG, Wu Z (2008) Ranked centroid projection: a data visualization approach with self-organizing maps. IEEE Trans Neural Netw 19(2):245–259
19. Bogdan M, Rosenstiel W (2001) Detection of cluster in self-organizing maps for controlling a prostheses using nerve signals. In: Proceedings of 9th European symposium on artificial neural networks. ESANN, pp 131–136
20. Berglund E, Sitte J (2006) The parameterless self-organizing map algorithm. IEEE Trans Neural Netw 17(2):305–316
21. Berglund E (2010) Improved PLSOM algorithm. Appl Intell 32(1):122–130
22. Hameed AA, Karlik B, Salman MS, Eleyan G (2019) Robust adaptive learning approach to self-organizing maps. Knowl-Based Syst 171:25–36
23. Hameed AA, Karlik B, Salman MS (2016) Back-propagation algorithm with variable adaptive momentum. Knowl Based Syst 114:79–87

Detection of Abnormal Flights Using Fickle Instances in SOM Maps

Marie Cottrell[1], Cynthia Faure[1,3], Jérôme Lacaille[2],
and Madalina Olteanu[1,4(✉)]

[1] SAMM, EA 4543, Panthéon-Sorbonne University, 90 rue de Tolbiac,
75013 Paris, France
madalina.olteanu@univ-paris1.fr
[2] Safran Aircraft Engines, Rond Point René Ravaud, Réau,
77550 Moissy Cramayel, France
[3] Aosis Consulting, 20 impasse Camille Langlade, 31100 Toulouse, France
[4] MaIAGE, INRA, Paris-Saclay University, Domaine de Vilvert,
78352 Jouy en Josas, France
http://samm.univ-paris1.fr, https://www.safran-aircraft-engines.com
http://www.aosis.net/, http://maiage.jouy.inra.fr

Abstract. For aircraft engineers, detecting abnormalities in a large dataset of recorded flights and understanding the reasons for these are crucial development and monitoring issues. The main difficulty comes from the fact that flights have unequal lengths, and data is usually high dimensional, with a variety of recorded signals. This question is addressed here by introducing a new methodology, combining time series partitioning, relational clustering and the stochasticity of the online self-organizing maps (SOM) algorithm. Our method allows to compress long and high-frequency bivariate time series corresponding to real flights into a sequence of categorical labels, which are next clustered using relational SOM. Eventually, by training SOM with a large number of initial configurations and by taking advantage of the stability of the clusters, we are able to isolate the most atypical flights, and, thanks to discussions with experts, understand what makes a flight an "abnormal" data.

1 Introduction

This present paper is a part of joint work with the Health Monitoring Department of Safran Aircraft Engines Company. The Pronostic Health Monitoring consists in a set of methods to proactively detect any abnormal behavior with the goal of optimizing and planning the maintenance operations.

In an aircraft, sensors are installed on board to record multivariate time series which describe the behavior of the engines. However analyzing this important amount of data is a difficult task impossible to achieve manually. Even if the experts have a very thorough knowledge about the engine operation data, they need some help from algorithmic methods, as mentioned in [1–3].

© Springer Nature Switzerland AG 2020
A. Vellido et al. (Eds.): WSOM 2019, AISC 976, pp. 120–129, 2020.
https://doi.org/10.1007/978-3-030-19642-4_12

In this paper, the flights are considered as a whole and represented by a sequence of labels. Clustering these sequences leads to highlight some groups of similar flights whilst putting to evidence some unclassifiable flights which are very interesting to study and are good candidates for "abnormality".

The data are initially constituted by 549 flights with 8 different engines, with a mean duration of 2.8 h per flight. The acquisition frequency is 8 Hz. No assumption is done about the observed time series, but one of the component is supposed to be a *key variable*, which strongly influences the behavior of the rest.

For the sake of simplicity, the bivariate case only is presented here, but the method can be easily extended to higher dimensional data.

The paper is organized as follows: Sect. 2 presents the data, Sects. 3 and 4 define the two-levels clustering which leads to represent each flight by a sequence of labels. In Sect. 5, the dissimilarity matrix of all the flights is defined and computed. Section 6 shows how to use the relational SOM to cluster the flights and identify the abnormal ones. Section 7 is a short conclusion.

2   The Data

One flight F is represented by a multivariate time series Z_t, with $1 \leq t \leq T$ and $Z_t \in \mathbb{R}^d$. The components of Z_t are the variables recorded by the on board sensors, for example the fan speed, the temperature inside the motor, the plane speed, the oil temperature, etc.

As we take $d = 2$ in this contribution, we only consider $Z_t = (X_t, Y_t), 1 \leq t \leq T$, where the key variable X_t is the fan speed, and Y_t is the temperature inside the engine.

V flights of different lengths are recorded. For each v, $1 \leq v \leq V$, the flight F_v is thus denoted by $Z_t^v = (X_t^v, Y_t^v), 1 \leq t \leq T_v$.

For each flight v, the time series X_t^v is split into phases which can be increasing transient, decreasing transient or stables, by using a rupture detection algorithm such as the PELT algorithm ([4]). The methodology is described in [5,6]. These phases have different lengths and the number of phases per flight varies.

3   First Level of Labeling

To overcome the difficulty of dealing with phases of different lengths, each increasing or decreasing X-phase is substituted by a fixed-length vector composed of its relevant numerical features, as lengthy, midpoint value, median, variance, variances of the two halves, means of the two halves, ...).

Then any clustering algorithm may be used on these vectors. Here the procedure consists in a SOM map training, combined with a hierarchical agglomerative clustering (HAC) applied to the code-vectors computed by SOM ([7]). We group the increasing phases of all the series X_t^v into clusters, denoted by $CA_1, CA_2, \ldots, CA_I$. The same holds for the decreasing phases grouped into clusters denoted by $CD_1, CD_2, \ldots, CD_J$. The set of the stable phases is denoted by CS. These clusters are called "level-1 clusters".

At this step, each flight is labeled by a sequence of labels, which are elements of the set $\{A_1, A_2, \ldots, A_I, D_1, D_2, \ldots, D_J, S\}$, according to the nature of the successive phases of time series X_t^v: A_i if the phase belongs to cluster CA_i, D_j if the phase belongs to cluster CD_j, S for the stable phases of CS.

4   Two-Levels Clustering and Resulting Labels

To take into account the second variable Y, we define an embedded level 2 clustering: each cluster CA_i or CD_j is split into a partition formed by the clusters $CA_{i,k}, k = 1, \ldots, K(i)$ or $CD_{j,l}, l = 1, \ldots, L(j)$, which are built according to the second variable Y_t^v. In the same way as for the level-1 clustering, each Y-phase is summarized by its numerical features to make possible the use of classical clustering algorithms.

This two-levels allows us to assign a two-indexes label $A_{i,k}, D_{j,l}$ or label S, to any bi-dimensional phase of any flight, so that all of the data is now summarized into V label sequences denoted by F_v - for the sake of simplicity we use the same notation for a flight and for its sequence of labels.

Table 1 presents the computed values of I, J, $K(i), i = 1, \ldots, I$, $L(j), j = 1, \ldots, J$.

We obtain $I = 7$ level-1 clusters of the increasing phases and $J = 8$ level-1 clusters of the decreasing phases.

Table 1. Number of clusters.

I	1	2	3	4	5	6	7
Number of level-2 clusters	4	6	6	6	6	10	4

J	1	2	3	4	5	6	7	8
Number of level-2 clusters	6	8	10	6	10	6	6	8

The number of labels of the 549 labeled sequences resulting from the two-levels clustering, is 22,5 on average, with a minimum of 10 and a maximum of 35. Figure 1 shows the distribution of the 20 most frequent labels (outside the S label).

The more frequent labels are the label $A_{1,1}$ (684 occurrences) which is a taxi-phase label during which the plane rolls on the tarmac, followed by other taxi-phases labels $(A_{1,1}, A_{1,4}, D_{7,3}, A_{3,6}, A_{3,3})$ and some descent phases $(D_{6,2}, D_{2,6})$.

5   Dissimilarity Matrix and Relational SOM

As these labeled sequences are not known by numerical features, we have to use the Relational SOM defined in [8], which is a generalization of the original SOM

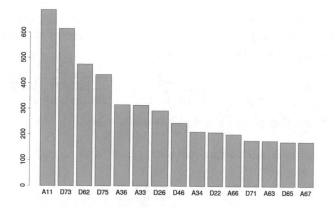

Fig. 1. Distribution of the 20 most frequent labels (outside the S label)

algorithm defined for numerical data. It only requires as input a dissimilarity matrix between the data. Hence, this method may be applied to any complex data (time series, graphs, texts, etc.) as long as a dissimilarity matrix can be computed.

The dissimilarities are defined according to the Optimal Matching method [9], borrowed to biology and to genetic algorithms and which is based on the computation of transition costs from a label to another one.

Several cases have to be distinguished:

- Substitution costs: two labels are exchanged
- Deletion costs: a label is deleted
- Adding costs: a label is added

5.1   Substitutions Costs

Let us define the substitution costs for which we have to consider several cases:

- *Between increasing phases labels substitution costs*

Let A_{ik} and $A_{i'k'}$ be two different labels of i phases.

If $i = i'$, the two level-2 clusters belong to the same level-1 cluster A_i, and we define the cost function c by:

$$c(A_{ik}, A_{ik'}) = \frac{\|\overline{CA_{ik}} - \overline{CA_{ik'}}\|}{\max_{s,s'} \|\overline{CA_{is'}} - \overline{CA_{is'}}\|}$$

where $\overline{CA_{xy}}$ is the bidimensional mean vector of the cluster CA_{xy} and $\| \cdot \|$ is the Euclidean distance in the numerical features space.

If $i \neq i'$, one has to take into account the distance between the level-1 clusters CA_i and CA'_i and also the distance between the level-2 clusters CA_{ik} and $CA_{i'k'}$. So the substitution cost is defined by:

$$c(A_{ik}, A_{i'k'}) = \frac{\|\overline{CA_i} - \overline{CA_{i'}}\|}{\max_{s,s'} \|\overline{CA_s} - \overline{CA_{s'}}\|} + \frac{\|\overline{CA_{ik}} - \overline{CA_{i'k'}}\|}{\max_{s,s'} \|\overline{CA_{is'}} - \overline{CA_{i's'}}\|}$$

– *Between decreasing phases labels substitution costs*

The substitution cost between decreasing phase labels D_{jl} and $D_{j'j'}$ is defined in the same way by:

$$c(D_{jl}, D_{jl'}) = \frac{\|\overline{CD_{jl}} - \overline{CD_{jl'}}\|}{\max_{s,s'} \|\overline{CD_{js'}} - \overline{CD_{js'}}\|}$$

if $j = j'$,

and

$$c(D_{jl}, D_{j'l'}) = \frac{\|\overline{CD_j} - \overline{CD_{j'}}\|}{\max_{s,s'} \|\overline{CD_s} - \overline{CD_{s'}}\|} + \frac{\|\overline{CD_{jl}} - \overline{CD_{j'l'}}\|}{\max_{s,s'} \|\overline{CD_{js'}} - \overline{CD_{j's'}}\|}$$

if $j \neq j'$.

– *Between increasing and decreasing phases labels substitution costs*

According to the definition of increasing and decreasing phases, these substitution costs have to take large values. We take all these costs equal to

$$\alpha \max \left(\max_{i,k,i',k'} c(A_{ik}, A_{i'k'}), \max_{j,l,j',l'} c(D_{jl}, D_{j'l'}) \right)$$

where α is a positive number chosen by the user.

– *Between increasing or decreasing phases labels and S labels substitution costs*

These costs are defined in such a way that they are larger than all the substitution costs between increasing phases or those between decreasing phases, therefore

$$c(A_{ik}, S) = \max_{s,u,s',u'} c(A_{su}, A_{s'u'})$$

and

$$c(D_{jl}, S) = \max_{s,u,s',u'} c(D_{su}, D_{s'u'})$$

5.2   Adding Costs and Deletion Costs

These costs are also defined to be very high equal to

$$\beta \max \left(\max_{i,k,i',k'} c(A_{ik}, A_{i'k'}), \max_{j,l,j',l'} c(D_{jl}, D_{j'l'}) \right)$$

where β is a positive number chosen by the user.

Table 2 shows a part of the substitution costs matrix. We observe that the substitution costs between two labels beginning by A_1 are smaller than those between one label beginning by A_1 and other one beginning by A_2, as desired.

Table 2. Partial representation of the substitution cost matrix.

	$A11$	$A12$	$A13$	$A14$	$A21$	$A22$	$A23$	$A24$	$A25$
$A11$	0,00	0,14	0,62	0,92	1,29	1,95	1,97	1,78	1,20
$A12$	0,14	0,00	0,65	1,00	1,28	1,65	1,59	1,78	1,42
$A13$	0,62	0,65	0,00	1,00	1,77	1,78	1,76	1,01	1,59
$A14$	0,92	1,00	1,00	0,00	1,79	1,00	1,70	1,14	1,59
$A21$	1,29	1,28	1,77	1,79	0,00	0,02	0,18	0,03	0,29
$A22$	1,95	1,65	1,78	1,00	0,02	0,00	0,16	0,05	0,27
$A23$	1,97	1,59	1,76	1,70	0,18	0,16	0,00	0,21	0,24
$A24$	1,78	1,78	1,01	1,14	0,03	0,05	0,21	0,00	0,30
$A25$	1,20	1,42	1,59	1,59	0,29	0,27	0,24	0,30	0,00

Let us denote by Δ the dissimilarity matrix, where $\Delta(v, v')$ is the dissimilarity between the labeled sequences F_v and $F_{v'}$, defined as the minimal value of the sum of the required changes costs to exchange F_v and $F_{v'}$.

The distribution of the dissimilarities is illustrated at Fig. 2.

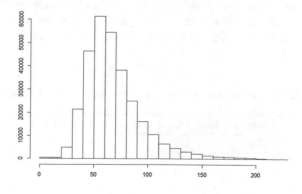

Fig. 2. Dissimilarity distribution

Figure 3 presents the most representative flight, determined as that one which minimizes the sum of all the dissimilarities between it and all the others. It has a "normal" behavior (taxi, take-off, climb, cruise, descent, landing)!

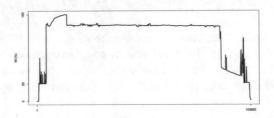

Fig. 3. Variable fan speed of the representative flight

6   Clustering the Labeled Sequences and Identifying Fickle Flights

We use a 10×10 Kohonen map and the relational SOM algorithm trained on the dissimilarity matrix Δ to get a clustering of the flights, that is of the labeled sequences.

If we consider several runs (at least 50) of the SOM algorithm, for a given size of the map and for a given data set, we observe that most of the pairs of flights are almost always or almost never in the same cluster. But there are also pairs of flights whose associations look random. These pairs of flights are called *fickle* pairs. This question was addressed by [10] in a bootstrap framework and used for text mining in [11, 12].

After having identified the fickle pairs, we define the fickle flights as being those which belong to an important number of fickle pairs (greater than a certain threshold).

The most fickle flights are then identified and are good candidates for expertise in order to detect anomalies.

After 100 runs of the relational SOM algorithm, the percentages of attractive, repulsive, fickle pairs are computed (see Table 3). Figure 4 shows the labeled sequences re-ordered according to their fickleness, i.e. the number of fickle pairs they belong to.

Table 3. Computed percentages of attractive, repulsive, fickle pairs

Attractive pairs	Repulsive pairs	Fickle pairs
30.02	58.09	11.80

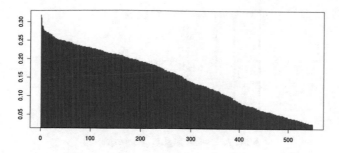

Fig. 4. Fickleness

Then it is possible to seek the most fickle labeled sequences that is the most fickle flights. They are mainly on the edges of the Kohonen maps. The following figures represent 4 fickle flights. The abscissa is the time, the left ordinate is the value of the X variable which is the fan speed, the right ordinate in red is the altitude represented to facilitate the interpretation.

The first fickle flight (Fig. 5) looks like the representative flight of Fig. 3, however the climb phase is very long and there is an unusual variation during the climb.

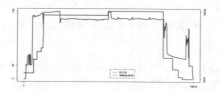

Fig. 5. Fickle flight (Example 1)

Next example (Fig. 6) is a very short flight with an atypical behavior of the fan speed variable.

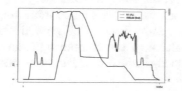

Fig. 6. Fickle flight (Example 2)

In Fig. 7, there is an inconsistency between the fan speed and the altitude: it can be a measurement error of the altitude, that has to be confirmed by the experts.

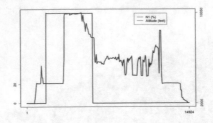

Fig. 7. Fickle flight (Example 3)

For the last example (Fig. 8), the altitude has several levels but seems to be normal, whilst the fan speed is chaotic!

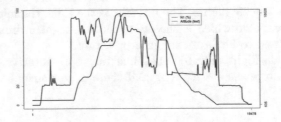

Fig. 8. Fickle flight (Example 4)

All these examples are illustrations of quite atypical flights, which have to be analyzed and characterized by the specialized experts.

7   Conclusion

The transformation of the flights represented by bidimensional time series into sequences of labels makes possible their clustering, in order to identify groups of similar flights, but overall to highlight some atypical flights which are the fickle flights computed after repeated runs of SOM.

This methodology is an interesting tool to mine very complex data and discover abnormal or atypical individuals. The generalization to multidimensional data is straightforward, although the computing time could be increasing with the number of the embedded clustering which are necessary to define the labels.

References

1. Bellas A, Bouveyron C, Cottrell M, Lacaille J (2014) Anomaly detection based on confidence intervals using SOM with an application to health monitoring. In: Villmann T, Schleif F-M, Kaden M, Lange M (eds) Advances in self-organizing maps and learning vector quantization. Proceedings of the 10th international workshop, (WSOM 2014), AISC. Springer, Mittweida, Germany, pp 145–155

2. Rabenoro T, Lacaille J, Cottrell M, Rossi F (2014) Anomaly detection based on indicators aggregation. In: International joint conference on neural networks (IJCNN 2014), pp 2548–2555, Beijing, China, July 2014
3. Lacaille J, Gerez V (2011) Online abnormality diagnosis for real-time implementation on turbofan engines and test cells. In: Annual conference of the prognostics and health management society 2011, vol 2
4. Killick R, Eckley I (2012) Optimal detection of changepoints with a linear computational cost. JASA 107(500):1590–1598
5. Bardet J-M, Faure C, Lacaille J, Olteanu M (2016) Comparison of three algorithms for parametric change-point detection. In: Verleysen M (ed) European symposium on artificial neural networks (ESANN 2016). Computational Intelligence and Machine Learning, Bruges, Belgium, pp 2–7
6. Faure C, Lacaille J, Bardet J-M, Olteanu M (2016) Indexation of bench test and flight data. In: Third Dety 2016. PHM Society
7. Faure C, Bardet J-M, Olteanu M, Lacaille J (2017) Design aircraft engine bivariate data phases using change-point detection method and self-organizing maps. In: Conference: ITISE - international work-conference on time series, Granada, Spain, September 2017. University of Granada
8. Olteanu M, Villa-Vialaneix N (2015) On-line relational and multiple relational SOM. Neurocomputing 147(1):15–30
9. Abbott A, Forrest J (1986) Optimal matching methods for historical sequences, journal of interdisciplinary history. Neural Netw 16(3):471–494
10. de Bodt E, Cottrell M, Verleysen M (2002) Statistical tools to assess the reliability of self-organizing maps. Neural Netw 15(8–9):967–978
11. Bourgeois N, Cottrell M, Deruelle B, Lamassé S, Letrémy P (2015) How to improve robustness in kohonen maps and display additional information in factorial analysis: application to text mining. Neurocomputing 147:120–135
12. Bourgeois N, Cottrell M, Lamasse S, Olteanu M (2015) Search for meaning through the study of co-occurrences in texts. In: Rojas I, Joya G, Catala A (eds) Advances in computational intelligence, IWANN 2015, Part II. Lecture Notes in Computer Science, vol 9095. Springer, Palma de Mallorca, Spain, June 2015, pp 578–591

LVQ-type Classifiers for Condition Monitoring of Induction Motors: A Performance Comparison

Diego P. Sousa[1], Guilherme A. Barreto[1(✉)], Charles C. Cavalcante[1], and Cláudio M. S. Medeiros[2]

[1] Federal University of Ceará - Department of Teleinformatics Engineering, Campus of Pici, Center of Technology, Fortaleza, Ceará, Brazil
diegoperdigao@gmail.com, {gbarreto,charles}@ufc.br
[2] Federal Institute of Ceará, Department of Industry, Laboratory of Energy Processing, Fortaleza, Ceará, Brazil
claudiosa@ifce.edu.br

Abstract. In this paper, we introduce a design methodology for prototype-based classifiers, more specifically the well-known LVQ family, aiming at improving their accuracy in fault detection/classification tasks. A laboratory testbed is constructed to generate the datasets which are comprised of short-circuit faults of different impedance levels, in addition to samples of the normal functioning of the motor. The generated data samples are difficult to classify as normal or faulty ones, especially if the faults are of high impedance (usually misinterpreted as non-faulty samples). Aiming at reducing misclassification, we use K-means and cluster validation techniques for finding an adequate number of labeled prototypes and their correct initialization for the efficient design of LVQ classifiers. By means of comprehensive computer simulations, we compare the performances of several LVQ classifiers in the aforementioned engineering application, showing that the proposed methodology eventually leads to high classification rates.

Keywords: Learning vector quantization ·
Prototype-based classifiers · Fault detection · Induction motors ·
Condition monitoring

1 Introduction

The family of learning vector quantization (LVQ) algorithms is comprised of prototype-based neural network (NN) models which have been used as alternatives to more traditional approaches (e.g. MLP and RBF networks). LVQ classifiers present classification accuracies at least as high as that of any other NN algorithm [1] and are simpler to interpret due to the local nature of the prototypes, which are positioned at representative regions (Voronoi cells) of the data.

© Springer Nature Switzerland AG 2020
A. Vellido et al. (Eds.): WSOM 2019, AISC 976, pp. 130–139, 2020.
https://doi.org/10.1007/978-3-030-19642-4_13

The finest feature of standard LVQ algorithms, that of interpretability due to the local nature of the prototype vectors, is also their main drawback. That is, the performance of an LVQ classifier is highly dependent on the prespecified number of labeled prototypes. Previous works have been developed trying to make the set of prototypes either adaptive [2] or optimally determined by means of evolutionary algorithms [3]. However, in the vast majority of the applications, that number is set by trial and error or exhaustive grid search (see [4,5] for excellent surveys on LVQ-based and other prototype-based classifiers).

From the exposed, in this paper, we aim to introduce a systematic methodology for finding a suitable number of prototypes and their reliable initialization. Roughly speaking, instead of inserting and/or removing prototypes on the fly as did by adaptive LVQ classifiers, we resort to clustering strategies to find the optimum number of prototypes per class. For the sake of simplicity, we use the K-means and well-known cluster validation indices, but any other clustering methodology can be used as well.

For assessing the proposed methodology we evaluate the state of the art in LVQ models on a fault detection/classification dataset obtained from 3-phase AC induction motors. Our target task is the identification of inter-turn short-circuit faults in the stator winding, which we have been investigating lately using standard powerful nonlinear classifiers, such as the MLP and the SVM [6] and SOM techniques [7,8]. For this purpose, we built a lab scale testbed for simulating faults of different impedance levels with different degrees of severity.

The remainder of the paper is divided as follows: in Sect. 2, the basics of cluster validation techniques are presented; in Sect. 3.1, LVQ-based classifiers are described; in Sect. 4, the experimental data acquisition is explained; in Sect. 4 our proposal is introduced and the results are shown and discussed; finally, in Sect. 5, the conclusions are made.

2 Basics of Cluster Validation Techniques

Techniques for cluster validation are used *a posteriori* to evaluate the results of a given clustering algorithm. It should be noted, however, that each cluster validation technique has its own set of assumptions, so that the final results may vary across the chosen techniques.

2.1 Cluster Validity Indices

Some well-known indices available in the clustering literature are described next. We denote K as the number of clusters, K_{max} is the maximum allowed number of clusters, d as the number of features, $\bar{\mathbf{x}}$ as the centroid of the $d \times N$ data matrix $\mathbf{X}$, n_i as the number of objects in cluster C_i, $\mathbf{c}_i$ as the centroid of cluster C_i, and $\mathbf{x}_l^{(i)}$ as the l-th feature vector, $l = 1, \ldots, n_i$, belonging of the cluster C_i.

(*i*) The *Davies-Bouldin* (DB) index [9] is a function of the ratio of the sum of within-cluster scatter to between-cluster separation, and it uses the clusters' centroids for this purpose. Initially, we need to compute the scatter within the i-th cluster and the separation between the i-th and j-th clusters, respectively, as

$$S_i = \left[\frac{1}{n_i} \sum_{l=1}^{n_i} \|\mathbf{x}_l^{(i)} - \mathbf{c}_i\|^2 \right]^{1/2} \quad \text{and} \quad d_{ij} = \|\mathbf{c}_i - \mathbf{c}_j\| \tag{1}$$

where $\| \cdot \|$ is the Euclidean norm. Finally, the DB index is defined as

$$DB(K) = \frac{1}{K} \sum_{i=1}^{K} R_i, \quad \text{where} \quad R_i = \max_{j \neq i} \left\{ \frac{S_i + S_j}{d_{ij}} \right\}. \tag{2}$$

The value of K leading to the smallest $DB(K)$ value is chosen as the optimal number of clusters.

(*ii*) The *Dunn* index [10] is represented generically by the following expression:

$$Dunn(K) = \frac{\min_{i \neq j} \{\delta(C_i, C_j)\}}{\max_{1 \leq l \leq k} \{\Delta(C_l)\}}, \tag{3}$$

where

$$\delta(C_i, C_j) = \min_{\mathbf{x} \in C_i, \mathbf{y} \in C_j} \{d(\mathbf{x}, \mathbf{y})\}, \quad \text{and} \quad \Delta(C_i) = \max_{\mathbf{x}, \mathbf{y} \in C_i} \{d(\mathbf{x}, \mathbf{y})\}, \tag{4}$$

with $d(\cdot, \cdot)$ denoting a dissimilarity function (e.g. Euclidean distance) between vectors. Note that, while $\delta(C_i, C_j)$ is a measure of separation between clusters C_i and C_j, $\Delta(C_i)$ is a measure of the dispersion of data within the cluster C_i. The value of K resulting in the largest $Dunn(K)$ value is chosen as the optimal number of clusters.

(*iii*) The *Calinski-Harabasz* (CH) index [11] is a function defined as

$$CH(K) = \frac{trace(\mathbf{B}_K)/(K-1)}{trace(\mathbf{W}_K)/(N-K)} \tag{5}$$

where $\mathbf{B}_K = \sum_{i=1}^{K} n_i (\mathbf{c}_i - \bar{\mathbf{x}})(\mathbf{c}_i - \bar{\mathbf{x}})^T$ is the between-group scatter matrix for data partitioned into K clusters, $\mathbf{W}_K = \sum_{i=1}^{K} \sum_{l=1}^{n_i} (\mathbf{x}_l^{(i)} - \mathbf{c}_i)(\mathbf{x}_l^{(i)} - \mathbf{c}_i)^T$ is the within-group scatter matrix for data clustered into K clusters. The $trace(\cdot)$ operator computes the sum of the elements on the main diagonal of a square matrix. The value of K resulting in the largest $CH(K)$ value is chosen as the optimal number of clusters.

(*iv*) The *Silhouette* (Sil) index [12] is defined as

$$Sil(K) = \sum_{i=1}^{N} S(i)/N, \quad S(i) = [b(i) - a(i)]/\max\{a(i), b(i)\}, \tag{6}$$

with $a(i)$ representing the average dissimilarity of the i-th feature vector to all other vectors within the same cluster (except i itself), and $b(i)$ denoting the lowest average dissimilarity of the i-th feature vector to any other cluster of which it is not a member. The silhouette can be calculated with any dissimilarity metric, such as the Euclidean or Manhattan distances. The value of K producing the largest $Sil(K)$ value is chosen as the optimal number of clusters.

3   Prototype-Based Classifiers

Let us consider a set of training input-output patterns $\{(\mathbf{x}_l, y_l)\}_{l=1}^N$, where $\mathbf{x}_l \in \mathbb{R}^p$ denotes the l-th input pattern and $y_l \in \mathcal{C}$ denotes its corresponding class label. Note that y_l is a discrete variable (of either numerical or nominal nature) which may assume only one out of K values in the finite set $\mathcal{C} = \{c_1, c_2, \ldots, c_K\}$.

Given a set of labeled prototype vectors $\mathbf{m}_i \in \mathbb{R}^p$, $i = 1, \ldots, M$, for all the prototype-based classifiers to be described in this section, class assignment for a new input pattern $\mathbf{x}(t)$ is based on the following decision criterion:

$$\text{Class of } \mathbf{x}(t) = \text{Class of } \mathbf{m}_c(t), \quad \text{where} \quad c = \arg \min_{i=1,\ldots,M} d(\mathbf{x}(t), \mathbf{m}_i(t)), \quad (7)$$

in which $d(\cdot, \cdot)$ denotes a dissimilarity measure specific to the extension of LVQ and c is the index of the nearest prototype among the M ones available. In the following paragraphs, we briefly described the learning rules for finding the positions of the prototypes $\mathbf{m}_i$, $i = 1, \ldots, M$ in the data space.

Minimum Distance-to-Centroid (MDC) Classifier [13]: For this classifier, we have $M = K$, i.e. the number of prototypes (M) is equal to the number of classes (K). In this case, the prototype of the i-th class is computed as the centroid of class i as $\mathbf{m}_i = \frac{1}{n_i} \sum_{\mathbf{x} \in c_i} \mathbf{x}$, $i = 1, \ldots, K$, where n_i is the number of training examples of class i.

3.1   LVQ Classifiers

For the whole family of LVQ classifiers, we have $M > K$, i.e. the number of prototypes (M) is higher than the number of classes (K). As a consequence, different prototypes may share the same label.

LVQ1 [14]: Let c be defined as in Eq. (7) for a new input pattern $\mathbf{x}(t)$. Then, the prototype $\mathbf{m}_c$ is updated as follows

$$\mathbf{m}_c(t+1) = \mathbf{m}_c(t) + s(t)\alpha(t)[\mathbf{x}(t) - \mathbf{m}_c(t)], \quad (8)$$

where $s(t) = +1$ if the classification is correct, and $s(t) = -1$ if the classification is wrong, and α is the learning rate.

LVQ2.1 [14]: In this algorithm, two prototypes $\mathbf{m}_i$ and $\mathbf{m}_j$ that are the nearest neighbors to $\mathbf{x}(t)$ are now updated simultaneously. One of them ($\mathbf{m}_i$, for example) must belong to the correct class and the other to the wrong class, respectively. Thus, the learning rules of the LVQ2.1 algorithm are given by

$$\mathbf{m}_i(t+1) = \mathbf{m}_i(t) + \alpha(t)[\mathbf{x}(t) - \mathbf{m}_i(t)], \quad (9)$$
$$\mathbf{m}_j(t+1) = \mathbf{m}_j(t) - \alpha(t)[\mathbf{x}(t) - \mathbf{m}_j(t)], \quad (10)$$

where $\mathbf{x}(t)$ must satisfy the following condition:

$$\min\left(\frac{d_i}{d_j}, \frac{d_j}{d_i}\right) > s, \quad \text{where } s = \frac{1-w}{1+w}, \quad (11)$$

where d_i and d_j are the Euclidean distances of $\mathbf{x}(t)$ from $\mathbf{m}_i$ and $\mathbf{m}_j$, respectively. A relatively 'window' w width from 0.2 to 0.3 is recommended.

LVQ3 [14]: For scenarios in which $\mathbf{x}(t)$, the nearest prototypes $\mathbf{m}_i$ and $\mathbf{m}_j$ belong to the same class, the following updating rule is applicable:

$$\mathbf{m}_k(t+1) = \mathbf{m}_k(t) + \epsilon s(t)\alpha(t)[\mathbf{x}(t) - \mathbf{m}_k(t)], \tag{12}$$

for $k \in \{i, j\}$, with $\mathbf{x}$ falling into the 'window'. In a series of experiments carried out in [14], feasible values of ϵ ranging from 0.1 to 0.5 were found, relating to $w = 0.2$ or 0.3. The optimal values for ϵ seems to depend on the size of the window, being smaller for narrower windows. An important feature of the LVQ3 algorithm is that it is self-stabilizing, in the sense that the optimal placements of the prototypes do not change in continued learning.

GLVQ [15]: The cost function is defined as follows:

$$E_{GLVQ} = \sum_{i=1}^{N} \phi(\mu(\mathbf{x})), \quad \mu(\mathbf{x}) = \frac{d^+ - d^-}{d^+ + d^-}, \tag{13}$$

where $\phi(\cdot)$ is the identity function for linear GLVQ or the logistic function for logistic GLVQ, and μ is the relative distance difference, and $d^+ = d(\mathbf{x}, \mathbf{m}^+)$ is the squared Euclidean distance of the input pattern $\mathbf{x}(t)$ to its closest prototype $\mathbf{m}^+(t)$ having the same label, and $d^- = d(\mathbf{x}, \mathbf{m}^-)$ is the squared Euclidean distance of the input pattern $\mathbf{x}(t)$ to its closest prototype $\mathbf{m}^-(t)$ having a different class label. Considering these scenarios, the following updating rules are applicable:

$$\mathbf{m}^+(t+1) = \mathbf{m}^+(t) + \alpha(t)\phi'(\mu(\mathbf{x}))[4d^-/(d^+ + d^-)^2][\mathbf{x}(t) - \mathbf{m}^+(t)], \tag{14}$$

$$\mathbf{m}^-(t+1) = \mathbf{m}^+(t) - \alpha(t)\phi'(\mu(\mathbf{x}))[4d^+/(d^+ + d^-)^2][\mathbf{x}(t) - \mathbf{m}^-(t)]. \tag{15}$$

GRLVQ [16,17]: In this algorithm, the cost function is defined as follows:

$$E_{GRLVQ} = \sum_{i=1}^{N} \phi(\mu(\mathbf{x})), \quad \mu(\mathbf{x}) = \frac{d_\lambda^+ - d_\lambda^-}{d_\lambda^+ + d_\lambda^-}, \tag{16}$$

where $\phi(\cdot)$ is the logistic function, and $d_\lambda^+ = d(\mathbf{x}, \mathbf{m}^+)$ is the squared Euclidean weighted distance of input pattern $\mathbf{x}(t)$ to its closest prototype $\mathbf{m}^+(t)$ having the same label, and $d_\lambda^- = d(\mathbf{x}, \mathbf{m}^-)$ is the squared Euclidean weighted distance of the input pattern $\mathbf{x}(t)$ to its closest prototype $\mathbf{m}^-(t)$ having a different class label. The relevance factors can be determined by the gradient descent as

$$\boldsymbol{\lambda}(t+1) = \boldsymbol{\lambda}(t) - \epsilon_\lambda(t)\phi'\left(\frac{(\mathbf{x} - \mathbf{m}^+)^2 d_\lambda^- - (\mathbf{x} - \mathbf{m}^-)^2 d_\lambda^+}{(d_\lambda^+ + d_\lambda^-)^2}\right), \tag{17}$$

where ϵ_λ is the gain factor, and $\lambda_a \geq 0$ for $a = 1, \ldots, p$, and numerical instabilities are avoided by applying the normalization $\sum_{a=1}^{p} \lambda_a = 1$. Finally, the updating rules are shown below:

$$\mathbf{m}^+(t+1) = \mathbf{m}^+(t) + \alpha(t)\phi'(\mu(\mathbf{x}))[4d_\lambda^-/(d_\lambda^+ + d_\lambda^-)^2][\mathbf{x}(t) - \mathbf{m}^+(t)], \tag{18}$$

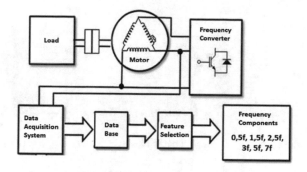

Fig. 1. Modules of the laboratory test bed and the data acquisition system.

$$\mathbf{m}^-(t+1) = \mathbf{m}^+(t) - \alpha(t)\phi'(\mu(\mathbf{x}))[4d_\lambda^+/(d_\lambda^+ + d_\lambda^-)^2][\mathbf{x}(t) - \mathbf{m}^-(t)]. \qquad (19)$$

4   Results and Discussion

In this section, we evaluate the proposed methodology to find the number of prototypes and its locations for the 3 types of classes existing in the available dataset, which are represented by the labels N (normal), H (high impedance) and L (low impedance). The dataset is comprised of 294 6-dimensional labeled feature vectors, in which the attribute values represent the FFT values for the chosen 6 harmonics of the fundamental frequency of the converter drive.

A 3-phase squirrel-cage induction motor built by WEG[1] industry is used in this study. Its main characteristics are 0.75 kW (power), 220/380 V (nominal voltage), 3.02/1.75 A (nominal current), 79.5% (efficiency), 1720 rpm (nominal rotational speed), Ip/In = 7.2 (peak to nominal current ratio), and 0.82 (power factor). The dataset is generated with this motor operating in different working conditions. The modules of the laboratory scale testbed are shown in Fig. 1.

The task of interest can be approached either as a 3-class problem (ternary classification) or as a 2-class problem (binary classification). For the ternary classification, the distribution of samples per class is as follows: normal condition (with 42 samples), high impedance fault (with 126 samples) or low impedance fault (also with 126 samples). As a binary classification problem, we merge the samples from classes H and L and label them simply as *Faulty*. Further details on the construction of this dataset and the experimental apparatus built for its generation is given in [6–8].

For each classifier, 100 independent turns of training and testing are carried out. For each run, the four steps of the proposed methodology are executed: (i) the holdout (division of the data set into training and validation data sets); (ii) determination of the K_{opt} and prototypes' initialization via application of clustering and cluster validity techniques per data class; and, (iv) LVQ training and testing. At the end of each run, the accuracy rate of each classifier is determined.

---

[1] http://www.weg.net/institutional/BR/en/.

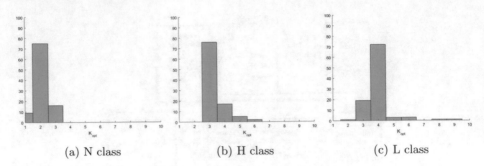

$$\text{(a) N class} \qquad \text{(b) H class} \qquad \text{(c) L class}$$

Fig. 2. Histograms of K_{opt} obtained by applying the majority voting scheme to the cluster validity techniques along 100 independent runs.

The 2nd step of the methodology is comprised of two stages. In the first stage, we apply the K-means algorithm on each class individually, for $K = 2, 3, \cdots, K_{max} = 10$ prototypes. For each value of K, we execute 10 independent runs of the K-means algorithm and choose the set of prototypes $\{\mathbf{p}_j\}_{j=2}^{K}$ that produces the lowest MSQE[2] for each class. Using these selected sets, we compute the corresponding values of the cluster validity indices in order to choose the optimal number of prototypes per class. We use the majority voting of these cluster validation techniques to make this choice. In the 2nd stage, we initialize the LVQ classifiers using the selected K_{opt} prototypes.

The histograms of the suggested optimal number of prototypes per class resulting from the majority voting scheme along the 100 independent turns are shown in Fig. 2. By analyzing this figure, it can be seen that the setup with $K_N = 2$, $K_H = 3$, and $K_L = 4$ is the most frequent one.

The performance of each LVQ-based classifier is shown in Fig. 3a for ternary classification and in Fig. 3b for binary classification. For the binary setting, we merge the L and H classes into a single faulty class. A closer look at these figures reveals that the linear GLVQ, logistic GLVQ, and GRLVQ classifiers performed better than the other three. Then, we choose GRLVQ and the linear GLVQ (due to the smaller computational cost) for further analyses. As shown in Table 1 the accuracy metrics of these two classifiers were basically the same.

It should be pointed out that as important as maximizing the overall accuracy rate is minimizing the occurrence of certain types of errors. As observed in [6,8], since the classification task is unbalanced (there are 252 samples of faulty conditions and only 42 samples of normal conditions), trained algorithms are biased toward classifying normal samples as faulty ones. This means that the false alarm rate is undesirably high in this case and should be minimized, because false alarms force unexpected downtime for maintenance purposes and, hence, cause an increase in production costs.

---

[2] Mean squared quantization error: $MSQE = \frac{1}{n_k} \sum_{\forall \mathbf{x} \in c_k} \|\mathbf{x} - \mathbf{m}_c^k\|^2$, where n_k is the number of data samples of the class c_k and $\mathbf{m}_c^k$ is the nearest prototype belonging to class c_k.

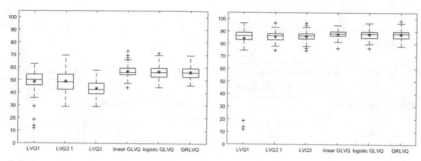

(a) Accuracy of ternary classification.   (b) Accuracy of binary classification.

Fig. 3. Boxplots of accuracy rates achieved by the evaluated LVQ classifiers.

Table 1. Accuracy rates for the ternary and binary classification tasks achieved by linear GLVQ and GRLVQ classifiers.

		Min	Max	Median	Std
Ternary	GLVQ	44.068%	72.881%	55.932%	5.575
	GRLVQ	45.763%	69.491%	55.932%	5.602
Binary	GLVQ	76.271%	96.610%	88.136%	3.606
	GRLVQ	77.966%	98.305%	88.136%	3.574

Table 2. Best ternary confusion matrix (GLVQ).

Linear GLVQ		Actual C.		
		N	H	L
Predicted class	N	2	0	1
	H	2	14	9
	L	0	13	18

Table 3. Best ternary confusion matrix (GRLVQ).

GRLVQ		Actual C.		
		N	H	L
Predicted class	N	1	1	0
	H	0	21	4
	L	0	13	19

In order to evaluate the linear GLVQ and GRLVQ in this regard, we compared their confusion matrices for the best case scenario, i.e. the one leading to the largest classification rate within the 100 independent runs. We observe in the 1st row of Tables 2 and 3 that both classifiers erroneously classified one faulty sample as a normal one (GLVQ: true L $\Rightarrow$ N, GRLVQ: true H $\Rightarrow$ N). However, only the GLVQ classified a normal sample as a faulty one (GLVQ: true N $\Rightarrow$ H) as we can see in the 1st column of those same tables. Similar behaviors are inferred for the binary task by analyzing Tables 4 and 5.

From the reported results, we can infer that the GRLVQ classifier is the best suited for the fault classification task of interest among all the evaluated LVQ variants. A nice feature of the GRLVQ is that we can check the relevance weights in order to have a clear notion which input attributes have more influence on

Table 4. Best binary confusion matrix (linear GLVQ).

Linear GLVQ		Actual class	
		Normal	Faulty
Predicted class	Normal	2	1
	Faulty	2	54

Table 5. Best binary confusion matrix (GRLVQ).

GRLVQ		Actual class	
		Normal	Faulty
Predicted class	Normal	1	1
	Faulty	0	57

the performance of the GRLVQ classifier. By analyzing the relevance weights' vector given by

$$\boldsymbol{\lambda} = [\lambda_{0.5\hat{f}_c} \mid \lambda_{1.5\hat{f}_c} \mid \lambda_{2.5\hat{f}_c} \mid \lambda_{3\hat{f}_c} \mid \lambda_{5\hat{f}_c} \mid \lambda_{7\hat{f}_c}]$$
$$= [0.1373|0.0952|0.1757|0.2277|0.1689|0.1952], \tag{20}$$

we can see that the 4th attribute $(\lambda_{3\hat{f}_c})$ is the most relevant one, while the 2nd attribute $(\lambda_{1.5\hat{f}_c})$ is the least relevant one.

5   Conclusions and Further Work

In this paper, we introduced a clustering-based methodology for building efficient LVQ-based classifiers. Our motivation had its origin in a complex fault classification task in which we have been working now for some years. The target task of detecting inter-turn short-circuit faults is challenging (even for human experts) because of the high probability of misinterpretation of high impedance faults as normal ones. Previous experience with powerful supervised neural network based classifiers, such as the MLP and RBF networks, has challenged us to apply much simpler prototype-based classifiers and get acceptable performances on the fault classification task of interest. We succeeded in reporting high accuracy rates, comparable to those achieved in previous works of our research group.

Currently, we are investigating the performances of kernelized versions of LVQ classifiers on the same fault classification task. Our ultimate goal is to develop an embedded software application capable of monitoring induction motors in an online fashion with high accuracy rates.

Acknowledgments. This study was financed in part by the Coordenação de Aperfeiçoamento de Pessoal de Nível Superior - Brasil (CAPES) - Finance Code 001. The authors also thank CNPq (grant 309451/2015-9) for the financial support and IFCE for the infrastructure of the Laboratory of Energy Processing.

References

1. Kohonen T (1990) Improved versions of learning vector quantization. In: 1990 IJCNN international joint conference on neural networks. IEEE, pp 545–550
2. Albuquerque RF, de Oliveira PD, Braga APdS (2018) Adaptive fuzzy learning vector quantization (AFLVQ) for time series classification. In: North American fuzzy information processing society annual conference. Springer, pp 385–397
3. Soares Filho LA, Barreto GA (2014) On the efficient design of a prototype-based classifier using differential evolution. In: 2014 IEEE symposium on in differential evolution (SDE). IEEE, pp 1–8
4. Biehl M, Hammer B, Villmann T (2016) Prototype-based models in machine learning. WIREs Cognit Sci 7(2):92–111
5. Nova D, Estévez PA (2014) A review of learning vector quantization classifiers. Neural Comput Appl 25(3–4):511–524
6. Coelho DN, Barreto GA, Medeiros CMS, Santos JDA (2014) Performance comparison of classifiers in the detection of short circuit incipient fault in a three-phase induction motor. In: Proceedings of the 2014 IEEE symposium on computational intelligence for engineering solutions (CIES 2014), pp 42–48
7. Sousa DP, Barreto GA, Medeiros CMS Efficient selection of data samples for fault classification by the clustering of the SOM, pp 1–12. http://cbic2017.org/papers/cbic-paper-71.pdf
8. Coelho DN, Medeiros CMS (2013) Short circuit incipient fault detection and supervision in a three-phase induction motor with a SOM-based algorithm. In: Advances in self-organizing maps. Springer pp 315–323
9. Davies DL, Bouldin DW (1979) A cluster separation measure. IEEE Trans Pattern Anal Mach Intell 2:224–227
10. Dunn JC (1973) A fuzzy relative of the ISODATA process and its use in detecting compact well-separated clusters
11. Caliński T, Harabasz J (1974) A dendrite method for cluster analysis. Commun Stat Theory Meth 3(1):1–27
12. Rousseeuw PJ (1987) Silhouettes: a graphical aid to the interpretation and validation of cluster analysis. J Comput Appl Math 20:53–65
13. Duda RO, Hart PE, Stork DG (2006) Pattern classification, 2nd edn. Wiley, Hoboken
14. Kohonen T (1990) Improved versions of learning vector quantization. In: Proceedings of the 1990 international joint conference on neural networks (IJCNN 1990), vol 1, pp 545–550
15. Sato A, Yamada K (1996) Generalized learning vector quantization. In: Advances in neural information processing systems, pp 423–429
16. Hammer B, Villmann T (2002) Generalized relevance learning vector quantization. Neural Netw 15(8–9):1059–1068
17. Hammer B, Strickert M, Villmann T (2005) On the generalization ability of GRLVQ networks. Neural Process Lett 21(2):109–120

When Clustering the Multiscalar Fingerprint of the City Reveals Its Segregation Patterns

Madalina Olteanu[1,2]([⊠]) and Jean-Charles Lamirel[3]

[1] MaIAGE, INRA, Université Paris Saclay, Domaine de Vilvert, 78352 Jouy en Josas, France
[2] SAMM, EA 4543, Université Paris 1 Panthéon Sorbonne, Paris, France
madalina.olteanu@univ-paris1.fr
[3] LORIA, Equipe Synalp, Bâtiment B, 54506 Vandoeuvre Cedex, France
lamirel@loria.fr

Abstract. The complexity of urban segregation challenges researchers to develop powerful and complex mathematical tools for assessing it. With more and more fine-grained and massive data becoming available these last years, individual-based models are now made possible in practice. Very recently, a mathematical object called *multiscalar fingerprint* [1], containing all possible and all scale individual trajectories in a city, was introduced. Here, we use clustering combined with specific measures for assessing features contributions to clusters, to explore this complex object and to single out *hotspots* of segregation. We illustrate how clustering allows to see where, how and to which extent segregation occurs.

Keywords: Spatial analysis · Segregation · Clustering · Feature importance

1 Introduction

Assessing the complexity and the multiple facets of urban segregation continues to challenge interdisciplinary research, and these last years even more so, with more and more fine-grained massive data becoming available. In the extensive literature on segregation, common measures are usually quantifying the differences between local concentrations of different groups [2]. The existing indices aim at emphasizing the uneven spatial distribution of groups, their concentration, clustering, centralization or exposure to each other. At a closer look, most of these indices – see [3,4] for a review – have two limitations: they are dependent on the arbitrary definition of local neighborhoods or spatial units, leading to the well-known modifiable areal unit problem [5], and they are scalar quantities, which means that one single number is supposed to summarize the entire information and complexity in the data.

These last ten years, new segregation measures aimed at taking into account its multiscalar behavior have been introduced [6–9]. Most of them are however

© Springer Nature Switzerland AG 2020
A. Vellido et al. (Eds.): WSOM 2019, AISC 976, pp. 140–149, 2020.
https://doi.org/10.1007/978-3-030-19642-4_14

limited to selecting an arbitrary number of scales, either by aggregating arbitrary numbers of spatial units, or by building circular neighborhoods of arbitrary radii around a starting point. Nonetheless, the major breakthrough of most of these approaches is the idea of building *egocentric profiles* or *egocentric signatures*, that is individual trajectories in the city. The limit on the arbitrary number of scales, as well as that on their arbitrary values, have been recently removed in [1], where trajectories are built using all possible scales in the data. Once the complex object represented by the set of all possible individual trajectories is computed, the next research challenge is to properly explore and exploit it, so that the segregation patterns encoded within are unveiled.

In a very recent previous work [10], the concepts of *focal distances* and *distortion coefficients* have been introduced as new measures for quantifying the perceived segregation at individual level. In the present manuscript, we use clustering as a complementary approach, able to single out and draw boundaries around *hotspots* of segregation. In order to make the manuscript more readable, we chose to mix technical details with experimental results, and we gradually introduce the methodology: in Sect. 2 we recall the definition of multiscalar individual trajectories as introduced in [1] and compute them on a dataset containing the social housing rate for the city of Paris; in Sect. 3 we compute the *distortion coefficients* as defined in [10] to assess the *hotspots* of segregation; in Sect. 4 we use clustering with recent indices for assessing features importance [11] to identify abnormal regions in the city, and characterize these abnormalities in terms of positive or negative segregation. We empirically prove that the two approaches are complementary and provide a fine-grained image of where, how and to which extent segregation occurs.

2 Building a Multiscalar Fingerprint of the City

Segregation being a complex phenomenon, one should ideally take into account a multiscalar and individual perspective when modeling and assessing it. Indeed, segregation patterns arise from a myriad of individual situations and perceptions, those experienced by every individual in the city while living in it. Using individual experiences to reveal spatial patterns builds upon the idea that the longer an individual has to go from his own home to seize what the city as a whole looks like, the more cut-off from the rest of the city he will feel [1]. With this in mind, one can compute individual trajectories encoding the perception of the walker while visiting first his direct neighborhood, then the next closest one, and so on, gradually, until having visited the entire city.

Suppose the residential coordinates of the N individuals in a city are known, as well as them belonging to one of two groups, A and B, with a proportion p_0 of group A in the whole city. For a given individual i, one may sequentially aggregate the rest of the population, using for instance a nearest neighbor rule. At step k, after having aggregated $k - 1$ individuals around the initial one, the proportion of group A *seen* by the i-th individual is $\hat{p}_{i,1:k}$. Hence, for each individual i, one computes a trajectory of proportions, $(\hat{p}_{i,1:k})_{k=1,...,N}$, which eventually converges to p_0. But the way it reaches the city average, more or less rapidly, encodes for the degree of segregation of the starting point.

We illustrate this on the distribution of social housing in the city of Paris, in 2014. The data comes from the French National Statistical Institute and, as most census data, the information is not available at housing level, but comes as already aggregated units: the approx. 1,150,000 housings in Paris are summarized by 987 spatial units. As one may see in Fig. 1(left), social housing is unevenly distributed in the city: while the actual average is 17,86% (close to the 20% official target), the median is equal to 7% only, and the rates in each spatial unit go from 0% to 97%, with a concentration of the high-rate units on the borders.

At this point, we may see that combining statistical data with geographical information and computing all possible individual trajectories in the city allows one to create a powerful mathematical object, a *fingerprint* containing all the information about the variable of interest, at all scales and from the finest possible available point of view. However, the size and the complexity of this object make it difficult to use as such for large datasets. One needs to summarize these trajectories into new indices and features, which will contain a maximum amount of information and bring to light patterns of perceived segregation.

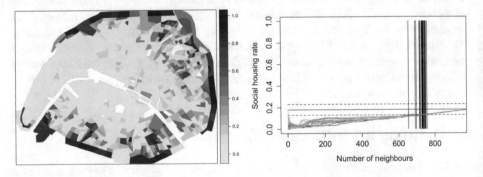

Fig. 1. Left: social housing rate per spatial unit. Right: social housing rate trajectories, for the units in Paris 8th district. The solid flat line is the city average, the dashed lines correspond to ±5% around it. Solid vertical lines correspond to *focal distances*.

3   Focal Distances and Distortion Coefficients

For quantifying the segregation perceived at a given location i, one may look at his trajectory $(\hat{p}_{i,1:k})_{k=1,...,N}$ and extract the *instant* of convergence $\tau_{i,\delta} = \arg\min_{k=1,...,N}\{|\hat{p}_{i,1:\tilde{k}} - p_0| \leq \delta \ \forall \tilde{k} \geq k\}$ for a fixed threshold $\delta > 0$. This instant, also called *focal distance* and first used in [1], translates the distortion in the perception of the city that an individual living in unit i has: the smaller the *focal distance*, the less "cut off" from the city he is; the larger the *focal distance*, the more persistent the feeling of segregation. We illustrate this concept in Fig. 1(right), for one specific Paris district, the Champs-Elysées and their surrounding area, and for a threshold δ fixed at 5%. All trajectories converge

very slowly and from below the city average, the 8th district being the "core" of
the rich neighborhoods of Paris, with almost no social housing.

Focal distances allow to draw an already significant picture of individual segregation patterns, but their definition is weakened by the arbitrary choice of
a convergence threshold. For removing this arbitrariness, one may let δ vary
between 0 and the maximum possible value $\delta_{\max} = \max\{|\hat{p}_{i,1:\tilde{k}} - p_0|, i =
1, ..., N; k = 1, ..., N\}$, on a very fine grid, and then integrate the resulting *focal
distances* over δ. For a spatial unit i, the *distortion coefficient* [10] measures how
distorted the city looks from the i-th location:

$$\Delta_i = \int_0^{\delta_{\max}} \tau_{\delta,i} d\delta \; . \tag{1}$$

Let us point out that the definitions of focal distances and distortion coefficients
may be easily extended to a more general context like multi-group or continuous
distributions, by replacing $\hat{p}_{i,1:\tilde{k}}$ with the aggregated distribution $\hat{f}_{i,1:\tilde{k}}$, p_0 with
the distribution in the whole city f_0, and $|\hat{p}_{i,1:\tilde{k}} - p_0|$ with any distance between
probability distributions, such as the Kullback-Leibler (KL) divergence. In the
following, we illustrate the results for Paris social housing rate using distortion
coefficients computed from the KL divergence trajectories. This is mainly for
stressing the advantage of using a more general framework as mentioned above,
but also because the two approaches (proportions trajectories and KL-divergence
trajectories) lead to similar convergence results.

Figures 2 and 3 illustrate the normalized (with respect to the theoretical
maximum) distortion coefficients for the trajectories of the social housing rate.
As one may see in Fig. 2, most spatial units have low distortion coefficients,
with a median value of 2.5% and 75% of the units below 5% (the normalized
distortion coefficients take values between 0 and 1, and may be viewed as the %
of theoretically maximum possible segregation). The tail of the distribution is
however rather heavy, and when looking into detail, one may identify a *hotspot*
of segregation around the Champs-Elysées (see Fig. 3). Furthermore, the spatial
unit with the highest distortion coefficient, 10%, hence maximum segregation in
the actual city configuration, is situated Place Vendôme, while the spatial unit
reaching the lowest distortion coefficient, 0.1%, is situated in the 10th district of
the city, the Hôpital Saint Louis neighborhood, a historically well mixed area.
One may also note the two orders of magnitude between the two.

Normalized distortion coefficients prove to be a powerful index for quantifying segregation within a city, both on an individual level, and on a multiscalar
level. However, we argued at the end of Sect. 2 for the need of various indices
summarizing the trajectories and allowing further analysis. One may be interested, for example, in assessing whether the segregation is an individual choice
or an imposed constraint, whether there are similar behaviors, and whether the
differences between trajectories are induced by all their instants, or by some
more *important* ones. In the next section, we aim at bringing some first answers
to these questions using a clustering procedure.

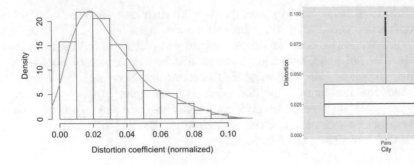

Fig. 2. Normalized distortion coefficients computed from the KL divergence trajectories on the social housing rate.

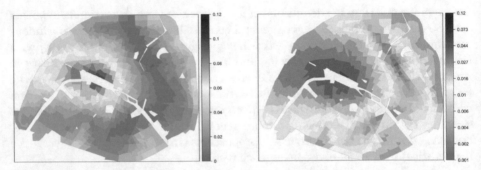

Fig. 3. Maps of the normalized distortion coefficients computed from the KL divergence trajectories on the social housing rate. Left: linear color scale; right: log color scale.

4   Clustering Trajectories

Since we aim at identifying similar patterns in the perception of segregation, the next step of our analysis is to cluster the trajectories into homogeneous groups. Furthermore, we are interested in extracting the *instants* in the trajectories which discriminate these patterns, and assessing where the differences between them occur in the trajectories, whether on a local or larger scale. This is obtained by quantifying the importance of each *instant*, using a feature importance criterion introduced in [12]. We briefly recall its definition, before describing the results on social housing in Paris.

4.1   Defining Contrasts and Indices of Features Importance

We consider a partition C obtained after having clustered a dataset D represented by a set of numerical positive features F. We define the F-measure $FF_c(f)$ of a feature f in cluster c as the harmonic mean of the feature recall $FR_c(f)$ and of the feature predominance $FP_c(f)$,

$$FF_c(f) = 2\left(\frac{FR_c(f) \cdot FP_c(f)}{FR_c(f) + FP_c(f)}\right) , \qquad (2)$$

where

$$FR_c(f) = \frac{\Sigma_{d\in c} w_d^f}{\Sigma_{c\in C}\Sigma_{d\in c} w_d^f} \; , \quad FP_c(f) = \frac{\Sigma_{d\in c} w_d^f}{\Sigma_{f'\in F_c, d\in c} w_d^{f'}} \; . \tag{3}$$

In the equations above, w_d^f represents the value of feature f for input d, and F_c is the set of all features present in the data assigned to cluster c. A feature is said to be present in a cluster c if its values in the cluster are strictly positive. Feature predominance measures the ability of f to *describe* the cluster c. In a complementary way, feature recall allows to characterize f according to its ability to *discriminate* c from other clusters.

Next, we define the set of features characteristic to a given cluster c:

$$S_c = \{f \in F_c \mid FF_c(f) > \overline{FF}(f) \text{ and } FF_c(f) > \overline{FF}_D\} \tag{4}$$

where

$$\overline{FF}(f) = \Sigma_{c'\in C}\frac{FF_{c'}(f)}{|C_{/f}|} \text{ and } \overline{FF}_D = \Sigma_{f\in F}\frac{\overline{FF}(f)}{|F|} \; . \tag{5}$$

$C_{/f}$ represents the subset of C in which the feature f occurs. The set of all selected features is then $S_C = \cup_{c\in C}S_c$.

The F-measures may be used to define the contrast of feature f in cluster c:

$$G_c(f) = \frac{FF_c(f)}{\overline{FF}(f)}. \tag{6}$$

The active features of a cluster are those for which the contrast is larger than 1. Moreover, the higher the contrast of a feature for one cluster, the better its performance in describing the cluster content. Conversely, the passive features in a cluster are those present in the data, but with a contrast less than unity. With the notations above, we may define the average contrast in all clusters as

$$\overline{G} = \frac{1}{|C|}\Sigma_{c\in C}\frac{1}{|S_c|}\Sigma_{f\in S_c}G_c(f) \; . \tag{7}$$

By comparing the contrast of a feature in a cluster to the average one, we define the set of features with an abnormal positive contrast in cluster c, as well as the amount and the size of the positive abnormality:

$$S_{c,+} = \{f \in S_c \mid G_c(f) > \overline{G}\} \; , \tag{8}$$

$$A_{c,+} = \Sigma_{f\in S_{c,+}}G_c(f) \; , \quad L_{c,+} = |S_{c,+}|/|F| \; . \tag{9}$$

In a similar fashion, we also define the set of features with abnormal negative contrasts in cluster c, the amount and the size of the negative abnormality:

$$S_{c,-} = \left\{f \in S_{c'}, c' \neq c \mid \frac{1}{G_c(f)} > \overline{G}\right\} \; , \tag{10}$$

$$A_{c,-} = \Sigma_{f\in S_{c,-}}\frac{1}{G_c(f)} \; , \quad L_{c,-} = |S_{c,-}|/|F| \; . \tag{11}$$

In the following, a *neutral cluster* is a cluster in which the cumulated sizes of positive and negative abnormalities are less than 5% of the total number of features.

Eventually, contrasts as previously defined allow to define the EC index [12], which can be used for selecting an optimal number of clusters:

$$EC(C) = \frac{1}{C} \sum_{c=1}^{C} \left(\frac{|S_c| \sum_{f \in S_c} G_c(f) + |\overline{S_c}| \sum_{h \in \overline{S_c}} \frac{1}{G_c(h)}}{|S_c| + |\overline{S_c}|} \right) , \qquad (12)$$

where $|S_c|$ is the number of active features in c, and $|\overline{S_c}|$, the number of passive features in the same cluster, whereas the set $\overline{S_c}$, representing the set of passive features in c, is equivalent to the complementary set of S_c in S_C. One of the main reasons for using the EC index in our procedure is its ability of producing reliable results with real data in cases ranging from low dimensional to high dimensional contexts, with a low computational time, conversely to all usual clustering indices, as it is shown in [12].

4.2   Five *hotspots* of Segregation for the City of Paris

We trained two different clustering algorithms on the social-housing rate trajectories in Paris, namely k-means [13], a winner-takes-all method, and growing neural gas (GNG) [14], a winner-takes-most method, with Hebbian learning trained with the usual Fritzke parameters. The GNG algorithm proved to be superior to k-means, because of the fact that (altogether) the Hebbian, incremental and winner-takes-most learning processes provide better independence to initial conditions and avoid producing degenerated clustering results. This kind of results have already been obtained in former experiments [12]. Several models with a number of clusters varying from 2 to 30 were trained, and the optimal number of clusters was selected by maximizing the EC index. As illustrated in Figs. 4 and 5, the optimal partitioning contains 24 clusters.

While contrast-based indices defined above could have been used in various ways, we chose to map the clusters using two criteria related to the ratio between abnormality amount and length: a first map with the average abnormal negative contrast computed by cluster, and a second map with the average abnormal positive contrast. The intensity of the negative and/or positive contrasts are represented in Fig. 6, where neutral clusters are colored in white. According to these, the abnormal clusters mainly correspond to the already detected *hotspot* around the Champs-Elysées, and also to the north-east, south, and south-east borders of the city, which contain spatial units with a high rate of social housing.

Furthermore, when comparing the sizes of positive or negative abnormalities, one may identify a cluster as being rather positively abnormal, or negatively abnormal. We impose this supplementary constraint in the representation in Fig. 7, and are then able to identify five distinct regions in the city: on the one hand, a large area in the central and western part, very similar to the *hotspot* detected by the distortion coefficients analysis, and, on the other hand,

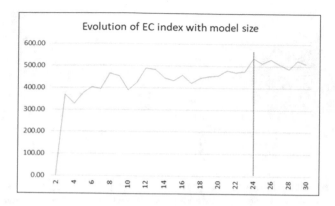

Fig. 4. Evolution of the EC index for various clustering configurations.

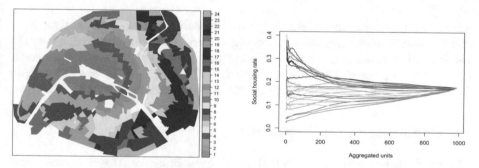

Fig. 5. Final clustering mapping (left) and average trajectories within clusters (right).

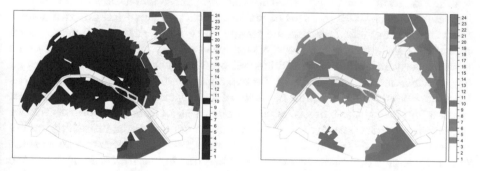

Fig. 6. Clusters with negative (left) and positive (right) contrasts; neutral clusters are colored in white. The color intensity is proportional to the average value of the contrast.

four smaller regions, all situated near the *Périphérique*, and where the density of social housing is very important. At a closer look, these four regions were already visible on the distortion coefficients map in Fig. 3, in the logarithmic scale representation. But one of them was contiguous and impossible to differentiate from the large *hotspot* next to it, and, furthermore, the distortion coefficients

Fig. 7. Clusters with a majority of negative (left) and positive (right) contrasts; neutral clusters are colored in white. The intensity is proportional to the average value of the contrast.

map did not provide the information on the *quality* of the *hotspots*. Clustering combined with features contrasts shows that there is a strong correlation between the type of contrast and the variable at study: *hotspots* with negative contrasts correspond to areas with a below average social housing rate, while *hotspots* with positive contrasts correspond to areas with an above average social housing rate.

5   Conclusion and Perspectives

The very recent notion of individual trajectories in a city [1] is a complex and powerful mathematical object which encodes all patterns of segregation at all scales. Extracting meaningful information from these trajectories is a challenging research issue, both in terms of mathematical modeling and practical exploration, since the data may become rapidly very large. After having computed *distortion coefficients* [10], we used clustering combined with recent measures for assessing features importance and identified which of the clusters were abnormal and in which sense. The results of the two approaches proved to be complementary, by singling out the same *hotspots* of segregation, however with different information brought to light. Clustering appears as an interesting perspective for mining trajectories, although preprocessing the data and dimensionality reduction could be a serious issue that we did not address here. Furthermore, clustering analysis may also allow to extract the instants in the trajectories, hence the scales, where abnormalities occur and which are thus critical for characterizing the implicit scales at which the segregation phenomenon occurs.

Acknowledgments. The authors wish to thank W. Clark (UCLA) and J. Randon-Furling (Université Panthéon Sorbonne) for the many discussions on the topics of spatial segregation and individual trajectories analysis.

References

1. Randon-Furling J, Olteanu M, Lucquiaud A (2018) From urban segregation to spatial structure detection. Urban Analytics and City Science, Environment and Planning B
2. Reardon SF, Firebaugh G (2002) Measures of multigroup segregation. Sociol Methodol 32(1):33–67
3. Kramer MR, Cooper HL, Drews-Botsch CD, Waller LA, Hogue CR (2010) Do measures matter? comparing surface-density-derived and census-tract-derived measures of racial residential segregation. Int J Health Geogr 9(1):29
4. Hong S, O'Sullivan D, Sadahiro Y (2014) Implementing spatial segregation measures in R. PLOS ONE 9:1–18
5. Openshaw S (1984) The modifiable areal unit problem. University of East Anglia
6. Reardon SF, Matthews SA, O'Sullivan D, Lee BA, Firebaugh G, Farrell CR, Bischoff K (2008) The geographic scale of metropolitan racial segregation. Demography 45(3):489–514
7. Clark WAV, Andersson E, Östh J, Malmberg B (2015) A multiscalar analysis of neighborhood composition in Los Angeles, 2000–2010: a location-based approach to segregation and diversity. Ann Assoc Am Geogr 105(6):1260–1284
8. Spielman SE, Logan JR (2013) Using high-resolution population data to identify neighborhoods and establish their boundaries. Ann Assoc Am Geogr 103(1):67–84
9. Fowler C (2016) Segregation as a multiscalar phenomenon and its implications for neighborhood-scale research: the case of south seattle 1990–2010. Urban Geogr 37(1):1–25
10. Olteanu M, Randon-Furling J, Clark W (2019, to appear) Spatial analysis in high resolution geo-data. In: ESANN Proceedings. Preprint available on demand
11. Lamirel J-C, Dugue N, Cuxac P (2016) New efficient clustering quality indexes. In: International joint conference on neural networks, pp 3649–3657
12. Lamirel J-C, Mall R, Cuxac P, Safi G (2011) Variations to incremental growing neural gas algorithm based on label maximization. In: International joint conference on neural networks, pp 956–965
13. James M (1967) Some methods for classification and analysis of multivariate observations. In: Proceedings of the fifth Berkeley symposium on mathematical statistics and probability, vol 1, pp 281–297
14. Fritzke B (1995) A growing neural gas network learns topologies. In: Advances in neural information processing systems, pp 625–632

Using Hierarchical Clustering to Understand Behavior of 3D Printer Sensors

Ashutosh Karna[✉] and Karina Gibert[✉]

Knowledge Engineering and Machine Learning Group (KEMLG),
Intelligent Data Science and Artificial Intelligence Research Center,
Universitat Politécnica de Catalunya – BarcelonaTech, Barcelona, Spain
{ashutosh.karna,karina.gibert}@upc.edu

Abstract. 3D Printing is one of the latest industrial revolutions massively disrupting the manufacturing value of chain and will deeply impact in the new context of Industry 4.0, with new ways of products manufacturing, delivery and maintenance. The 3D Printing process is heavily reliant on the power of data both coming from the physical OEM (original equipment manufacturer) and print files. The Jet fusion technology of 3D printing can take hours to produce and occurs on minuscule scales. Thus, it is necessary to develop new data-driven techniques to actively prevent possible issues (both hardware failure as well as part defects). As this is a relatively new field, researchers are still actively studying various sensors and their impact on printing process and the outcome itself. By appropriate profiling of printing sensors, one can reduce the post processing effort to a minimum while ensuring the desired part quality. In this work, the authors are studying some specific sensors and their behaviour while the machine is printing a job to understand relationships among them and how they overall govern the printing process. Also, attempts are being made to create print profiles by appropriately applying clustering techniques and using visual inspection.

Keywords: Hierarchical · Clustering · 3D printing · Data preprocessing ·
Part quality profile · Class Panel Graph

1 Introduction

1.1 3D Printing Overview

Additive manufacturing [1], or professional 3D printing, refers to several technologies that produce parts in an additive way. The printing process starts with a digital 3D model of the job which is sliced in thin layers by specific software. The build process begins by laying down a thin layer of powdered material across the working area. Then the material recoater carriage scans from top-to-bottom. Next, the printing and fusing carriage scans from right-to-left across the working area. The leading energy source preheats the working area immediately before printing to provide consistent and accurate temperature control of each layer as it is printed. The printheads then print functional parts by firing fusing and detailing agents in precise locations onto the material to define the part's geometry and its properties. The printing and fusing

© Springer Nature Switzerland AG 2020
A. Vellido et al. (Eds.): WSOM 2019, AISC 976, pp. 150–159, 2020.
https://doi.org/10.1007/978-3-030-19642-4_15

carriage then returns left-to-right to fuse the areas that were just printed. At the ends of the scans, supply bins refill the recoater with fresh material and service stations can test, clean, and service the printheads on the printing and fusing carriage as needed to ensure reliable operation. The process continues layer-by-layer until a complete part or set of parts is built in the Build unit. In Fig. 1 [2], we can see the powder deposition and fusion process in every step. Figure 1 shows the step by step process of powder deposition on bed in Multi Jet fusion printer developed at HP Inc.

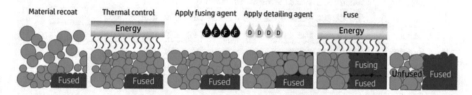

Fig. 1. Jet Fusion Process in various stages

To support the optimal printing behavior throughout the job, dozens of sensors placed on various internal subsystems capture pressure, temperature, humidity and breakage at periodic intervals. The data coming from the sensors is sent to a cloud storage through unstructured log files which could be useful to detect patterns and diagnose issues related to specific types of print job failures of part abnormalities.

1.2   Data Collection and Parsing

The analyzed data were collected from eight anonymized HP 3D Printers over 300 different printing sessions, generating raw log files of more than 35 GB in size, containing over one hundred sensor values representing the temperature, humidity, pressure, etc. of different subsystems.

For data analysis, Python 3.6 was used along with the popular machine learning libraries Scikit-Learn, Pandas, Numpy and SciPy. Vector quantization and parallel processing was heavily used while parsing the large log files containing sensor data in order to speed up the process. More than 562,000 records of sensor behavior representing the 3D Printer's behavior during different task – Warming-Up, Printing and Annealing were parsed.

For this work, we focused on the **Print** phase to understand how sensors react with one another while the part is being manufactured additively. From over one hundred sensors that are located throughout the machine, the printer sends periodic recordings of 57 sensors to the cloud that are monitored during various phases of job workflow. Thus, this study will focus on just these 57 sensors. Further, there were over 46,000 records observed from the original dataset that represents **Print** task by the printer.

Table 1. List of Sensors during printing

Variable	Description
TimeStamp	Timestamp of the sensor recording
Sensor_1	To measure pressure in Air release system
Sensor_2	To measure Ambient temperature
Sensor_3	To measure temperature in cooling system-1
Sensor_4	To measure temperature in cooling system-2
Sensor_5	To measure temperature in cooling system-3
Sensor_6	To detect glass breakage on left fusing system
Sensor_7	To detect glass breakage on right fusing lamp
Sensor_8	To measure temperature in carriage back
Sensor_9	To measure temperature in carriage front
Sensor_10	To measure temperature in carriage middle
Sensor_11	Internal camera reading
Sensor_12	To measure the reference temperature in subsystem_x – back
Sensor_13	To measure the reference temperature in subsystem_x – front
Sensor_14	To measure the reference temperature in subsystem_x – middle
Sensor_15	To measure the temperature in the subsystem_x – back
Sensor_16	To measure the temperature in the subsystem_x – front
Sensor_17	To measure the temperature in the subsystem_x – middle
Sensor_18	To measure the reference temperature in subsystem_y – back
Sensor_19	To measure the reference temperature in subsystem_y – front
Sensor_20	To measure the reference temperature in subsystem_y – middle
Sensor_21	To measure the temperature in the subsystem_y – back
Sensor_22	To measure the temperature in the subsystem_y – front
Sensor_23	To measure the temperature in the subsystem_y – middle
Sensor_24	To check obstruction in pressure system-left
Sensor_25	Temperature coefficient sensor
Sensor_26	Temp. coefficient for fusing system1 – left
Sensor_27	Temp. coefficient for fusing system1 – right
Sensor_28	Temp. coefficient for fusing system2 – left
Sensor_29	Temp. coefficient for fusing system2 – right
Sensor_30	Temp. coefficient for camera system
Sensor_31	Temp. coefficient for Cooling left air exit
Sensor_32	Temp. coefficient for Top heating
Sensor_33	Temp. coefficient for right air exit
Sensor_34	Humidity sensor for subsystem_xx
Sensor_35	Temperature sensor for subsystem_xx
Sensor_36	Connectivity check sensor for fusing system-left
Sensor_37	Connectivity check sensor for fusing system – right
Sensor_38	To check obstruction in pressure system-right
Sensor_39	Sensor in subsystem_yy
Sensor_40	Temperature Sensor in subsystem_z1
Sensor_41	Temperature Sensor in subsystem_z2

1.3   Data Preprocessing

The first step of the analysis is the generation of descriptive statistics and preprocessing of data [3, 4].

While doing an initial exploratory analysis, 16 out of original 57 sensors were discarded based on the reliability of measurements, poor domain interpretation and presence of faulty measurements. Now, this leaves us with 41 sensor variables. Table 1 represents the list of all sensors finally considered in the study. They mainly capture temperature, humidity and pressure in different subsystems of 3D Printer. The names of the sensors have been anonymized to maintain the confidentiality of the subsystems. All these sensor variables go through the following phases in the job lifetime: **Parsing, Processing, Warming-Up, Printing, Annealing, Cooling, Preparing and Unpacking.**

All the analyzed variables taken are numerical in nature but with varying scale and unit.

2   Statistical Clustering Method

Since 3D Printing is a recent technological advancement, the literature investigating how various sensors behave is limited. A neural network-based approach to predict the quality of electronic parts and shape accuracy has been discussed in [5] in 3D Inkjet printing process. The literature on the additive manufacturing technology space and statistical analysis is still very limited.

This paper focuses on ***Multi-Jet Fusion*** based 3D printing technology and attempts to discover the underlying patterns and structure of sensor measurements describing the process. Clustering techniques have been used in this work to identify typical scenarios in the printing jobs. Although K-means is a popular technique, it suffers from the specific drawbacks lack of established methods to determine the most adequate number of clusters in advance and even the inadequacy to handle a large number of patterns [6]. Hierarchical clustering methods allow to determine the number of clusters as an output, by analyzing the dendrogram and visualizing significant variability in the data. Hence, the current work employs hierarchical clustering using Ward's method and normalized Euclidean distance, since the sensor variables are recorded in different scales and units. This allows their comparison at a uniform scale with balanced impact in the cluster formation.

As hierarchical clustering is of $O(n^2)$ time complexity, it is computationally very expensive even for relatively small datasets. Keeping this in mind, the study starts with taking a random sample of 10,000 records drawn without replacement. In future work this will be scaled up by using a CURE-like strategy [7] that helps bringing hierarchical clustering strategies to big datasets.

The dendrogram (see Fig. 2) showed four potential clusters (as Calinski-Harabasz index also indicates) that need to be analyzed to appropriately profile them. Conditional distributions of all 41 numerical sensors with regards to each individual cluster were analyzed and insights were used in subsequent analysis.

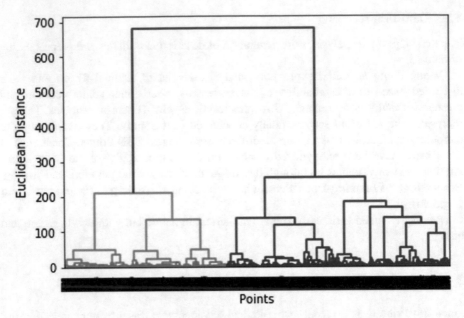

Fig. 2. Dendrogram of sensor readings

3 Interpretation of Clusters

3.1 Analysis of Conditional Distributions Versus Classes

The Class Panel Graph (CPG) [8] synthetizes the conditional distributions of the variables against the clusters. The CPG of the sensor variables versus the clusters was built to understand the behavior of sensory fluctuation. Additionally, illustrative variables (that were not used for the clustering) were used to enrich interpretation of the patterns. They are added as new columns on the CPG (Fig. 3). These news variables are:

- 'Status_message': it represents the final status of the print job (success, cancelled, errors etc.)
- 'Fail_reason': reports the primary reason in case of job failure.

Figure 3 shows a part of the CPG containing only some of the active and illustrative variables. In particular, it includes variables that show differences through the clusters and can potentially explain the cluster patterns.

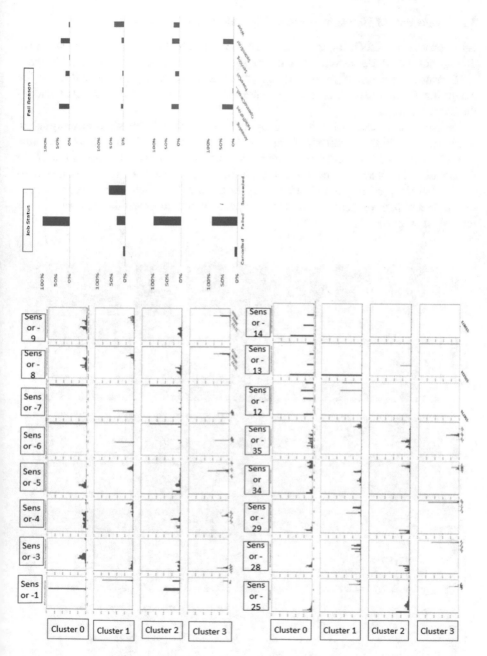

Fig. 3. Class Panel Graph of some relevant sensors

3.2   Detection of Non-informative and Redundant Variables

By observing the CPG, it can be seen that some of the sensors take a constant value throughout the whole dataset while others show similar behavior for all the clusters.

Sensor-24 and Sensor-38 are constant at zero reading throughout the dataset, which means there is no obstruction in the pressure system and is correctly connected in all of the analyzed jobs.

Sensor-2 and sensor-3 show similar distribution in all the clusters (correlation coefficient = 0.9992). Similarly, Sensor – 9 and Sensor-10 also exhibit the same behavior throughout the clusters (correlation coefficient = 0.9439). Also, similarity in distribution is observed between Sensor-26 and Sensor-28 (correlation coefficient = 0.9975) and might add redundancy to the dataset. Thus, it is enough to keep one of each pair of sensors for the analysis, as the information they provide is redundant (Fig. 4).

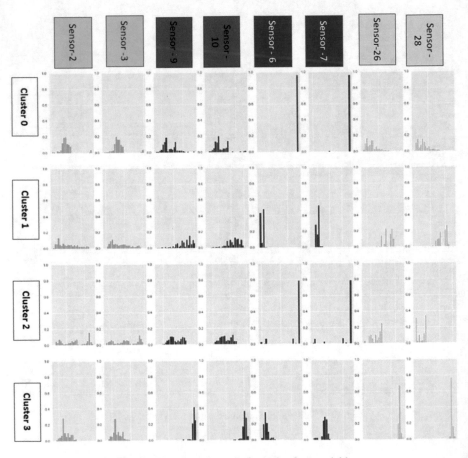

Fig. 4. Class panel graph for redundant variables

3.3   Pattern Conceptualization

The CPG for the whole dataset helps reveal some useful insights about the 3D printer sensors and the behavior of the printer during the printing phase.

Cluster 0 $\rightarrow$ In this cluster we can see that the Air Release System does not reach recommended pressure to start printing (Sensor-1). Temperature in another cooling system and carriage is lower than recommended (Sensor-3, Sensor-4 and Sensor-5, Sensor-9). Also, carriage movements are slower than recommended. Finally, Sensors 6 and 7 indicate that fusing lamps are too close to a certain component and printing is too risky. Projecting the illustrative variable 'Job Status' in this cluster, it can be seen that all the jobs in this category have failed and main reasons are *"Failed to Print"* and *"System Error"*.

Cluster 1 $\rightarrow$ Many sensors show acceptable values in this cluster. Fusing lamps are in the right position (Sensor-6 and Sensor-7) and allow printing. Carriage (Sensor-5, with sufficiently high value) indicates that carriage movement is fine. Subsystem X has correct temperature (Sensor – 12). Temperature coefficient is correct as well (Sensor-25) and so is Airlung pressure (Sensor-1). However, very low *"air-in"* and *"air-out"* temperatures are registered of Sensor – 25, which is associated to bad part quality as well as to system errors in the machine. Projecting the illustrative variable 'Job Status' in this cluster, we see that more than half of the jobs succeeded and some failed. Those were basically related with "Wiper". This is quite uncommon and points to a certain subsystems of the printer.

Cluster 2 $\rightarrow$ This cluster represents a situation in which the pressure target has not been reached (Sensor 1) and thus printing cannot start, or the machine would trigger system error. This is sufficient for job failure. Also, lamps are too close to a certain internal component (Sensor-6, Sensor-7). Internal cooling is too hot (Sensor-8), but this would not have much effect as the printing process will not finish successfully anyhow. Sensor-34 also indicates overheating. This cluster is again formed by failed jobs and the reasons of failure are heterogeneous, including "System Error", "power cut", "wiper" or "failure to print". Some of the sensors show a clear bimodal distribution, thus indicating that the cluster might contain jobs from two different printers. Further analysis will help us to understand if this cluster requires further subdivision.

Cluster 3 $\rightarrow$ In this cluster sufficient pressure is reached (Sensor-1) to start printing process. Here, the problem is that Sensor-9 is running at very high temperature while Sensor-3, which is located near the subsystem where sensor 9 is placed, is not properly matching the behavior of Sensor-9. This might cause imbalance in internal cooling, and thereby generate system alert. Projecting the CPG with illustrative variables, we can see that most of the jobs in this clusters failed either by "System Error" or "Failed to Print".

4   Conclusion

This study is the first part of a wider research to establish real time data science methodologies for the extraction of conceptual information from the sensors providing on-line readings from 3D printers. In this paper, off-line hierarchical clustering has been applied on a limited sample of data to show the potential of the clustering process in understanding different printing scenarios. The case study has been quite helpful in getting an idea of how sensors can contribute to understand the underlying behavior in 3D Printers. Some specific interpretation-oriented tools (like Class Panel Graphs) have been used to profile the obtained clusters. On the one hand, this permitted to identify redundant sensors that provide similar information even if they are placed in different components of the 3D printers. On the other hand, the CPG allowed a quick identification of sensors that were possibly significant with regards to the structure of clusters. Significance has been verified by the appropriate statistical tests and the interpretation of significant differences among clusters permitted to build the profiles described in previous section. Four different profiles were identified, from which three of them correspond to unsuccessful jobs. From these, two were associated with insufficient pressure even to start printing, and the third profile is associated with the mismatch between two sensors in a certain subsystem that reports unbalance in the internal cooling. Cluster 0 is associated with problems in carriage and fusing lamps, whereas cluster 3 shows overheating problems on top of that and failure can have many different origins. Finally, cluster 1 contains successful jobs and some failures due to "wiper". This seems to be associated with very low temperature in air-in and air-out. This work has shown the promising added value of standardizing a sensor data-intensive data science process to better understand 3D printing processes and identify improvements in the 3D printers design that allow job quality improvement. The analysis presented in this work has raised some hypothesis to be confirmed with through further future research and suggests that quality of the job (failure/success) may not be wholly associated to individual components, but also the interaction between different subsystems.

References

1. Wong KV, Hernandez A (2012) A review of additive manufacturing. ISRN Mech Eng
2. HP Inc. Technical Whitepaper. http://images.engage.hp.com/Web/HPInc/%7B12bd7228-a080-4361-889c-bc30e7d7633a%7D_Technical_white_paper_4AA5-5472ENW_March_2018.pdf. Accessed 28 Feb 2019
3. Gibert K, Sànchez-Marrè M, Codina V (2010) Choosing the right data mining technique: classification of methods and intelligent recommenders. In: Proceedings of the iEMSs fifth biennial meeting: international congress on environmental modelling and software, vol 1, pp 2448–2453
4. Gibert K, Spate J, Sànchez-Marrè M, Athanasiadis IN, Comas J (2008) Chapter twelve data mining for environmental systems. In: Developments in integrated environmental assessment, pp 205–228

5. Tourloukis G, Stoyanov S, Tilford T, Bailey C (2015) Data driven approach to quality assessment of 3D printed electronic products. In: 38th international spring seminar on electronics technology (ISSE). IEEE, pp 300–305
6. Jain AK (2010) Data clustering: 50 years beyond K-means. Pattern Recogn Lett 31(8):651–666
7. Guha S, Rastogi R, Shim K (1998) CURE: an efficient clustering algorithm for large databases. In: ACM sigmod record, vol 27, no 2. ACM, pp 73–84
8. Gibert K, Nonell R, Velarde JM, Colillas MM (2005) Knowledge discovery with clustering: impact of metrics and reporting phase by using Klass. Neural Netw World 15(4):319

A Walk Through Spectral Bands: Using Virtual Reality to Better Visualize Hyperspectral Data

Henry Kvinge[1], Michael Kirby[1(✉)], Chris Peterson[1], Chad Eitel[2], and Tod Clapp[2]

[1] Department of Mathematics, Colorado State University, Fort Collins, CO, USA
kirby@math.colostate.edu
[2] Department of Biomedical Sciences, Colorado State University, Fort Collins, CO, USA

Abstract. One of the basic challenges of understanding hyperspectral data arises from the fact that it is intrinsically 3-dimensional. A diverse range of algorithms have been developed to help visualize hyperspectral data trichromatically in 2-dimensions. In this paper we take a different approach and show how virtual reality provides a way of visualizing a hyperspectral data cube without collapsing the spectral dimension. Using several different real datasets, we show that it is straightforward to find signals of interest and make them more visible by exploiting the immersive, interactive environment of virtual reality. This enables signals to be seen which would be hard to detect if we were simply examining hyperspectral data band by band.

Keywords: Hyperspectral imaging · Virtual reality ·
Data visualization · Chemical signal visualization

1 Introduction

The visualization of large data sets in high dimensions still poses a major challenge. The ultimate goal is to convert these observed, or simulated, phenomenon into rules, laws or principles. This is the process of knowledge discovery from data. Recent advances in virtual reality (VR), coupled with enhancements in computational power, present the possibility of a new paradigm for data exploration. This will involve the combination of an expert user interacting with mathematical algorithms that characterize information in the observations. The final objective is the guided navigation of high-dimensions in order to extract fundamental insights into the data.

Unlike RGB (red-green-blue) data which has 3-bands and which can be visualized in 2-dimensions by exploiting our trichromacy (i.e. our color vision), most hyperspectral data is inescapably 3-dimensional by virtue of having many more spectral bands (commonly more than 150). In order to overcome the challenge of

© Springer Nature Switzerland AG 2020
A. Vellido et al. (Eds.): WSOM 2019, AISC 976, pp. 160–165, 2020.
https://doi.org/10.1007/978-3-030-19642-4_16

analyzing such data, algorithms and software have been developed which either allow the user to scan through the data and selectively display different bands [1], selects the bands automatically to try to maximize some information criteria [2,3], or treats band selection as a dimensionality reduction problem [4–6].

In this paper we take a different approach. Instead of trying to develop methods which condense the information of a hyperspectral image into a digestible 2-dimensional format, we instead embrace the 3-dimensionality of hyperspectral data by choosing to use VR where we can visualize the data cubes in full. We propose this framework not only because it fully leverages our innate 3-dimensional visual abilities, but also because it provides for an immersive experience where the user's full attention can be directed to exploring the data. The idea of VR has a history dating back to 1965 [7]. It has only been in the last 15 years however that hardware capabilities have improved to the point that many of the envisioned applications of VR have started to become a reality. In general, VR is a good tool for exploring data where the addition of a third spatial dimension for exploration provides significant improvements in understanding. A few examples of applications of VR include [8–13].

We see at least two advantages to visualizing hyperspectral data using VR. The most obvious has already been pointed out; in VR, researchers would be exploring the data in its natural dimension, without any reductions necessary. The difference here is equivalent to the difference between giving a physician a collection of 2-dimensional images of slices of a human heart versus giving the physician the ability to look at that stack of images formed into a single 3-dimensional representation of that heart. In VR we expect explorations of hyperspectral data to not only be more thorough but also vastly more efficient and accessible. The second benefit to moving the analysis of hyperspectral data to the virtual realm is that it will make it easier for researchers to collaborate in real time regardless of whether they are in the same location or not [14].

2   Background

2.1   Hyperspectral Data

The hyperspectral data that we use in this paper comes in the form of a *data cube*, that is an $n \times m \times b$ array of numbers where n, m, b are positive integers. The first two coordinates correspond to the *spatial dimensions* while the values of n and m denote the *spatial resolution* of the data cube. The third coordinate corresponds to the *spectral dimension* with b denoting the number of *spectral bands*. For each k with $1 \leq k \leq b$, the kth band is represented by an $n \times m$ array that captures the intensity of energy in that particular wavelength across the scene.

2.2   Virtual Reality for Data Visualization

Virtual reality utilizes a combination of hardware and software to create an immersive and explorable environment in which a user can interact with data.

Its immersive environment affords users a unique and intuitive perspective of structural relationships and scale.

We used BananaVision software developed at Colorado State University to load and visualize our hyperspectral data in VR. BananaVision is a networked, multi-user virtual reality tool that renders both human anatomical learning models as well as medical images (CT, PET/CT, MRI, MEG, NIfTII) for the creation of instructional content and clinical evaluation. The BananaVision software takes "stacks" of 2-dimensional images and volumizes them to give a 3-dimensional representation that can be explored in virtual reality. The volumetric data is colored by the positive scalar value attached to each position in the data cube. In the case of hyperspectral data, the coloring is thus determined by the spectral intensity values at each point in the cube.

Without manipulation, a data cube would be presented in VR as a solid object and we would only be able to study its exterior (the edges of each band as well as the first and last band). In order to begin to segment out interesting structures within the cube we need to: slice the cube to expose intensity values in the interior, make certain locations in the cube more or less transparent, and choose a coloring scheme (based on intensity) which helps us to discriminate between different structures.

3   Example Visualizations

In this section we describe and provide pictures of the application of VR to two hyperspectral data sets. These explorations were carried out using a machine running an Nvidia 1080 Ti graphics card.

3.1   Indian Pines

The Indian Pines data set consists of 220 spectral bands at a resolution of 145×145 pixels (that is, a $145 \times 145 \times 220$ data cube). We use a corrected data set that has bands from the region of water absorption removed, resulting in 200 remaining spectral bands.[1] The data was collected with an airborne visible/infrared imaging spectrometer (AVIRIS) at the Indian Pines test site in Indiana [15]. The scene contains various classes, such as woods, grass, corn, alfalfa, and buildings.

In Fig. 1(a) we show what this data looks like with the default visualization settings in VR. As described above, the cube is colored by the intensity at a location in the cube. As can be seen, without adjusting the display settings, the only parts of the data cube that we can examine are the boundary faces. In Fig. 1(b) we have removed certain bands, thresholded out certain intensity levels, and chosen to make other intensity values more transparent. The result is that we can essentially carve out the spectral information related to some of the

---

[1] The data is available at: http://www.ehu.eus/ccwintco/index.php/Hyperspectral_Remote_Sensing_Scenes.

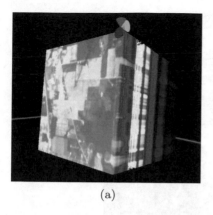

(a)   (b)

Fig. 1. The Indian Pines hyperspectral data cube as seen in virtual reality. (a) Shows how the cube appears when it is first loaded (that is, in the default display settings). The yellow rectangle is the first band. One can see the outline of agricultural fields. (b) Shows what the cube looks like after we have removed certain bands, made completely transparent positions in the cube with intensity values above or below certain thresholds, and selectively colored and assigned transparency values to the remaining intensity values. As can be seen, using these relatively simple operations we can begin to chisel out figures of interest in the cube. The agricultural fields in (a) for example, now stand out in 3-dimensions as plateau-like structures in (b).

features from the cube. For example, at least some of the plateau-like structures seen in (b) correspond to agricultural fields with different kinds of crops. The structures that resemble trees at the far top corner of the cube do not correspond to a labeled ground truth class. This is an example of a feature that stands out in VR that might warrant further examination.

3.2   Chemical Plume Detection

In this section we give an example of VR applied to hyperspectral data containing a chemical simulant release. The data set that we use is the Johns Hopkins Applied Physics Lab FTIR-based longwave infrared sensor hyperspectral dataset [16]. This dataset consists of several hyperspectral videos showing a chemical simulant release. We will consider two cubes from a video showing the release of the chemical simulant sulfur hexafluoride (SF_6). One cube was taken before the release of the chemical simulant, the other was taken after and contains the chemical signal. Each data cube consists of 129 spectral bands with spatial resolution 128×320 (for a data cube of size $128 \times 320 \times 129$). In order to make the comparison between these with both visible simultaneously, we concatenate them along the spectral dimension. In Fig. 2(a) we show what the two concatenated data cubes look like in VR with default settings. The left cube in both (a) and (b) is the one that contains the chemical release. While this is not immediately obvious in Fig. 2(a), with some straight-forward intensity thresholding, color shifting, and cube slicing the chemical release becomes apparent as

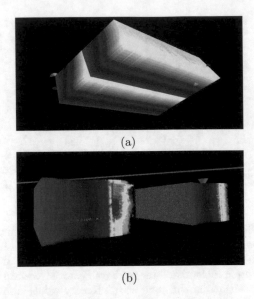

(a)

(b)

Fig. 2. Two cubes from the Applied Physics Lab FTIR-based longwave infrared sensor hyperspectral dataset, concatenated along the spectral dimension. One cube contains a chemical release of the simulant SF_6 while the other does not. It is not obvious from (a) which cube contains the chemical release. After some straight-forward intensity thresholding, color shifting, and cube slicing, the chemical release becomes apparent (b) as the blue nodule in the left cube (the spectral dimension from this viewpoint runs horizontally).

seen in Fig. 2(b). The chemical plume can be seen in the left cube taking the form of the blue nodule.

That we can find the chemical plume relatively easily in VR for this example is significant. It is non-trivial to find this plume by scanning through bands in 2-dimensions. We believe that our results here point toward hyperspectral data being a promising domain in which to apply VR.

4   Conclusion

In this paper we proposed the use of virtual reality as a tool to explore hyperspectral data cubes. We showed that virtual reality allows one to leverage human's innate 3-dimensional spatial ability to explore this type of data.

We end by noting some features that we believe would be useful in future VR software designed specifically for the purpose of studying hyperspectral imagery. These include: the incorporation hyperspectral specific data within the VR environment, the ability to extract spectral curves from within the VR environment, the ability to load and independently move multiple data cubes at once, and the ability to apply selected algorithms (such as wavelet or Fourier transforms) to the data cube within VR.

We speculate that the future development of geometric and topological algorithms for data analysis, such as Self-Organizing Mappings, may benefit from the ability to visualize high-dimensions using VR as a tool.

References

1. Biehl L, Landgrebe D (2002) MultiSpec—a tool for multispectral–hyperspectral image data analysis. Comput Geosci 28(10):1153–1159
2. Le Moan S, Mansouri A, Voisin Y, Hardeberg JY (2011) A constrained band selection method based on information measures for spectral image color visualization. IEEE Trans Geosci Remote Sens 49(12):5104–5115
3. Demir B, Celebi A, Erturk S (2009) A low-complexity approach for the color display of hyperspectral remote-sensing images using one-bit-transform-based band selection. IEEE Trans Geosci Remote Sens 47(1):97–105
4. Zhu Y, Varshney PK, Chen H (2007) Evaluation of ICA based fusion of hyperspectral images for color display. In: 2007 10th international conference on information fusion, pp 1–7. IEEE
5. Tyo JS, Konsolakis A, Diersen DI, Olsen RC (2003) Principal-components-based display strategy for spectral imagery. IEEE Trans Geosci Remote Sens 41(3):708–718
6. Du Q, Raksuntorn N, Cai S, Moorhead RJ (2008) Color display for hyperspectral imagery. IEEE Trans Geosci Remote Sens 46(6):1858–1866
7. Sutherland IE (1965) The ultimate display. In: Multimedia: from wagner to virtual reality, pp 506–508
8. Gamito P, Oliveira J, Coelho C, Morais D, Lopes P, Pacheco J, Brito R, Soares F, Santos N, Barata AF (2017) Cognitive training on stroke patients via virtual reality-based serious games. Disabil Rehabil. 39(4):385–388
9. Mujber TS, Szecsi T, Hashmi MS (2004) Virtual reality applications in manufacturing process simulation. J Mater Proces Technol 155:1834–1838
10. Meola A, Cutolo F, Carbone M, Cagnazzo F, Ferrari M, Ferrari V (2017) Augmented reality in neurosurgery: a systematic review. Neurosurg Rev 40(4):537–548
11. Heise N, Hall HA, Garbe BA, Eitel CM, Clapp TR (2018) A virtual learning modality for neuroanatomical education. FASEB J **32**. 635.10
12. Salvadori A, Del Frate G, Pagliai M, Mancini G, Barone V (2016) Immersive virtual reality in computational chemistry: applications to the analysis of QM and MM data. Int J Quant Chem. 116(22):1731–1746
13. Basantes J, Godoy L, Carvajal T, Castro R, Toulkeridis T, Fuertes W, Aguilar W, Tierra A, Padilla O, Mato F et al (2017) Capture and processing of geospatial data with laser scanner system for 3D modeling and virtual reality of amazonian caves. In: Ecuador technical chapters meeting (ETCM), 2017 IEEE. IEEE, pp 1–5
14. Zyga L (2009) Virtual worlds may be the future setting of scientific collaboration. https://phys.org/news/2009-08-virtual-worlds-future-scientific-collaboration.html
15. Baumgardner MF, Biehl LL, Landgrebe DA (2015) 220 band AVIRIS hyperspectral image data set: 12 June 1992 Indian Pine test site 3. https://doi.org/10.4231/R7RX991C. https://purr.purdue.edu/publications/1947/1
16. Broadwater JB, Limsui D, Carr AK (2011) A primer for chemical plume detection using LWIR sensors. Technical Paper, National Security Technology Department, Las Vegas, NV

Incremental Traversability Assessment Learning Using Growing Neural Gas Algorithm

Jan Faigl$^{(\boxtimes)}$ and Miloš Prágr

Department of Computer Science, Faculty of Electrical Engineering,
Czech Technical University in Prague, Technická 2, 166 27 Prague 6, Czech Republic
{faiglj,pragrmi1}@fel.cvut.cz
https://comrob.fel.cvut.cz/

Abstract. In this paper, we report early results on the deployment of the growing neural gas algorithm in online incremental learning of traversability assessment with a multi-legged walking robot. The addressed problem is to incrementally build a model of the robot experience with traversing the terrain that can be immediately utilized in the traversability cost assessment of seen but not yet visited areas. The main motivation of the studied deployment is to improve the performance of the autonomous mission by avoiding hard to traverse areas and support planning cost-efficient paths based on the continuously collected measurements characterizing the operational environment. We propose to employ the growing neural gas algorithm to incrementally build a model of the terrain characterization from exteroceptive features that are associated with the proprioceptive based estimation of the traversal cost. Based on the reported results, the proposed deployment provides competitive results to the existing approach based on the Incremental Gaussian Mixture Network.

Keywords: Terrain characterization · Multi-legged walking robot

1 Introduction

The problem studied in this paper is arising from robotic data collection missions where an autonomous robot is requested to repeatably visit a set of locations to measure some phenomena of interest [3]. During such a mission, the robot not only collects the requested data measurements, but it also experiences the terrain. Hence, the experience can be exploited in finding more efficient paths through the environment, and thus improve the mission performance. This might be especially suitable for multi-legged robots that can traverse rough terrains [1], but they suffer from low stability and high energy requirements when traversing difficult terrains [6]. Therefore, we aim to incrementally learn a model of the traversability assessment that can be instantly utilized in the evaluation of the

© Springer Nature Switzerland AG 2020
A. Vellido et al. (Eds.): WSOM 2019, AISC 976, pp. 166–176, 2020.
https://doi.org/10.1007/978-3-030-19642-4_17

Fig. 1. Example of the traversability assessment model used in path planning for a hexapod walking robot. The gray areas are parts of the operational environment mapped so far, and the traversability cost is visualized by the respective color of the areas, where hard to traverse parts are in red and low-cost parts are in blue.

seen but not yet visited areas, to support planning cost-efficient paths. Our motivational deployment is visualized in Fig. 1.

Existing approaches to terrain characterization can be categorized into terrain classification and traversability assessment using a continuous function. The traversability assessment for path planning is of our particular interest; however, we consider assessment only of passable terrains, and thus we rely on prior binary classification to traversable and untraversable areas. The estimation of the motion cost can be based on locally observed properties [14] and modeling of such a spatial phenomenon can be performed by the Gaussian Process (GP) based regression, e.g., to create continuous occupancy [9] or elevation maps [16]. However, GPs are computationally demanding to be directly utilized in online decision-making and incremental model learning. Therefore, in our previous work [12], we employed the Incremental Gaussian Mixture Network Model (IGMN) [10] to learn a cost of transport model that is then utilized in the cost prediction using exteroceptive sensing of robot surroundings.

Although the IGMN provides a satisfiable performance in the cost estimation suitable for path planning [12,13], we aim to explore other possibilities to combine terrain classification with terrain traversability learning in a computationally efficient way. Motivated by recent advancements in the application of the Growing Neural Gas (GNG) algorithms in online labeling [2], time series classification [8], online anomaly detection [15] using motion and appearance features [17] and clustering data streams [5], we tackled the studied problem using the original GNG algorithm [4] proposed by Bernd Fritzke in 1994. The only modification is in adding a new node if the current winning unit is not close enough [11] instead of growing every fixed number of adaptations.

In the rest of the paper, we specify the problem context and report on our early evaluation results and comparison of the GNG with the IGMN.

2   Problem Specification

The studied problem is to estimate the traversal cost based on the experience of the robot with terrain traversing. The considered terrain characterization is a combination of exteroceptive signals, which allows predicting the cost from range measurements, with proprioceptive measurements characterizing the robot experience with the terrain. The robot operational environment is modeled as the 2.5D elevation map to store the elevation and RGB color information that is utilized to compute the exteroceptive part of the terrain descriptor. In particular, we utilize a terrain feature descriptor that consists of three shape features [7] and two appearance features. The shape features s_1, s_2, and s_3 are defined as

$$s_1 = \frac{\lambda_1}{\lambda_3}, \quad s_2 = \frac{\lambda_2 - \lambda_1}{\lambda_3}, \quad s_3 = \frac{\lambda_3 - \lambda_2}{\lambda_3}, \tag{1}$$

where $\lambda_1 \leq \lambda_2 \leq \lambda_3$ are the eigenvalues of the covariance matrix of the elevation in a particular area of interest. The appearance part of the descriptor consists of the ab channel means of Lab color space denoted a_1 and a_2. Finally, the traversal cost is characterized by the experienced stability cost c determined as the square root of the robot roll variance for 10 s period measured by the onboard attitude heading reference system running at 400 Hz. The model descriptor is thus six dimensional vector $d = (s_1, s_2, s_3, a_1, a_2, c)$ and the traversability assessment is based on the model inference to predict c using $d_{sa} = (s_1, s_2, s_3, a_1, a_2)$.

During the mission, the robot builds a map of its surroundings, traverse the terrain, and its experience with the terrain can be represented as a sequence of n descriptors that is further called trail $\mathcal{T}$, i.e., $\mathcal{T} = (d(1), \ldots, d(n))$. Every single descriptor $d(k)$ is utilized to incrementally update the model $\mathcal{M}(k)$

$$\mathcal{M}(k) \leftarrow \text{learn}(\mathcal{M}(k-1), d(k)) \tag{2}$$

that can be immediately used to assess a set of descriptors characterizing seen but not yet visited areas, e.g., organized in a grid map, $\mathcal{G} = \{(x_i, y_j, d_{sa}(i, j)) \mid 1 \leq i \leq w, 1 \leq j \leq h\}$, where x_i and y_j are the spatial coordinates of the corresponding exteroceptive measurements $d_{sa}(i, j)$ for the $w \times h$ large grid map. Since the measurements might not be available for every cell of the map, the number of descriptors $m = |\mathcal{G}|$ can be $m \leq w \cdot h$.

The evaluation of the learned model and its generalization to other environments can be based on measuring the difference between the predicted values of c for the grid map $\mathcal{G}$ and the ground truth values. However, it is nearly impossible to establish the ground truth, because it would require a precise and complete traversing of all areas of the particular environment, but most importantly the measured experience depends on many factors, and it is generally a random variable. Therefore, we consider a reference value of the predicted cost determined by the computationally demanding GPs using the whole particular trail $\mathcal{T}$ denoted as $\mathcal{M}_{\text{GP}}^{\mathcal{T}}$. For each k-th descriptor of the trail, $\mathcal{M}(k)$ is used to assess $\mathcal{G}$, and we measure the performance of the incrementally learned model as the evolution of the root-mean-square error (RMSE) to the GP-based predictor

$$RMSE(k) = \sqrt{\frac{\sum\limits_{\boldsymbol{d}_{sa} \in \mathcal{G}} \left(\text{predict}(\mathcal{M}(k), \boldsymbol{d}_{sa}) - \text{predict}(\mathcal{M}_{\text{GP}}^{\mathcal{T}}, \boldsymbol{d}_{sa})\right)^2}{m}}. \tag{3}$$

The model inference provides a prediction of the traversal cost as a continuous variable, and its particular value depends on many factors. Therefore, models learned by different techniques would unlikely provide the identical value of the predicted traversal cost as the reference GP-based model. Hence, we can take advantage of the learned GP-based model that provides the variance of the learned random variables, and we can consider a model is well approximating the GP-based reference if the predicted value is close to the predicted mean value of the GP-based model. Thus, for each particular descriptor $\boldsymbol{d}_{sa} \in \mathcal{G}$, we can estimate the mean $\mu(\boldsymbol{d}_{sa}) = \text{predicted}(\mathcal{M}_{\text{GP}}^{\mathcal{T}}, \boldsymbol{d}_{sa})$ and its variance $\sigma^2(\boldsymbol{d}_{sa})$. Then, we can consider that the predicted value by the model $\mathcal{M}$ is correct with respect to the reference model $\mathcal{M}_{\text{GP}}^{\mathcal{T}}$ if its distance from $\mu(\boldsymbol{d}_{sa})$ is shorter than two times of the standard deviation, i.e., the predicted value fits about 95% values of the corresponding distribution represented by the GP. Based on this idea, the model correctness quality indicator R_c can be defined as the ratio of the number of the correctly estimated traversal costs to the total number of the descriptors in $\mathcal{G}$

$$R_c(\mathcal{M}) = \frac{|\{\boldsymbol{d}_{sa} | \boldsymbol{d}_{sa} \in \mathcal{G} \text{ and } |\,\text{predict}(\mathcal{M}, \boldsymbol{d}_{sa}) - \mu(\boldsymbol{d}_{sa})| \leq 2\sigma(\boldsymbol{d}_{sa})\}|}{|\mathcal{G}|} \cdot 100\%. \tag{4}$$

Finally, we can further exploit explicitly labeled terrains and learn individual GP-based model for each particular terrain type using only the corresponding parts of the trails for the specific (human labeled) terrain types. In the evaluation of the testing grid map $\mathcal{G}$, we can use all learned models to predict the traversal cost, but the value with the lowest variance (i.e., with the highest confidence of the predicted value) is considered to be the reference traversal cost for the particular descriptor. Such a compound model of individual GPs for particular terrain types is denoted $\mathcal{M}_{\text{GP}}^{tt}$ and it can be used in (4) as the reference GP-based model. Note that such an evaluation is possible only if the explicit labels of the terrain types for the specific parts of the trails are available, which is not the case of the incremental learning in the motivational deployment, but labels are available for evaluation of the examined learning methods.

3   Evaluation Results

The experimental evaluation of the GNG in terrain assessment learning has been performed for a hexapod walking robot in a set of laboratory terrains that consists of flat ground, black fabric, artificial turf, wooden blocks, and wooden stairs. Each terrain type has been traversed four times and a single $\mathcal{T}_{\text{all}}$ of all concatenated trails has 827 descriptors. The evaluation is performed for a grid $\mathcal{G}$ created for a different setup with slightly modified terrain types, see Fig. 2.

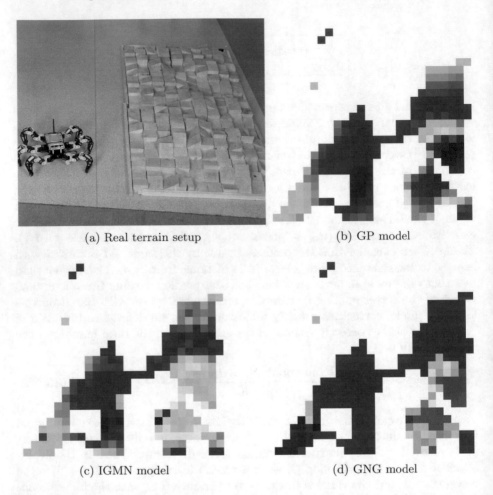

(a) Real terrain setup   (b) GP model

(c) IGMN model   (d) GNG model

Fig. 2. A snapshot of the robot on wooden blocks and visualization of the assessment of the reference $\mathcal{G}$ using the full GP-based model and learned IGMN and GNG for $\mathcal{T}_{\text{all}}$.

The GNG algorithm [4] has been implemented in C++, and both the learning and inference take a fraction of millisecond, and it is practically negligible. The utilized parameters according to notation in [4] are $\epsilon_b = 0.2$, $\epsilon_n = 0.1$, $a_{max} = 10$, $\alpha = 0.5$, $d = 0.995$, and new node is added if the Euclidean distance of the new measurement d to the nearest unit exceeds 0.15.

In addition to the GP-based model, we compare the GNG with the IGMN [10] that is supervised approximation of the EM algorithm that incrementally constructs the Gaussian mixture model, adjusts its components and parameters based on each presented training sample. New components are inserted during the learning process and must prove their relevance by accumulating sufficient posterior probability to be retained by the mixture. The IGMN has been utilized in our previous work [12] and here, we use the same setup with the maximal number of com-

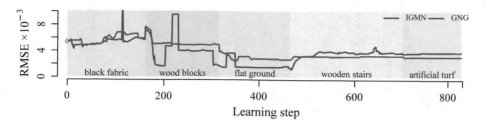

Fig. 3. Evolution of $RMSE(k)$ for particular learning step – each learning step the whole grid map $\mathcal{G}$ is assessed using the currently learned model.

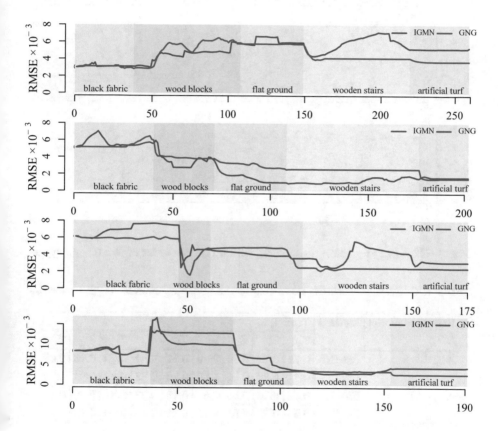

Fig. 4. Evolution of $RMSE(k)$ for individual trails $\mathcal{T}_1$–$\mathcal{T}_4$ from top to bottom.

ponents limited to ten, the grace period $v_{\min} = 100$, minimal accumulated posterior $sp_{\min} = 3$, and scaling factor $\delta = 1$, but with terrain descriptor $\boldsymbol{d}$. Due to its implementation in Python, it is a bit more computationally demanding, and it operates in a fraction of second, which is, however, still satisfiable for deployment in online path planning.

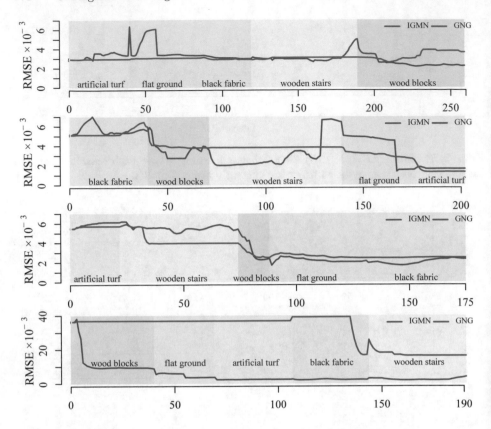

Fig. 5. $RMSE(k)$ for the trails $\mathcal{T}_{r_1}$–$\mathcal{T}_{r_4}$ with a random order of the terrain types.

Table 1. Correctness ratio R_c according to the reference models $\mathcal{M}_{\mathrm{GP}}^{\mathcal{T}_{all}}$ and $\mathcal{M}_{\mathrm{GP}}^{tt}$

Reference model	Method	Trail								
		$\mathcal{T}_{all}$	$\mathcal{T}_1$	$\mathcal{T}_2$	$\mathcal{T}_3$	$\mathcal{T}_4$	$\mathcal{T}_{r_1}$	$\mathcal{T}_{r_2}$	$\mathcal{T}_{r_3}$	$\mathcal{T}_{r_4}$
$\mathcal{M}_{\mathrm{GP}}^{\mathcal{T}_{all}}$	IGMN	0.72	0.70	0.86	0.99	0.87	0.82	0.99	0.98	0.88
	GNG	0.71	0.82	0.91	1.00	0.96	0.82	0.82	0.94	0.82
$\mathcal{M}_{\mathrm{GP}}^{tt}$	IGMN	0.52	0.16	0.53	0.56	0.18	0.23	0.31	0.62	0.18
	GNG	0.66	0.53	0.61	0.71	0.26	0.47	0.68	0.73	0.24

The performance of the predictors has been evaluated using (3) for the whole trail $\mathcal{T}_{all}$, but also for four individual trails $\mathcal{T}_1$, $\mathcal{T}_2$, $\mathcal{T}_3$, and $\mathcal{T}_4$ with the terrain sequence of black fabric, wooden blocks, flat, stairs, and artificial turf. Besides, we consider four additional trails with shuffled terrain types denoted $\mathcal{T}_{r_i}$. The evolution of the RMSE is depicted in Figs. 3, 4, and 5. In addition to the RMSE, we consider the correctness ratio R_c defined in (4) for evaluation of the final

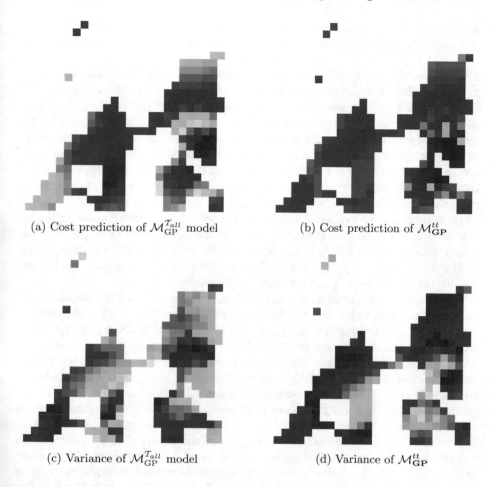

(a) Cost prediction of $\mathcal{M}_{\mathrm{GP}}^{\mathcal{T}_{all}}$ model

(b) Cost prediction of $\mathcal{M}_{\mathbf{GP}}^{tt}$

(c) Variance of $\mathcal{M}_{\mathrm{GP}}^{\mathcal{T}_{all}}$ model

(d) Variance of $\mathcal{M}_{\mathbf{GP}}^{tt}$

Fig. 6. Reference GP-based traversal cost model using the trails $\mathcal{T}_{all}$ with all learning data (right) and compound model based on known terrain types (left). The bottom row visualizes variances of the predicted traversal costs. The lower variances are shown in the dark blue, and it can be observed that the compound model estimates the traversal cost with the overall lower variances. The most unsure prediction is for wooden blocks, see Fig. 2a. A single GP-based model using $\mathcal{T}_{all}$ has the highest variances for the flat ground, and the wooden blocks with the high traversal cost are predicted with the relatively lower variance, but only a few regions have the lowest variance.

learned models per particular method and individual trails. Since the trail $\mathcal{T}_{all}$ can be considered as the most information-rich, the used reference GP-based model is learned from $\mathcal{T}_{all}$. The results are depicted in Table 1.

A similar evaluation can be performed for the compound reference model consisting of the individual GP-based models for the particular terrain types denoted $\mathcal{M}_{\mathrm{GP}}^{tt}$, see the two bottom rows in Table 1. In this case, the number of the correctly estimated traversal costs is overall noticeably smaller than for a single GP-based model using $\mathcal{T}_{all}$. The two reference GP-based models are compared in Fig. 6 regarding the predicted traversal cost and estimated variance.

Discussion – The assessments in Fig. 2 indicate that both the IGMN and GNG models partially learned the traversability assessment in comparison to the GP-based model. However, regarding path planning, it is important that the cost is sufficiently distinguishable to avoid difficult terrains, which is satisfied for both models. Regarding the evolution of the RMSE, the GNG performs a bit better in the particular case shown in Fig. 3. On the other hand, it is evident from the individual trails in Fig. 4 and especially with shuffled terrain types in Fig. 5 that particular sequence of the terrains can significantly affect the performance regarding the reference GP-based model. It is also not surprising that difficult terrains need several samples to improve the model, see Fig. 5 for the trail that starts with the wooden blocks. Although we employed the GNG algorithm in a very straightforward way, it provides the relatively competitive performance to the IGMN regarding (3), except the trail that starts at wooden blocks (see Fig. 5), which motivates us for further development.

In particular, the results indicate that terrain characterization purely based on a continuous function can provide sporadic results as the real performance of the robot can vary significantly. It is partially addressed in the evaluation using a compound model based on the individual GP-based model for each particular terrain type. Regarding the results visualized in Fig. 6, an advantage of the individual models of the traversal cost per particular terrain types is not clearly supported. Even though there are more parts with the lower variance of the predicted values, there is also the relatively unsure part corresponding to the wooden blocks, where the predicted cost is significantly lower than in Fig. 6a. It is most likely because the terrain descriptor of the wooden blocks is similar to the wooden stairs and considering individual terrain classes reduces the number of samples used in the model learning. Therefore, in our future work, we plan to consider identification of the terrain types and eventually combine the benefits of the both approaches to improve the overall traversal cost prediction. Thus, we aim to investigate techniques of unsupervised clustering to automatically identify possible terrain types and incrementally learn a model of the aggregated traversability cost for such identified terrain classes.

4   Conclusion

We presented evaluation results on a straightforward deployment of the GNG algorithm in incremental traversability assessment learning. We described the problem and evaluation challenges related to the nature of the incremental model learning and simultaneous usage of the model in decision-making for improving the mission performance by a more informed path planning. Although the presented results do not support the GNG is the most suitable technique for the addressed problem, its main benefit is in simplicity and computational efficiency, which allows modeling the traversability cost using tens and hundreds of units in comparison to the fixed number of components in the IGMN, where the size is limited to ten to get a reasonable performance. We consider the added value of this paper in reporting on evaluation results and introducing the methodology for comparing predictors in incremental traversability assessment learning.

Regarding the results, there are still several open questions, but also promising ideas. We aim to further work on combining the continuous traversability assessment function with more explicit terrain classification to improve the performance by recently proposed GNG for anomaly detection in data streams. Moreover, we also plan to consider the explicit sequence of the data measurements and support the terrain classification and traversal cost modeling by multi-dimensional time series.

Acknowledgments. This work was supported by the Czech Science Foundation under research project No. 18-18858S. The authors acknowledge the support of the OP VVV funded project CZ.02.1.01/0.0/0.0/16_019/0000765 "Research Center for Informatics".

References

1. Bartoszyk S, Kasprzak P, Belter D (2017) Terrain-aware motion planning for a walking robot. In: International workshop on robot motion and control (RoMoCo), pp 29–34
2. Beyer O, Cimiano P (2011) Online labelling strategies for growing neural gas. Intell Data Eng Autom Learn (IDEAL) 6936:76–83
3. Dunbabin M, Marques L (2012) Robots for environmental monitoring: significant advancements and applications. IEEE Robot Autom Maga 19(1):24–39
4. Fritzke B (1994) A growing neural gas network learns topologies. In: Neural information processing systems (NIPS), pp 625–632
5. Ghesmoune M, Lebbah M, Azzag H (2016) A new growing neural gas for clustering data streams. Neural Netw 78:36–50
6. Homberger T, Bjelonic M, Kottege N, Borges PVK (2016) Terrain-dependant control of hexapod robots using vision. In: International symposium on experimental robotics (ISER), pp 92–102
7. Kragh M, Jørgensen RN, Pedersen H (2015) Object detection and terrain classification in agricultural fields using 3D lidar data. In: International conference on computer vision systems (ICVS), vol 9163. Springer, pp 188–197
8. Nooralishahi P, Seera M, Loo CK (2017) Online semi-supervised multi-channel time series classifier based on growing neural gas. Neural Comput Appl 28(11):3491–3505
9. O'Callaghan S, Ramos FT, Durrant-Whyte H (2009) Contextual occupancy maps using Gaussian processes. In: IEEE international conference robotics and automation (ICRA), pp 1054–1060
10. Pinto R, Engel P (2015) A fast incremental Gaussian mixture model. PLOS, e0141942
11. Prudent Y, Ennaji A (2005) An incremental growing neural gas learns topologies. In: International joint conference on neural networks (IJCNN), vol 2, pp 1211–1216
12. Prágr M, Čížek P, Faigl J (2018) Cost of transport estimation for legged robot based on terrain features inference from aerial scan. In: IEEE/RSJ international conference intelligent robots and systems (IROS), pp 1745–1750

13. Prágr M, Čížek P, Faigl J (2018) Incremental learning of traversability cost for aerial reconnaissance support to ground units. In: Modelling & simulation for autonomous system (MESAS)
14. Stelzer A, Hirschmüller H, Görner M (2012) Stereo-vision-based navigation of a six-legged walking robot in unknown rough terrain. Int J Robot Res 31(4):381–402
15. Sun Q, Liu H, Harada T (2017) Online growing neural gas for anomaly detection in changing surveillance scenes. Patt Recogn 64:187–201
16. Vasudevan S, Ramos F, Nettleton E, Durrant-Whyte H, Blair A (2009) Gaussian process modeling of large scale terrain. In: IEEE international conference robotics and automation (ICRA), pp 1047–1053
17. Zhang Y, Lu H, Zhang L, Ruan X (2016) Combining motion and appearance cues for anomaly detection. Patt Recogn 51:443–452

Learning Vector Quantization:
Theoretical Developments

Investigation of Activation Functions for Generalized Learning Vector Quantization

Thomas Villmann[1]([✉]), Jensun Ravichandran[1], Andrea Villmann[1,2],
David Nebel[1,2], and Marika Kaden[1]

[1] Saxony Institute for Computational Intelligence and Machine Learning,
University of Applied Sciences Mittweida, Mittweida, Germany
thomas.villmann@hs-mittweida.de
[2] Schulzentrum Döbeln-Mittweida, Döbeln, Germany
https://www.institute.hs-mittweida.de/webs/sicim.html

Abstract. An appropriate choice of the activation function plays an important role for the performance of (deep) multilayer perceptrons (MLP) in classification and regression learning. Usually, these activations are applied to all perceptron units in the network. A powerful alternative to MLPs are the prototype-based classification learning methods like (generalized) learning vector quantization (GLVQ). These models also deal with activation functions but here they are applied to the so-called classifier function instead. In the paper we investigate whether successful candidates of activation functions for MLP also perform well for GLVQ. For this purpose we show that the GLVQ classifier function can also be interpreted as a generalized perceptron.

Keywords: Learning vector quantization · Classification ·
Activation function · ReLU · Swish · Sigmoid · Perceptron ·
Prototype-based networks

1 Introduction

Prototype-based classification learning like learning vector quantization (LVQ) was introduced by KOHONEN in [1] and belongs to robust and stable classification models in machine learning [2]. One of the most prominent variants is generalized learning vector quantization (GLVQ, [3]). The GLVQ cost function to be minimized by *stochastic gradient descent learning* (SGDL) is an approximation of the overall classification error. Further, GLVQ belongs to the family of margin optimizers for classification learning maximizing the hypothesis margin [4]. Margin optimizers ensure a certain level of model robustness.

Geometrically, LVQ algorithms distribute class-dependent prototypes in the data space during learning. After training new data are classified according to the nearest prototype principle (NPP) based on a data space dissimilarity measure, frequently chosen as the (squared) Euclidean distance. Yet, also adaptive parameterized dissimilarity are popular. Particularly, linear projection metrics are favored allowing an easy interpretation [5].

© Springer Nature Switzerland AG 2020
A. Vellido et al. (Eds.): WSOM 2019, AISC 976, pp. 179–188, 2020.
https://doi.org/10.1007/978-3-030-19642-4_18

Recently, the neural network perspective of LVQ gains an attraction, because it allows to combine LVQ models with techniques of deep learning [6,7]. In this perspective the LVQ prototypes are interpreted as weight vector of linear perceptrons, such that the maximum perceptron excitation by a data vector, here taken as (neural) stimulus, corresponds to minimum distance in NPP [8]. As shown in [2], GLVQ can be seen as a feedforward multilayer (neural) network (MLN) with a single projection layer and a subsequent competition layer.

Another class of those feedforward networks are multilayer perceptrons (MLP) consisting of several perceptron layers [9,10]. For deep networks, the number of layers is huge. The convergence behavior in SGDL as well as the final network performance of (deep) MLPs strongly depends on the activation function in use. If only linear functions are applied, the whole MLP network remains a linear classifier. For non-linear problems, non-linear activation functions are required. Here sigmoidal functions are standard. However, during the last years ReLU-activation (rectified linear units) became very popular because of better performance during training and higher recall quality of respective networks [11]. Yet, newest investigations show that ReLU-units can be further improved using more sophisticated activation function like *swish*, *max* and others [12–15].

For GLVQ, to our best knowledge, only linear and sigmoid activation functions were considered regarding the classification behavior whereas convergence behavior was not in the focus so far. In this context, (class) border sensitive learning was established using an optimized sigmoid activation function [16]. Therefore, the aim of this contribution is to investigate several prominent state-of-the-art activation functions of (deep) MLPs regarding their convergence behavior and resulting final classification performance when applied in GLVQ. For this purpose, we explicitly refer to the multilayer network perspective of GLVQ.

2   Generalized Learning Vector Quantization - A Multilayer Network Perspective

2.1   Basics of GLVQ

We start considering GLVQ from the geometric perspective.

In GLVQ, a set $W = \{\mathbf{w}_1, \mathbf{w}_2, ..., \mathbf{w}_N\}$ of prototypes $\mathbf{w}_k \in \mathbb{R}^p$ is assumed as well as (training) data $\mathbf{x} \in X \subseteq \mathbb{R}^n$ with class labels $c(\mathbf{x})$. Each prototype is uniquely responsible for a certain class $c_k = c(\mathbf{w}_k)$. The data are projected by means of a projection $\pi : \mathbb{R}^n \to \mathbb{R}^p$ into the prototype space $\mathbb{R}^p$ also denoted as projection space in this context. The vector quantization mapping $\mathbf{x} \mapsto \kappa(\mathbf{x})$ takes place as a winner-takes-all (WTA) rule according to

$$\kappa(\mathbf{x}, W) = \mathrm{argmin}_{k:\mathbf{w}_k \in W} \{d(\pi(\mathbf{x}), \mathbf{w}_k) | k = 1, 2, ...N\}. \tag{1}$$

realizing the nearest prototype principle with respect to a predefined dissimilarity measure d. The value $\kappa(\mathbf{x}, W)$ is called the index of the best matching (winner) prototype with respect to the set W. An unknown data vector $\mathbf{u}$ is assigned to the class $c_{\kappa(\mathbf{u})} = c(\mathbf{w}_{\kappa(\mathbf{u})})$.

Usually, the (squared) Euclidean distance

$$d_\pi \left(\mathbf{x}, \mathbf{w}_k \right) = \left\langle \pi \left(\mathbf{x} \right), \pi \left(\mathbf{x} \right) \right\rangle_E - 2 \left\langle \pi \left(\mathbf{x} \right), \mathbf{w}_k \right\rangle_E + \left\langle \mathbf{w}_k, \mathbf{w}_k \right\rangle_E \tag{2}$$

is applied, where $\left\langle \mathbf{z}, \mathbf{w}_k \right\rangle_E$ denotes the Euclidean inner product. Let $\mathbf{w}^+ \in W^+$ and $\mathbf{w}^- \in W^-$ be the best matching prototypes according to $W^+ \left(\mathbf{x} \right) = \{ \mathbf{w}_k | \mathbf{w}_k \in W \wedge c_k = c \left(\mathbf{x} \right) \} \subset W$ and $W^- \left(\mathbf{x} \right) = W \setminus W^+$, respectively. Local costs in GLVQ are defined as

$$l \left(\mathbf{x}, W, \gamma \right) = f \left(\frac{d_\pi^+ \left(\mathbf{x} \right) - d_\pi^- \left(\mathbf{x} \right)}{\eta \left(\mathbf{x} \right)} - \gamma \right) \tag{3}$$

with $d_\pi^\pm \left(\mathbf{x} \right) = d_\pi \left(\mathbf{x}, \mathbf{w}^\pm \right)$, the *GLVQ-activation function* f is a monotonically increasing and differentiable function, $\eta \left(\mathbf{x} \right) = d_\pi^+ \left(\mathbf{x} \right) + d_\pi^- \left(\mathbf{x} \right)$ is the local normalization, and $\gamma \in \mathbb{R}$ is a shifting variable frequently set to zero. The difference quantity $h \left(\mathbf{x} \right) = \frac{1}{2} | d_\pi^- \left(\mathbf{x} \right) - d_\pi^+ \left(\mathbf{x} \right) |$ is denoted as the (local) *hypothesis margin* [4], which is related to the *hypothesis margin vector*

$$\mathbf{h} \left(\mathbf{x} \right) = \mathbf{w}^- - \mathbf{w}^+ \tag{4}$$

via the triangle $\triangle \left(\mathbf{x}, \mathbf{w}^+, \mathbf{w}^- \right)$. The quantity

$$\mu \left(\mathbf{x}, \gamma \right) = \frac{d_\pi^+ \left(\mathbf{x} \right) - d_\pi^- \left(\mathbf{x} \right)}{\eta \left(\mathbf{x} \right)} \tag{5}$$

is the parametrized classification function yielding negative values only for correctly classified training samples iff $\gamma = 0$.

The cost function to be minimized by SGDL becomes

$$E_{GLVQ} \left(X, W \right) = \sum_{\mathbf{x} \in X} l \left(\mathbf{x}, W \right) \tag{6}$$

as explained in [3]. SGDL can be realized based on simple stochastic gradients schemes or more advanced approaches like the *Broyden–Fletcher–Goldfarb–Shanno–algorithm* (BFGS, [17,18]). The stochastic gradient of the cost function $E_{GLVQ} \left(X, W \right)$ from (6) is equal to the gradient of the local cost such that we get

$$\triangle \mathbf{w}^\pm \propto - \frac{\partial l \left(\mathbf{x}, W, \gamma \right)}{\partial \mathbf{w}^\pm} \tag{7}$$

for the descent step with

$$\frac{\partial l \left(\mathbf{x}, W, \gamma \right)}{\partial \mathbf{w}^\pm} = - \frac{\partial f \left(\mu \left(\mathbf{x}, \gamma \right) \right)}{\partial \mu \left(\mathbf{x}, \gamma \right)} \cdot \frac{\partial \mu \left(\mathbf{x}, \gamma \right)}{\partial d_\pi^\pm \left(\mathbf{x} \right)} \cdot \frac{\partial d_\pi^\pm \left(\mathbf{x} \right)}{\partial \mathbf{w}^\pm} \tag{8}$$

taking into account the derivative $\frac{\partial f \left(\mu \left(\mathbf{x}, \gamma \right) \right)}{\partial \mu \left(\mathbf{x}, \gamma \right)}$ of the activation function.

The choice $\pi \left(\mathbf{x} \right) = \mathbf{x}$ yields the standard GLVQ whereas for the linear mapping $\pi_\Omega \left(\mathbf{x} \right) = \mathbf{\Omega} \mathbf{x}$ the matrix variant GMLVQ is obtained, which reduces to relevance GLVQ (GRLVQ) for a diagonal matrix $\mathbf{\Omega}$ [5,19]. If $\pi \left(\mathbf{x} \right)$ is realized by a deep network, DeepGLVQ is resulted [6,7].

2.2   GLVQ - A Neural Network Perspective

In the following we reconsider GLVQ taking the neural network perspective. For this purpose, we rewrite the squared Euclidean distance $d_\pi (\mathbf{x}, \mathbf{w}_k)$ as

$$d_\pi (\mathbf{x}, \mathbf{w}_k) = -2 \langle \pi (\mathbf{x}), \mathbf{w}_k \rangle_E - b_k (\mathbf{x}) \tag{9}$$

which is a linear perceptron with weight vector $\mathbf{w}_k$ and the bias $b_k (\mathbf{x}) = \langle \pi (\mathbf{x}), \pi (\mathbf{x}) \rangle + \langle \mathbf{w}_k, \mathbf{w}_k \rangle_E$ regarding the projected input vector $\pi (\mathbf{x}) \in \mathbb{R}^p$ [20]. Thus we get for the local costs (3)

$$
\begin{aligned}
l (\mathbf{x}, W) &= f \left(-\frac{2 (\langle \pi (\mathbf{x}), \mathbf{w}^+ \rangle_E - \langle \pi (\mathbf{x}), \mathbf{w}^- \rangle_E)}{\eta (\mathbf{x})} + \frac{b^- (\mathbf{x}) - b^+ (\mathbf{x})}{\eta (\mathbf{x})} - \gamma \right) \\
&= f \left(-\frac{2 \langle \pi (\mathbf{x}), \mathbf{w}^+ - \mathbf{w}^- \rangle_E}{\eta (\mathbf{x})} + B^\pm (\mathbf{x}, \gamma) \right) \\
&= f \left(\langle \hat{\pi} (\mathbf{x}), \mathbf{h} (\mathbf{x}) \rangle_E + B^\pm (\mathbf{x}, \gamma) \right) \\
&= \Pi_f (\mathbf{x}, W)
\end{aligned}
$$

with the scaled data mapping $\hat{\pi} (\mathbf{x}) = \frac{2\pi(\mathbf{x})}{\eta(\mathbf{x})}$ and hypothesis margin vector $\mathbf{h} (\mathbf{x})$ taking the linearity of the inner product into account. Thus, the local loss can be seen as a linear perceptron with hypothesis margin vector $\mathbf{h} (\mathbf{x})$ as weight vector and parameterized bias $B^\pm (\mathbf{x}, \gamma)$ for the projected data $\hat{\pi} (\mathbf{x})$. In this sense, the GLVQ activation function can be interpreted as an activation function for the special *GLVQ-perceptron* $\Pi_f (\mathbf{x}, W)$. Note that the GLVQ-perceptron $\Pi_f (\mathbf{x}, W)$ delivers maximum local costs for maximum excitation, such that GLVQ-classification-learning relates to minimum excitation learning for GLVQ-perceptrons $\Pi_f (\mathbf{x}, W)$. In GLVQ, standard choices for the activation function are identity $\mathrm{id} (x) = x$ and the sigmoid $\mathrm{sgd} (x, \beta)$ with $\beta = 1$, see Table 1.

2.3   Activation Function for MLP and GLVQ-MLN

As we have explained in the previous subsection, the local loss in GLVQ can be described as a particular perceptron structure. Hence, the consideration of the regarding activation function becomes inevitable. Many considerations for (deep) MLPs have shown that the appropriate choice of activation functions is essentially for convergence behavior and final network performance [15]. Originally, sigmoid functions like *tangens-hyperbolicus* or standard sigmoid function $\mathrm{sgd} (x, \beta)$ with $\beta = 1$ (see Table 1) were preferred to ensure non-linearity and differentiability together with easy analytical computation of derivatives. Later, *Rectified linear Units* (ReLU) $\mathrm{ReLU} (x) = \max (0, x)$ became popular due to its performance and computational simplicity [11]. Yet, many other, frequently similar, variants were proposed during the last years showing improvements for particular tasks. The recently proposed investigation [12], systematically studied the behavior of MLPs if activation function are generated using a modular system of basis functions. It turns out that the *swish*-function $\mathrm{swish} (x, \beta) = x \cdot \mathrm{sgd} (x, \beta)$ introduced in [21] is in average most successful although not always being the best choice [12, 14]. In fact, swish (x, β) can be seen as an intermediate between ReLU (x) and the scaled identity id (x) according to *functional* limits

$$\text{Swish}(x, \beta) \xrightarrow[\beta \to \infty]{} \text{ReLU}(x) \text{ and Swish}(x, \beta) \xrightarrow[\beta \searrow 0]{} \text{id}(x) \tag{10}$$

respectively. Yet, other activation functions like $m(x, \beta) = \max(x, \text{sgd}(x, \beta))$ also perform very well for deep MLP as outlined in [13]. Further, the choice of the activation also affects the classification robustness [22].

As outlined above, optimization of the cost function in GLVQ depends on the gradients of the local loss and, hence, also on the GLVQ-activation function f via the derivative contained in (8). Therefore, the GLVQ-perceptron $\Pi_f(\mathbf{x}, W)$ interpretation of the local loss in GLVQ still containing the derivative of the GLVQ activation function f motivates to consider successful activation function candidates known for MLP also for GLVQ-MLN. Respective candidates according to [13] are collected in Table 1 together with their derivatives necessary for the calculation of the gradient $\frac{\partial l(\mathbf{x}, W, \gamma)}{\partial \mathbf{w}^{\pm}}$.[1] Note that for $\beta = 0$ the *Leaky ReLU* LReLU (x, β) introduced in [25] simply becomes ReLU (x), whereas swish$_\tau(x, \beta)$ and $m_\tau(x, \beta)$ are variants of swish (x, β) and $m(x, \beta)$ replacing the sigmoid sgd (x, β) by the tangens-hyperbolicus function $\tau(x, \beta)$.

3   Numerical Results for Activation Functions in GLVQ

We performed numerical investigations regarding the performance of the activation functions for four widely used standard data sets.

3.1   Data Sets

The Tecator Data Set consists of 215 spectra measured for several meat probes. The spectral range is 850–1050 nm with $D = 100$ spectral bands. The data set is labeled according to the fat content (high/low). The data set is provided as a training set ($N_{V_{train}} = 172$) and a test set ($N_{V_{test}} = 43$) [26].[2]

---

[1] The derivative of the maximum function $m(x, \beta)$ could be approximated using the quasi-max function $\mathcal{Q}_\alpha(x, \beta) = \frac{1}{\alpha} \log \left(e^{\alpha x} + e^{\alpha \cdot \text{sgd}(x, \beta)} \right)$ proposed by J.D. Cook [23] with $\alpha \gg 0$. The respective consistent derivative approximation is

$$\frac{d\,m(x, \beta)}{dx} \approx \frac{\left(\exp(\alpha x) + \frac{d\text{sgd}(x, \beta)}{dx} \cdot \exp(\alpha \cdot \text{sgd}(x, \beta)) \right)}{(\exp(\alpha x) + \exp(\alpha \cdot \text{sgd}(x, \beta)))} \tag{11}$$

as provided in [24]. Analogously, the quasi-max approximation $m_\tau(x, \beta) \approx \frac{1}{\alpha} \log (\exp(\alpha x) + \exp(\alpha \cdot \tau(x, \beta)))$ is valid with

$$\frac{d\,m_\tau(x, \beta)}{dx} \approx \frac{\left(\exp(\alpha x) + \frac{d\tau(x, \beta)}{dx} \cdot \exp(\alpha \cdot \tau(x, \beta)) \right)}{(\exp(\alpha x) + \exp(\alpha \cdot \tau(x, \beta)))} \tag{12}$$

as the derivative approximation.

[2] Tecator data set is available at *StaLib*: http://lib.stat.cmu.edu/datasets/tecator.

Table 1. Successful activation functions for MLP according to [13] together with their derivatives.

Activation function	Derivative
$\mathrm{sgd}\,(x,\beta) = \frac{1}{1+\exp(-\beta\cdot x)}$	$\frac{d\mathrm{sgd}(x,\beta)}{dx} = \beta\cdot\mathrm{sgd}\,(x,\beta)\cdot(1-\mathrm{sgd}\,(x,\beta))$
$\tau\,(x,\beta) = \tanh(x\cdot\beta)+1$	$\frac{d\tau(x,\beta)}{dx} = \beta\cdot\left(1-(\tanh(x,\beta))^2\right)$
$\mathrm{swish}\,(x,\beta) = x\cdot\mathrm{sgd}\,(x,\beta)$	$\frac{d\mathrm{swish}(x,\beta)}{dx} = \beta\cdot\mathrm{swish}\,(x,\beta)+\mathrm{sgd}\,(x,\beta)\cdot(1-\beta\cdot\mathrm{swish}\,(x,\beta))$
$\mathrm{swish}_\tau\,(x,\beta) = x\cdot\tau\,(x,\beta)$	$\frac{d\mathrm{swish}_\tau(x,\beta)}{dx} = x\cdot\beta+\tau\,(x,\beta)\cdot(1-\beta\cdot\mathrm{swish}_\tau\,(x,\beta))$
$\mathrm{LReLU}\,(x,\beta) = \max(0,\beta\cdot x)$	$\frac{\partial\mathrm{LReLU}(x)}{dx} = \begin{cases}0 & x<0\\ \beta & x>0\end{cases}$ or $\frac{\partial\mathrm{LReLU}(x)}{dx} = \beta\cdot\frac{d\mathrm{swish}(x,\alpha)}{dx}\approx\beta\cdot\frac{d\mathrm{swish}(x,\alpha)}{dx}$ for $\alpha\gg 0$ acc. (10)
$m\,(x,\beta) = \max(x,\mathrm{sgd}\,(x,\beta))$	$\frac{\partial m(x,\beta)}{dx} = \begin{cases}1 & x>\mathrm{sgd}\,(x,\beta)\\ \frac{d\mathrm{sgd}(x,\beta)}{dx} & x<\mathrm{sgd}\,(x,\beta)\end{cases}$ or $\frac{dm(x,\beta)}{dx}\approx\frac{\left(\exp(\alpha x)+\frac{d\mathrm{sgd}(x,\beta)}{dx}\cdot\exp(\alpha\cdot\mathrm{sgd}(x,\beta))\right)}{(\exp(\alpha x)+\exp(\alpha\cdot\mathrm{sgd}(x,\beta)))}$ for $\alpha\gg 0$ acc. to (11)
$m_\tau\,(x,\beta) = \max(x,\tau\,(x,\beta))$	$\frac{\partial m_\tau(x,\beta)}{dx} = \begin{cases}1 & x>\tau\,(x,\beta)\\ \frac{d\tau(x,\beta)}{dx} & x<\tau\,(x,\beta)\end{cases}$ or $\frac{dm_\tau(x,\beta)}{dx}\approx\frac{\left(\exp(\alpha x)+\frac{d\tau(x,\beta)}{dx}\cdot\exp(\alpha\cdot\tau(x,\beta))\right)}{(\exp(\alpha x)+\exp(\alpha\cdot\tau(x,\beta)))}$ for $\alpha\gg 0$ acc. to (12)
$\mathrm{cosxx}\,(\beta,x) = \beta\cdot x-\cos(x)$	$\frac{\partial\mathrm{cosxx}(\beta,x)}{dx} = \beta+\sin(x)$
$\mathrm{soft}+(\beta,x) = \log(1+\exp(\beta\cdot x))$	$\frac{\partial\mathrm{soft}+(\beta,x)}{dx} = \frac{\beta\cdot\exp(\beta\cdot x)}{1+\exp(\beta\cdot x)}$

The Indian Pine Data Set is a spectral data set from remote sensing.[3] It was generated by an AVIRIS sensor capturing an area corresponding to 145×145 pixels in the Indian Pine test site in the northwest of Indiana [27]. The spectrometer operates in the visible and mid-infrared wavelength range $(0.4\text{--}2.4\,\mu\text{m})$ with $D = 220$ equidistant bands. The area includes 16 different kinds of forest or other natural perennial vegetation and non-agricultural sectors, which are also denoted as background. These background pixels are removed from the data set as usual. Additionally, we remove 20 wavelengths, mainly affected by water content (around $1.33\,\mu\text{m}$ and $1.75\,\mu\text{m}$). Finally, all spectral vectors were normalized according to the l_2-norm. This overall preprocessing is usually applied to this data set [27]. Data classes with less than 100 samples were removed yielding a 12-class-problem.

WBCD and PIMA. The Wisconsin-Breast-Cancer-data (WBCD) and the Indian diabetes data set (PIMA) contain 562 and 768 data vectors with 32 and 8 data dimensions, respectively, and each divided into two classes (healthy/ill). A detailed description can be found in [28].

3.2 Results

The results reported here were obtained for the test data (30% of data). GLVQ always was applied with only one prototype per class.

Table 2. Results for the Tecator (left) and WBCD (right) data set

$f(x)$	β	acc (in %)	epochs	stability	β	acc (in %)	epochs	stability
τ	0.1	71.69 ± 0.42	5 ± 0.0	1.00	1	94.77 ± 0.0	8 ± 0.6	1.00
$swish_\tau$	1	71.53 ± 1.44	10 ± 5.0	0.80	1	97.69 ± 0.1	4 ± 2.0	0.88
m_τ	0.1	71.58 ± 0.88	5 ± 0.1	1.00	0.1	94.73 ± 0.1	7 ± 1.5	0.84
					1	94.77 ± 0.0	8 ± 0.5	1.00
$soft+$	100	100.00 ± 0.00	105 ± 16.3	1.00	10	98.84 ± 0.0	22 ± 1.8	1.00
					100	97.09 ± 0.0	233 ± 16.6	1.00
$Cosxx$	0.1	88.59 ± 1.50	16 ± 2.3	1.00	0.5	98.27 ± 0.1	19 ± 1.9	1.00
					1	94.77 ± 0.0	0.1 ± 0.6	1.00
sgd	1	62.27 ± 0.75	9 ± 0.7	1.00	1	94.19 ± 0.00	7 ± 0.5	1.00
	100	98.16 ± 7.23	73 ± 21.3	0.98	10	98.84 ± 0.0	6 ± 1.6	1.00
					100	98.77 ± 0.2	33 ± 14.2	1.00
id	–	62.04 ± 0.31	9 ± 0.5	1.00	–	93.60 ± 0.0	9 ± 0.4	1.00
$swish$	100	99.73 ± 0.55	30 ± 10.8	1.00	10	97.67 ± 0.0	92 ± 8.3	1.00
$ReLU$	–	97.00 ± 1.85	34 ± 10.1	0.46	–	97.26 ± 0.7	234 ± 120	0.65
$LReLU$	0.1	88.09 ± 0.78	87 ± 15.0	0.33	0.1	97.67 ± 0.0	30 ± 5.7	0.71

---

[3] The data set can be found at www.ehu.es/ccwintco/uploads/2/22/Indian_pines.mat.

For each data set and each activation function from Table 1 we performed 100 runs for several parameter values β according to a grid search for them where the averaged accuracy was the evaluation criterion. The obtained best results are depicted in Tables 2 and 3. Additionally, the value 'epochs' gives the averaged number of learning epochs until convergence. The value 'stability' refers to the probability of convergence within a maximum number of 10000 epochs.

Table 3. Results for the PIMA (left) and Indian Pine (right) data set

$f(x)$	β	acc (in %)	epochs	stability	β	acc (in %)	epochs	stability
τ	1	74.0 ± 0.0	14 ± 1.4	1.00	1	57.7 ± 0.1	111 ± 6.2	1.00
$swish_\tau$	1	77.7 ± 0.7	12 ± 4.0	0.74	1	63.3 ± 0.7	56 ± 14.3	0.93
m_τ	1	74.5 ± 0.0	70 ± 9.7	1.00	0.5	57.7 ± 0.1	80 ± 7.3	1.00
					1	57.4 ± 0.1	110 ± 6.0	1.00
$soft+$	10	$77.5 \pm 0.$	70 ± 9.7	1.00	10	72.5 ± 0.0	658 ± 125.9	1.00
	100	77.8 ± 0.2	336 ± 133.8	1.00	100	82.8 ± 0.7	2277 ± 307.0	1.00
$Cosxx$	0.1	77.6 ± 1.1	38 ± 9.1	1.00	0.1	65.8 ± 2.1	24 ± 4.6	1.00
	1	77.5 ± 0.0	14 ± 1.4	1.00	0.5	65.9 ± 0.2	87 ± 4.6	1.00
sgd	1	71.0 ± 0.00	13 ± 1.5	1.00	1	56.4 ± 0.22	50 ± 7.4	1.00
	10	78.4 ± 0.0	17 ± 2.1	1.00	1	56.4 ± 0.2	50 ± 7.4	1.00
					10	66.9 ± 0.0	297 ± 52.2	1.00
					100	77.7 ± 2.0	2386 ± 896.1	1.00
id	–	70.6 ± 0.1	16.3 ± 1.6	1.00	–	56.1 ± 0.1	6 ± 5.9	1.00
$swish$	1	77.5 ± 0.0	13 ± 1.3	1.00	10	75.2 ± 0.6	80 ± 13.0	1.00
					100	78.2 ± 1.0	166 ± 16.9	1.00
$ReLU$	–	67.0 ± 5.9	68 ± 7.7	0.46	–	75.2 ± 0.9	222 ± 23.5	1.00
$LReLU$	0.5	77.5 ± 0.0	34 ± 8.6	0.33	0.1	78.3 ± 0.7	201 ± 36.6	1.00

With respect to the averaged accuracy, $soft+ \approx 89.86$, $sgd \approx 88.28$ and $swish \approx 88.28$ perform best for appropriate choices of the β-parameter. $LReLU \approx 85.39$ is weaker compared to $ReLU \approx 84.06$. If we consider the averaged convergence speed ratio related to $ReLU$ we get trivially $ReLU = 1$ and $soft+ \approx 4.60$, $sgd \approx 3.29$, and $swish \approx 0.55$ whereas $LReLU \approx 1.02$.

We can conclude that $soft+$, sgd and $swish$ achieve similar accuracies for an appropriate parameter choice β. However, $swish$ clearly outperforms the others regarding the convergence speed. Both standard activation functions for GLVQ, $id \approx 70.58$ and $sgd \approx 70.96$ with $\beta = 1$, are significantly weaker and, hence, should be avoided.

4   Conclusions

In this paper we studied the influence of several MLP activation function candidates regarding their performance influence for GLVQ. Motivation for this investigation is the first main result of this paper is that the classifier function of GLVQ can be described as a generalized perceptron and, hence, GLVQ activation function plays the role of a perceptron activation. The numerical experiments have shown that $soft+$, sgd and $swish$ achieve the best accuracy performance as it is also the case for (deep) MLP networks [13]. Yet, regarding the convergence speed $swish$ has to be clearly favored. Summarizing these experiments we suggest to switch over to *Swish* for GLVQ activation.

References

1. Kohonen T (1988) Learning vector quantization. Neural Netw 1(Suppl 1):303
2. Villmann T, Saralajew S, Villmann A, Kaden M (2018) Learning vector quantization methods for interpretable classification learning and multilayer networks. In: Sabourin C, Merelo JJ, Barranco AL, Madani K, Warwick K (eds) Proceedings of the 10th international joint conference on computational intelligence (IJCCI), Sevilla. SCITEPRESS - Science and Technology Publications, Lda., Lisbon, pp 15–21. ISBN 978-989-758-327-8
3. Sato A, Yamada K (1996) Generalized learning vector quantization. In: Touretzky DS, Mozer MC, Hasselmo ME (eds) Advances in neural information processing systems 8, Proceedings of the 1995 conference. MIT Press, Cambridge, pp 423–429
4. Crammer K, Gilad-Bachrach R, Navot A, Tishby A (2003) Margin analysis of the LVQ algorithm. In: Becker S, Thrun S, Obermayer K (eds) Advances in neural information processing (Proceedings of NIPS 2002), vol 15. MIT Press, Cambridge, pp 462–469
5. Schneider P, Hammer B, Biehl M (2009) Adaptive relevance matrices in learning vector quantization. Neural Comput 21:3532–3561
6. de Vries H, Memisevic R, Courville A (2016) Deep learning vector quantization. In: Verleysen M (ed) Proceedings of the European symposium on artificial neural networks, computational intelligence and machine learning (ESANN 2016), Louvain-La-Neuve, Belgium, pp 503–508. i6doc.com
7. Villmann T, Biehl M, Villmann A, Saralajew S (2017) Fusion of deep learning architectures, multilayer feedforward networks and learning vector quantizers for deep classification learning. In: Proceedings of the 12th workshop on self-organizing maps and learning vector quantization (WSOM2017+). IEEE Press, pp 248–255
8. Kohonen T (1995) Self-organizing maps, vol 30. Springer series in information sciences. Springer, Heidelberg (Second Extended Edition 1997)
9. Haykin S (1994) Neural networks. A comprehensive foundation. Macmillan, New York
10. Hertz JA, Krogh A, Palmer RG (1991) Introduction to the theory of neural computation, vol 1. Santa Fe institute studies in the sciences of complexity: lecture notes. Addison-Wesley, Redwood City
11. Goodfellow I, Bengio Y, Courville A (2016) Deep learning. MIT Press, Cambridge
12. Ramachandran P, Zoph B, Le QV (2018) Swish: a self-gated activation function. Technical report arXiv:1710.05941v2, Google brain

13. Ramachandran P, Zoph B, Le QV (2018) Searching for activation functions. Technical report arXiv:1710.05941v1, Google brain
14. Eger S, Youssef P, Gurevych I (2018) Is it time to swish? comparing deep learning activation functions across NLP tasks. In: Proceedings of the 2018 conference on empirical methods in natural language processing (EMNLP), Brussels, Belgium. Association for computational linguistics, pp 4415–4424
15. Chieng HH, Wahid N, Pauline O, Perla SRK (2018) Flatten-T swish: a thresholded ReLU-swish-like activation function for deep learning. Int J Adv Intell Inform 4(2):76–86
16. Kaden M, Riedel M, Hermann W, Villmann T (2015) Border-sensitive learning in generalized learning vector quantization: an alternative to support vector machines. Soft Comput 19(9):2423–2434
17. Fletcher R (1987) Practical methods of optimization, 2nd edn. Wiley, New York. 2000 edition
18. LeKander M, Biehl M, de Vries H (2017) Empirical evaluation of gradient methods for matrix learning vector quantization. In: Proceedings of the 12th workshop on self-organizing maps and learning vector quantization (WSOM2017+). IEEE Press, pp 1–8
19. Hammer B, Villmann T (2002) Generalized relevance learning vector quantization. Neural Netw 15(8–9):1059–1068
20. Saralajew S, Holdijk L, Rees M, Kaden M, Villmann T (2018) Prototype-based neural network layers: incorporating vector quantization. Mach Learn Rep 12(MLR-03-2018):1–17. ISSN: 1865-3960, http://www.techfak.uni-bielefeld.de/~fschleif/mlr/mlr_03_2018.pdf
21. Elfwing S, Uchibe E, Doya K (2018) Sigmoid-weighted linear units for neural network function approximation in reinforcement learning. Neural Netw 107:3–11
22. Zhang H, Weng T-W, Chen P-Y, Hsieh C-J, Daniel L (2018) Efficient neural network robustness certification with general activation functions. In: Bengio S, Wallach H, Larochelle H, Grauman K, Cesa-Bianchi N, Garnett R (eds) Advances in neural information processing systems 31. Curran Associates, Inc., New York, pp 4944–4953
23. Cook J (2011) Basic properties of the soft maximum. Working paper series 70, UT MD Anderson cancer center department of biostatistics. http://biostats.bepress.com/mdandersonbiostat/paper70
24. Lange M, Villmann T (2013) Derivatives of l_p-norms and their approximations. Mach. Learn. Rep. 7(MLR-04-2013):43–59. ISSN: 1865-3960. http://www.techfak.uni-bielefeld.de/~fschleif/mlr/mlr_04_2013.pdf
25. Maas AL, Hannun AY, Ng AY (2013) Rectifier nonlinearities improve neural network acoustic models. In: Proceedings of ICML-workshop for on deep learning for audio, speech, and language processing, Proceedings of machine learning research, vol 28
26. Krier C, Rossi F, François D, Verleysen M (2008) A data-driven functional projection approach for the selection of feature ranges in spectra with ICA or cluster analysis. Chemometr Intell Lab Syst 91(1):43–53
27. Landgrebe DA (2003) Signal theory methods in multispectral remote sensing. Wiley, Hoboken
28. Asuncion A, Newman DJ: UC Irvine machine learning repository. http://archive.ics.uci.edu/ml/

Robustness of Generalized Learning Vector Quantization Models Against Adversarial Attacks

Sascha Saralajew[1(✉)], Lars Holdijk[1,2], Maike Rees[1], and Thomas Villmann[3]

[1] Dr. Ing. h.c. F. Porsche AG, Weissach, Germany
{sascha.saralajew,lars.holdijk,maike.rees}@porsche.de
[2] University of Groningen, Groningen, Netherlands
[3] Saxony Institute for Computational Intelligence and Machine Learning,
University of Applied Sciences Mittweida, Mittweida, Germany
thomas.villmann@hs-mittweida.de

Abstract. Adversarial attacks and the development of (deep) neural networks robust against them are currently two widely researched topics. The robustness of Learning Vector Quantization (LVQ) models against adversarial attacks has however not yet been studied to the same extent. We therefore present an extensive evaluation of three LVQ models: Generalized LVQ, Generalized Matrix LVQ and Generalized Tangent LVQ. The evaluation suggests that both Generalized LVQ and Generalized Tangent LVQ have a high base robustness, on par with the current state-of-the-art in robust neural network methods. In contrast to this, Generalized Matrix LVQ shows a high susceptibility to adversarial attacks, scoring consistently behind all other models. Additionally, our numerical evaluation indicates that increasing the number of prototypes per class improves the robustness of the models.

1 Introduction

The robustness against adversarial attacks of (deep) neural networks (NNs) for classification tasks has become one of the most discussed topics in machine learning research since it was discovered [1,2]. By making almost imperceptible changes to the input of a NN, attackers are able to force a misclassification of the input or even switch the prediction to any desired class. With machine learning taking a more important role within our society, the security of machine learning models in general is under more scrutiny than ever.

To define an adversarial example, we use a definition similar to [3]. Suppose we use a set of scoring functions $f_j : \mathcal{X} \to \mathbb{R}$ which assign a score to each class $j \in \mathcal{C} = \{1, \ldots, N_c\}$ given an input $\mathbf{x}$ of the data space $\mathcal{X}$. Moreover, the predicted class label $c^* (\mathbf{x})$ for $\mathbf{x}$ is determined by a winner-takes-all rule $c^* (\mathbf{x}) = \arg\max_j f_j (\mathbf{x})$ and we have access to a labeled data point $(\mathbf{x}, y)$ which is correctly classified as $c^* (\mathbf{x}) = y$. An adversarial example $\tilde{\mathbf{x}}$ of the sample $\mathbf{x}$ is defined as the minimal required perturbation of $\mathbf{x}$ by ϵ to find a point at the decision boundary or in the classification region of a different class than y, i.e.

© Springer Nature Switzerland AG 2020
A. Vellido et al. (Eds.): WSOM 2019, AISC 976, pp. 189–199, 2020.
https://doi.org/10.1007/978-3-030-19642-4_19

$$\min_{\epsilon} \|\epsilon\|, \text{ s.t. } f_j(\tilde{\mathbf{x}}) \geq f_y(\tilde{\mathbf{x}}) \text{ and } \tilde{\mathbf{x}} = \mathbf{x} + \epsilon \in \mathcal{X} \text{ and } j \neq y. \tag{1}$$

Note that the magnitude of the perturbation is measured regarding a respective norm $\|\cdot\|$. If $f_j(\tilde{\mathbf{x}}) \approx f_y(\tilde{\mathbf{x}})$, an adversarial example close to the decision boundary is found. Thus, adversarials are also related to the analysis of the decision boundaries in a learned model. It is important to define the difference between the ability to generalize and the robustness of a model [4]. Assume a model trained on a finite number of data points drawn from an unknown data manifold in $\mathcal{X}$. Generalization refers to the property to correctly classify an *arbitrary* point *from* the unknown data manifold (so-called on-manifold samples). The robustness of a model refers to the ability to correctly classify on-manifold samples that were *arbitrarily disturbed*, e.g. by injecting Gaussian noise. Depending on the kind of noise these samples are on-manifold or off-manifold adversarials (not located on the data manifold). Generalization and robustness have to be learned explicitly because the one does not imply the other.

Although Learning Vector Quantization (LVQ), as originally suggested by KOHONEN in [5], is frequently claimed as one of the most robust crisp classification approaches, its robustness has not been actively studied yet. This claim is based on the characteristics of LVQ methods to partition the data space into Vorono? cells (receptive fields), according to the best matching prototype vector. For the Generalized LVQ (GLVQ) [6], considered as a differentiable cost function based variant of LVQ, robustness is theoretically anticipated because it maximizes the hypothesis margin in the *input space* [7]. This changes if the squared Euclidean distance in GLVQ is replaced by adaptive dissimilarity measures such as in Generalized Matrix LVQ (GMLVQ) [8] or Generalized Tangent LVQ (GTLVQ) [9]. They first apply a projection and measure the dissimilarity in the corresponding *projection space*, also denoted as feature space. A general robustness assumption for these models seems to be more vague.

The **observations** of this paper are: **(1)** GLVQ and GTLVQ have a high robustness because of their hypothesis margin maximization in an appropriate space. **(2)** GMLVQ is susceptible to adversarial attacks and hypothesis margin maximization does not guarantee a robust model in general. **(3)** By increasing the number of prototypes the robustness *and* the generalization ability of a LVQ model increases. **(4)** Adversarial examples generated for GLVQ and GTLVQ often make semantic sense by interpolating between digits.

2   Learning Vector Quantization

LVQ assumes a set $\mathcal{W} = \{\mathbf{w}_1, \ldots, \mathbf{w}_{N_w}\}$ of prototypes $\mathbf{w}_k \in \mathbb{R}^n$ to represent and classify the data $\mathbf{x} \in \mathcal{X} \subseteq \mathbb{R}^n$ regarding a chosen dissimilarity $d(\mathbf{x}, \mathbf{w}_k)$. Each prototype is responsible for exactly one class $c(\mathbf{w}_k) \in \mathcal{C}$ and each class is represented by at least one prototype. The training dataset is defined as a set of labeled data points $X = \{(\mathbf{x}_i, y_i) | \mathbf{x}_i \in \mathcal{X}, y_i \in \mathcal{C}\}$. The scoring function for the class j yields $f_j(\mathbf{x}) = -\min_{k:c(\mathbf{w}_k)=j} d(\mathbf{x}, \mathbf{w}_k)$. Hence, the predicted class $c^*(\mathbf{x})$ is the class label $c(\mathbf{w}_k)$ of the closest prototype $\mathbf{w}_k$ to $\mathbf{x}$.

Generalized LVQ: GLVQ is a cost function based variant of LVQ such that stochastic gradient descent learning (SGDL) can be performed as optimization strategy [6]. Given a training sample $(\mathbf{x}_i, y_i) \in X$, the two *closest* prototypes $\mathbf{w}^+ \in \mathcal{W}$ and $\mathbf{w}^- \in \mathcal{W}$ with correct label $c(\mathbf{w}^+) = y_i$ and incorrect label $c(\mathbf{w}^-) \neq y_i$ are determined. The dissimilarity function is defined as the squared Euclidean distance $d_E^2(\mathbf{x}, \mathbf{w}_k) = (\mathbf{x} - \mathbf{w}_k)^T (\mathbf{x} - \mathbf{w}_k)$. The cost function of GLVQ is

$$E_{GLVQ}(X, \mathcal{W}) = \sum_{(\mathbf{x}_i, y_i) \in X} l(\mathbf{x}_i, y_i, \mathcal{W}) \tag{2}$$

with the local loss $l(\mathbf{x}_i, y_i, \mathcal{W}) = \varphi(\mu(\mathbf{x}_i, y_i, \mathcal{W}))$ where φ is a monotonically increasing differentiable activation function. The classifier function μ is defined as

$$\mu(\mathbf{x}_i, y_i, \mathcal{W}) = \frac{d^+(\mathbf{x}_i) - d^-(\mathbf{x}_i)}{d^+(\mathbf{x}_i) + d^-(\mathbf{x}_i)} \in [-1, 1] \tag{3}$$

where $d^\pm(\mathbf{x}_i) = d_E^2(\mathbf{x}_i, \mathbf{w}^\pm)$. Thus, $\mu(\mathbf{x}_i, y_i, \mathcal{W})$ is negative for a correctly classified training sample $(\mathbf{x}_i, y_i)$ and positive otherwise. Since $l(\mathbf{x}_i, y_i, \mathcal{W})$ is differentiable, the prototypes $\mathcal{W}$ can be learned by a SGDL approach.

Generalized Matrix LVQ: By substituting the dissimilarity measure d_E^2 in GLVQ with an adaptive dissimilarity measure

$$d_\Omega^2(\mathbf{x}, \mathbf{w}_k) = d_E^2(\Omega \mathbf{x}, \Omega \mathbf{w}_k), \tag{4}$$

GMLVQ is obtained [8]. The relevance matrix $\Omega \in \mathbb{R}^{r \times n}$ is learned during training in parallel to the prototypes. The parameter r controls the projection dimension of Ω and must be defined in advance.

Generalized Tangent LVQ: In contrast to the previous methods, GTLVQ [9] defines the prototypes as affine subspaces in $\mathbb{R}^n$ instead of points. More precisely, the set of prototypes is defined as $\mathcal{W}_T = \{(\mathbf{w}_1, \mathbf{W}_1), \ldots, (\mathbf{w}_{N_w}, \mathbf{W}_{N_w})\}$ where $\mathbf{W}_k \in \mathbb{R}^{n \times r}$ is the r-dimensional basis and $\mathbf{w}_k$ is the translation vector of the affine subspace. Together with the parameter vector $\boldsymbol{\theta} \in \mathbb{R}^r$, they form the prototype as affine subspace $\mathbf{w}_k + \mathbf{W}_k \boldsymbol{\theta}$. The tangent distance is defined as

$$d_T^2(\mathbf{x}, (\mathbf{w}_k, \mathbf{W}_k)) = \min_{\boldsymbol{\theta} \in \mathbb{R}^r} d_E^2(\mathbf{x}, \mathbf{w}_k + \mathbf{W}_k \boldsymbol{\theta}) \tag{5}$$

where r is a hyperparameter. Substituting d_E^2 in GLVQ with d_T^2 and redefining the set of prototypes to $\mathcal{W}_T$ yields GTLVQ. The affine subspaces defined by $(\mathbf{w}_k, \mathbf{W}_k)$ are learned by SGDL.

3   Experimental Setup

In this section adversarial attacks as well as robustness metrics are introduced and the setup of the evaluation is explained. The setup used here follows the one presented in [10] with a few minor modifications to the study of LVQ methods. All experiments and models were implemented using the KERAS framework in PYTHON on top of TENSORFLOW.[1] All evaluated LVQ models are made available as pretrained TENSORFLOW graphs and as part of the FOOLBOX ZOO[2] at https:// github.com/LarsHoldijk/robust_LVQ_models.

The FOOLBOX [11] implementations with default settings were used for the attacks. The evaluation was performed using the MNIST dataset as it is one of the most used datasets for robust model evaluation in the literature. Despite being considered by many as a solved 'toy' dataset with state-of-the-art (SOTA) deep learning models reaching close to perfect classification accuracy, the defense of adversarial attacks on MNIST is still far from being trivial [10]. The dataset consists of handwritten digits in the data space $\mathcal{X} = [0, 1]^n$ with $n = 28 \cdot 28$. We trained our models on the 60K training images and evaluated all metrics and scores on the *complete* 10K test images.

3.1   Adversarial Attacks

Adversarial attacks can be grouped into two different approaches, white-box and black-box, distinguished by the amount of knowledge about the model available to the attacker. White-box or gradient-based attacks are based on exploiting the interior gradients of the NNs, while black-box attacks rely only on the output of the model, either the logits, the probabilities or just the predicted discrete class labels. Each attack is designed to optimize the adversarial image regarding a given norm. Usually, the attacks are defined to optimize over L^p norms (or p-norms) with $p \in \{0, 2, \infty\}$ and, therefore, are called L^p-attacks.

In the evaluation, nine attacks including white-box and black-box attacks were compared. The white-box attacks are: Fast Gradient Sign Method (FGSM) [1], Fast Gradient Method (FGM), Basic Iterative Method (BIM) [12], Momentum Iterative Method (MIM) [13] and Deepfool [14]. The black-box attacks are: Gaussian blur, Salt-and-Pepper (S&P), Pointwise [10] and Boundary [15]. See Table 1 for the L^p definition of each attack. Note that some of the attacks are defined for more than one norm.

3.2   Robustness Metrics

The robustness of a model is measured by four different metrics, all based on the *adversarial distances* $\delta_A(\mathbf{x}, y)$. Given a labeled test sample $(\mathbf{x}, y)$ from a test set T and an adversarial L^p-attack A, $\delta_A(\mathbf{x}, y)$ is defined as: **(1)** zero if the data sample is misclassified $c^*(\mathbf{x}) \neq y$; **(2)** $\|\epsilon\|_p = \|\tilde{\mathbf{x}} - \mathbf{x}\|_p$ if A found an adversary $\tilde{\mathbf{x}}$ and $c^*(\mathbf{x}) = y$; **(3)** ∞ if no adversary was found by A and $c^*(\mathbf{x}) = y$.

For each attack A the *median-δ_A* score is defined as median $\{\delta_A(\mathbf{x}, y) \mid (\mathbf{x}, y) \in T\}$, describing an averaged δ_A over T robust to outliers.[3] The *median-δ_p^**

---

[1] TENSORFLOW: www.tensorflow.org; KERAS: www.keras.io.

[2] https://foolbox.readthedocs.io/en/latest/modules/zoo.html.

[3] Hence, *median-δ_A* can be ∞ if for over 50% of the samples no adversary was found.

score is computed for all L^p-attacks as the median $\{\delta_p^*(\mathbf{x}, y) \mid (\mathbf{x}, y) \in T\}$ where $\delta_p^*(\mathbf{x}, y)$ is defined as $\min\{\delta_A(\mathbf{x}, y) \mid A$ is a L^p-attack$\}$. This score is a worst-case evaluation of the median-δ_A, assuming that each sample is disturbed by the respective worst-case attack A_p^* (the attack with the smallest distance). Additionally, the threshold accuracies acc-A and acc-A_p^* of a model over T are defined as the percentage of adversarial examples found with $\delta_A(\mathbf{x}, y) \leq t_p$, using either the given L^p-attack A for all samples or the respective worst-case attack A_p^* respectively. This metric represents the remaining accuracy of the model when only adversaries under a given threshold are considered valid. We used the following thresholds for our evaluation: $t_0 = 12$, $t_2 = 1.5$ and $t_\infty = 0.3$.

Table 1. The results of the robustness evaluation. Attacks are clustered by their L^p class, the boxes denote the type of the attack (white- or black-box). Accuracies are given in percentages and the #prototypes is recorded per class. All scores are evaluated on the test set. For each model we report the clean accuracy (clean acc.), the median-δ_A (left value) and acc-A score (right value) for each attack and the worst-case (worst-c.) analysis over all L^p-attacks by presenting the median-δ_p^* (left value) and acc-A_p^* score (right value). Higher scores mean higher robustness of the model. The median-δ_A of the most robust model in each attack is highlighted in bold. Overall, the model with the best (highest) worst-case median-δ_p^* is underlined and highlighted.

			CNN		Madry		GLVQ				GMLVQ				GTLVQ			
#prototypes							1		128		1		49		1		10	
Clean acc.			99		99		83		95		88		93		95		97	
L^2	FGM	□	2.1	73	∞	96	∞	63	∞	76	0.6	7	0.8	15	∞	71	∞	81
	Deepfool	□	1.9	70	**5.5**	94	1.6	53	2.3	73	0.5	26	0.7	27	2.3	73	2.5	81
	BIM	□	1.5	50	**4.9**	94	1.5	50	2.1	68	0.6	6	0.7	8	2.1	68	2.3	77
	Gaussian	■	6.4	99	6.6	98	6.8	83	6.7	68	6.3	88	6.2	92	**7.1**	94	6.9	97
	Pointwise	■	4.2	96	2.1	80	4.5	79	5.4	92	1.6	54	2.4	78	5.5	92	**5.6**	95
	Boundary	■	1.9	76	1.5	52	2.1	61	**3.2**	76	0.6	7	0.8	7	2.8	78	3.1	86
	worst-c.		1.5	50	1.5	52	1.5	49	2.1	68	0.5	3	0.6	3	2.1	68	**2.2**	77
L^∞	FGSM	□	.17	7	**.52**	96	.17	11	.29	43	.04	0	.05	0	.22	18	.25	26
	Deepfool	□	.16	1	**.49**	95	.13	7	.22	21	.04	27	.05	19	.19	9	.22	19
	BIM	□	.12	0	**.41**	94	.12	3	.20	9	.04	0	.05	0	.17	3	.20	5
	MIM	□	.13	0	**.38**	93	.12	3	.19	9	.04	0	.05	0	.17	3	.20	5
	worst-c.		.12	0	**.38**	93	.11	2	.19	5	.03	0	.04	0	.17	3	.19	4
L^0	Pointwise	■	19	73	4	1	22	64	32	79	3	6	6	18	34	80	**35**	85
	S&P	■	65	94	17	63	126	77	**188**	92	8	37	17	61	155	91	179	95
	worst-c.		19	73	4	1	22	64	32	79	3	6	6	18	34	80	**35**	85

3.3 Training Setup and Models

All models, except the Madry model, were trained with the Adam optimizer [16] for 150 epochs using basic data augmentation in the form of random shifts by ± 2 pixels and random rotations by $\pm 15°$.

NN Models: Two NNs are used as baseline models for the evaluation. The first model is a convolutional NN, denoted as CNN, with two convolutional layers and two fully connected layers. The convolutional layers have 32 and 64 filters with a stride of one and a kernel size of 3×3. Both are followed by max-pooling layers with a window size and stride each of 2×2. None of the layers use padding. The first fully connected layer has 128 neurons and a dropout rate of 0.5. All layers use the ReLU activation function except for the final fully connected output layer which uses a softmax function. The network was trained using the categorical cross entropy loss and an initial learning rate of 10^{-4} with a decay of 0.9 at plateaus.

The second baseline model is the current SOTA model for MNIST in terms of robustness proposed in [17] and denoted as Madry. This model relies on a special kind of adversarial training by considering it as a min-max optimization game: before the loss function is minimized over a given training batch, the original images are partially substituted by perturbed images with $\|\epsilon\|_\infty \leq 0.3$ such that the loss function is *maximized* over the given batch. The Madry model was downloaded from https://github.com/MadryLab/mnist_challenge.

LVQ Models: All three LVQ models were trained using an initial learning rate of 0.01 with a decay of 0.5 at plateaus and with φ defined as the identity function. The prototypes (translation vectors) of all methods were class-wise initialized by k-means over the training dataset. For GMLVQ, we defined Ω with $n = r$ and initialized Ω as a scaled identity matrix with Frobenius norm one. After each update step, Ω was normalized to again have Frobenius norm one. The basis matrices $\mathbf{W}_k$ of GTLVQ were defined by $r = 12$ and initialized by a singular value decomposition with respect to each initialized prototype $\mathbf{w}_k$ over the set of class corresponding training points [9]. The prototypes were not constrained to $\mathcal{X}$ ('box constrained') during the training, resulting in possibly non-interpretable prototypes as they can be points in $\mathbb{R}^n$.[4]

Two versions of each LVQ model were trained: one with one prototype per class and one with multiple prototypes per class. For the latter the numbers of prototypes were chosen such that all LVQ models have roughly 1M parameters. The chosen number of prototypes per class are given in Table 1 by #prototypes.

4   Results

The results of the model robustness evaluation are presented in Table 1. Figure 1 displays adversarial examples generated for each model. Below, the four most notable observations that can be made from the results are discussed.

---

[4] A restriction to $\mathcal{X}$ leads to an accuracy decrease of less than 1%.

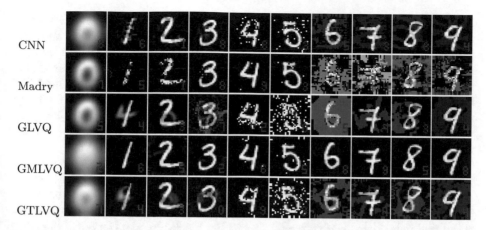

Fig. 1. For each model, adversarial examples generated by the attacks (from left to right): Gaussian, Deepfool (L_2), BIM (L_2), Boundary, Pointwise (L_0), S&P, FGSM, Deepfool (L_∞), BIM (L_∞) and MIM. For the LVQ models the version with more prototypes per class was used. The ten digits were randomly selected under the condition that every digit was classified correctly by all models. The original images are 0, 1, ..., 9 from left to right. The red digits in the lower right corners indicate the models prediction after the adversarial attack.

Hypothesis Margin Maximization in the Input Space Produces Robust Models (GLVQ and GTLVQ Are Highly Robust): Table 1 shows outstanding robustness against adversarial attacks for GLVQ and GTLVQ. GLVQ with multiple prototypes and GTLVQ with both one or more prototypes per class, outperform the NN models by a large difference for the L^0- and L^2-attacks while having a considerably lower clean accuracy. This is not only the case for individual black-box attacks but also for the worst-case scenarios. For the L^0-attacks this difference is especially apparent. A possible explanation is that the robustness of GLVQ and GTLVQ is achieved due to the input space hypothesis margin maximization [7].[5] In [7] it was stated that the hypothesis margin is a lower bound for the sample margin which is, *if defined in the input space*, used in the definition of adversarial examples (1). *Hence, if we maximize the hypothesis margin in the input space we guarantee a high sample margin and therefore, a robust model.* A first attempt to transfer this principle was made in [3] to create a robust NN by a first order approximation of the sample margin in the input space.

However, the Madry model still outperforms GLVQ and GTLVQ in the L^∞-attacks as expected. This result is easily explained using the manifold based definition of adversarial examples and the adversarial training procedure of the Madry model, which optimizes the robustness against $\|\epsilon\|_\infty \leq 0.3$. Considering the manifold definition, one could say that Madry augmented the original

---

[5] Note that the results of [7] hold for GTLVQ as it can be seen as a version of GLVQ with infinitely many prototypes learning the affine subspaces.

MNIST manifold to include small L^∞ perturbations. Doing so, Madry creates a new *training*-manifold in addition to the original MNIST manifold. In other words, the L^∞ robustness of the adversarial trained Madry model can be seen as its generalization on the new training-manifold (this becomes clear if one considers the high acc-A scores for L^∞). For this reason, the Madry model is only robust against off-manifold examples that are on the generated training-manifold. As soon as off-training-manifold examples are considered the accuracy will drop fast. This was also shown in [10], where the accuracy of the Madry model is significantly lower when considering a threshold $t_\infty > 0.3$.[6]

Furthermore, the Madry model has outstanding robustness scores for gradient-based attacks in general. We accredit this effect to potential obfuscation of gradients as a side-effect of the adversarial training procedure. While [18] was not able to find concrete evidence of gradient obfuscation due to adversarial training in the Madry model, it did list black-box-attacks outperforming white-box attacks as a signal for its occurrence.

Hypothesis Margin Maximization in a Space Different to the Input Space Does Not Necessarily Produce Robust Models (GMLVQ Is Susceptible for Adversarial Attacks): In contrast to GLVQ and GTLVQ, GMLVQ has the lowest robustness score across all attacks and all methods. Taking the strong relation of GTLVQ and GMLVQ into account [9], it is a remarkable result.[7] One potential reason is, that GMLVQ maximizes the hypothesis margin in a projection space which differ in general from the input space. The margin maximization in the projection space is used to construct a model with good generalization abilities, which is why GMLVQ usually outperforms GLVQ in terms of accuracy (see the clean accuracy for GLVQ and GMLVQ with one prototype per class). However, a large margin in the projection space does not guarantee a big margin in the input space. Thus, GMLVQ does not implicitly optimize the separation margin, as used in the definition of an adversarial example (1), in the input space. Hence, GMLVQ is a good example to show that a model, which generalizes well, is not necessarily robust.

Another effect which describes the observed lack of robustness by GMLVQ is its tendency to oversimplify (to collapse data dimensions) without regularization. Oversimplification may induce heavy distortions in the mapping between

---

[6] For future work a more extensive evaluation should be considered: including not only the norm for which a single attack was optimized but rather a combination of all three norms. This gives a better insight on the characteristics of the attack and the defending model. The L^0 norm can be interpreted as the number of pixels that have to change, the L^∞ norm as the maximum deviation of a pixel and the L^2 norm as a kind of average pixel change. As attacks are optimized for a certain norm, only considering this norm might give a skewed impression of their attacking capability. Continuing, calculating a threshold accuracy including only adversaries that are below all three thresholds may give an interesting and more meaningful metric.

[7] GTLVQ can be seen as localized version of GMLVQ with the constraint that the Ω matrices must be orthogonal projectors.

input and projection space, potentially creating dimensions in which a small perturbation in the input space can be mapped to a large perturbation in the projection space. These dimensions are later used to efficiently place the adversarial attack. This effect is closely related to theory known from metric learning, here oversimplification was used by [19] to optimize a classifier over d_Ω^2, which *maximally collapses (concentrates) the classes to single points* (related to the prototypes in GMLVQ). It is empirically shown that this effect helps to achieve a good generalization.

To improve the robustness of GMLVQ penalizing the collapsing of dimensions may be a successful approach. A method to achieve this is to force the eigenvalue spectrum of the mapping to follow a uniform distribution, as proposed in [20]. This regularization technique would also strengthen the transferability between the margin in the projection and input space. Unfortunately, it requires the possibly numerical instable computation of the derivative of a determinant of a product of Ω which makes it impossible to train an appropriate model for MNIST using this regularization so far. The fact that GTLVQ is a constrained version of GMLVQ gives additional reason to believe that regularizations/constraints are able to force a model to be more robust.

Increasing the Number of Prototypes Improves the Ability to Generalize and the Robustness: For all three LVQ models the robustness improves if the number of prototypes per class increases. Additionally, increasing the number of prototypes leads to a better ability to generalize. This observation provides empirical evidence supporting the results of [4]. In [4] it was stated that generalization and robustness are not necessarily contradicting goals, which is a topic recently under discussion.

With multiple prototypes per class, the robustness of the GLVQ model improves by a significantly larger margin than GTLVQ. This can be explained by the high accuracy of GTLVQ with one prototype. The high accuracy with one prototype per class indicates that the data manifold of MNIST is almost flat and can therefore be described with one tangent such that introducing more prototypes does not improve the model's generalization ability. If we add more prototypes in GLVQ, the prototypes will start to approximate the data manifold and with that implicitly the tangent prototypes used in GTLVQ. With more prototypes per class, the scores of GLVQ will therefore most likely converge towards those of GTLVQ.

GLVQ and GTLVQ Require Semantically Correct Adversarial Examples: Figure 1 shows a large semantic difference between the adversarial examples generated for GLVQ / GTLVQ and the other models. A large portion of the adversarial examples generated for the GLVQ and GTLVQ models look like interpolations between the original digit and another digit.[8] This effect is especially visible for the Deepfool, BIM and Boundary attacks. In addition to this,

---

[8] A similar effect was observed in [10] for k-NN models.

the Pointwise attack is required to generate features from other digits to fool the models, e.g. the horizontal bar of a two in the case of GLVQ and the closed ring of a nine for GTLVQ (see digit four). In other words, for GLVQ and GTLVQ some of the attacks generate adversaries that closer resemble on-manifold samples than off-manifold. For the other models, the adversaries are more like off-manifold samples (or in the case of Madry, off-training-manifold).

5   Conclusion

In this paper we extensively evaluated the robustness of LVQ models against adversarial attacks. Most notably, we have shown that there is a large difference in the robustness of the different LVQ models, even if they all perform a hypothesis margin maximization. GLVQ and GTLVQ show high robustness against adversarial attacks, while GMLVQ scores the lowest across all attacks and all models. The discussion related to this observation has lead to four important **conclusions**: **(1)** For (hypothesis) margin maximization to lead to robust models the space in which the margin is maximized matters, this must be the same space as where the attack is placed. **(2)** Collapsed dimensions are beneficial for the generalization ability of a model. However, they can be harmful for the model's robustness. **(3)** It is possible to derive a robust model by applying a fitting regularization/constraint. This can be seen in the relation between GTLVQ and GMLVQ and is also studied for NNs [21]. **(4)** Our experimental results with an increased number of prototypes support the claim of [4], that the ability to generalize and the robustness are principally not contradicting goals.

In summary, the overall robustness of LVQ models is impressive. Using only one prototype per class and no purposefully designed adversarial training, GTLVQ is on par with SOTA robustness on MNIST. With further research, the robustness of LVQ models against adversarial attacks can be a valid reason to deploy them instead of NNs in security critical applications.

References

1. Goodfellow I, Shlens J, Szegedy C (2015) Explaining and harnessing adversarial examples. In: International conference on learning representations
2. Szegedy C, Zaremba W, Sutskever I, Bruna J, Erhan D, Goodfellow I, Fergus R (2014) Intriguing properties of neural networks. In: International conference on learning representations
3. Elsayed G, Krishnan D, Mobahi H, Regan K, Bengio S (2018) Large margin deep networks for classification. In: Advances in neural information processing systems, pp 850–860
4. Stutz D, Hein M, Schiele B (2018) Disentangling adversarial robustness and generalization. arXiv preprint arXiv:1812.00740
5. Kohonen T (1988) Learning vector quantization. Neural networks, 1(Supplement 1)
6. Sato A, Yamada K (1996) Generalized learning vector quantization. In: Advances in neural information processing systems, pp 423–429

7. Crammer K, Gilad-Bachrach R, Navot A, Tishby N (2003) Margin analysis of the LVQ algorithm. In: Advances in neural information processing systems, pp 479–486
8. Schneider P, Biehl M, Hammer B (2009) Adaptive relevance matrices in learning vector quantization. Neural Comput 21(12):3532–3561
9. Saralajew S, Villmann T (2016) Adaptive tangent distances in generalized learning vector quantization for transformation and distortion invariant classification learning. In: 2016 international joint conference on neural networks (IJCNN). IEEE, pp 2672–2679
10. Schott L, Rauber J, Bethge M, Brendel W (2019) Towards the first adversarially robust neural network model on MNIST. In: International conference on learning representations
11. Rauber J, Brendel W, Bethge M (2017) Foolbox: a python toolbox to benchmark the robustness of machine learning models. arXiv preprint arXiv:1707.04131
12. Kurakin A, Goodfellow I, Bengio S (2016) Adversarial examples in the physical world. arXiv preprint arXiv:1607.02533
13. Dong Y, Liao F, Pang T, Su H, Zhu J, Hu X, Li J (2018) Boosting adversarial attacks with momentum. In: Proceedings of the IEEE conference on computer vision and pattern recognition, pp 9185–9193
14. Moosavi-Dezfooli S-M, Fawzi A, Frossard P (2016) DeepFool: a simple and accurate method to fool deep neural networks. In: Proceedings of the IEEE conference on computer vision and pattern recognition, pp 2574–2582
15. Brendel W, Rauber J, Bethge M (2018) Decision-based adversarial attacks: reliable attacks against black-box machine learning models. In: Proceedings of the 6th international conference on learning representations
16. Kingma DP, Ba JL (2015) Adam: a method for stochastic optimization. In: Proceedings of the international conference on learning representations, pp 1–13
17. Madry A, Makelov A, Schmidt L, Tsipras D, Vladu A (2018) Towards deep learning models resistant to adversarial attacks. In: International conference on learning representations
18. Athalye A, Carlini N, Wagner D (2018) Obfuscated gradients give a false sense of security: circumventing defenses to adversarial examples. In: Proceedings of the 35th international conference on machine learning
19. Globerson A, Roweis S (2006) Metric learning by collapsing classes. In: Advances in neural information processing systems, pp 451–458
20. Schneider P, Bunte K, Stiekema H, Hammer B, Villmann T, Biehl M (2010) Regularization in matrix relevance learning. IEEE Trans Neural Netw 21(5):831–840
21. Croce F, Andriushchenko M, Hein M (2018) Provable robustness of ReLU networks via maximization of linear regions. arXiv preprint arXiv:1810.07481

Passive Concept Drift Handling via Momentum Based Robust Soft Learning Vector Quantization

Moritz Heusinger[✉], Christoph Raab[✉], and Frank-Michael Schleif[✉]

University of Applied Sciences Würzburg-Schweinfurt,
Sanderheinrichsleitenweg 20, Würzburg, Germany
{moritz.heusinger,christoph.raab,frank-michael.schleif}@fhws.de

Abstract. Concept drift is a change of the underlying data distribution which occurs especially with streaming data. Besides other challenges in the field of streaming data classification, concept drift should be addressed to obtain reliable predictions. The Robust Soft Learning Vector Quantization has already shown good performance in traditional settings and is modified in this work to handle streaming data. Further, momentum-based stochastic gradient descent is applied to tackle concept drift passively due to increased learning capabilities. The proposed work is tested against common benchmark algorithms and streaming data in the field.

Keywords: Stream classification · Concept drift ·
Robust Soft Learning Vector Quantization

1   Introduction

A key concept in machine learning is to separate the training step of the model and the evaluation phase. However, this is not applicable in the domain of stream classification. In this field, it is assumed that data arrive continuously, making the storage of data in memory unfeasible. Further, not all training data is available at training time, which raises the need for constantly updating the model, e.g. online or incremental. Stream classification algorithms [2] tackle these requirements.

However, stream classifiers are prone to concept drift of streaming data, which is a change of underlying distribution and could lead to a collapse in prediction performance. There are various types of concept drift, i.e. incremental, abrupt, gradual and reoccurring, and as a consequence a variety of approaches addressing these issues have been proposed [10]. In general, these strategies are separated into active and passive [16]. Active adaptation changes a model noticeable. The passive ones use no explicit detection strategy, but are continually updating the model, without awareness of concept drift.

The family of prototype-based classification algorithms, the learning vector quantization (LVQ), receives much attention as a potential stream classification algorithm due to online learning capabilities [21]. However, objective driven

A. Vellido et al. (Eds.): WSOM 2019, AISC 976, pp. 200–209, 2020.
https://doi.org/10.1007/978-3-030-19642-4_20

LVQs learn by Stochastic Gradient Descent (SGD), which we assume as inappropriate in adapting the model if a concept drift occurs. Therefore, we propose in Sect. 4, a streaming robust soft learning vector quantization (RSLVQ) which maximizes the objective function with momentum based Stochastic Gradient Ascent (SGA). This is applied to increase learning speed and rapid adaptation to occurring concept drifts. The rest of the paper contains a discussion of related work in Sect. 2 and the comparison of our work to baseline RSLVQ and standard algorithms on common data streams in the field in Sect. 5.

2   Related Work

In the field of streaming data, different kinds of algorithms successfully apply passive drift handling to evolving data streams, which are summarized as lazy learning algorithms.

In [16] a modern approach of a self-adjusting memory (SAM) version of the K-Nearest Neighbor (KNN) which is called SAM-KNN is proposed. SAM-KNN is developed to handle various types of concept drift, using biologically inspired memory models and their coordination. The basic idea of it is to have dedicated models for current and former concepts used according to the demands of the given situation [16].

The Adaptive Windowing algorithm (ADWIN) [3] is a drift detector and works by keeping updated statistics of a variable sized window, so it can detect changes and perform cuts in its window to better adapt the learning algorithms. Note that most ensembles do active drift detection because an additional drift detection algorithm is used.

For evolving data stream classification often tree-based algorithms, e.g. the famous Hoeffding Tree (HT) [7], are used due to their good performance [5]. To address the problem of concept drift, an adaptive HT with ADWIN as drift detector was published in [4] and showed better prediction performance on evolving data streams as the classical HT.

Also, ensemble models are used to combine multiple classifiers [11] in the streaming domain. The OzaBagging (OB) [17] algorithm is an online ensemble model that uses different base classifiers. For the use of concept drift detection this ensemble model can be again combined with the ADWIN algorithm to the OzaBagging ADWIN (OBA) algorithm. For a comprehensive data stream description see [15].

Further, the family of LVQ algorithms first introduced by [13] has received attention as potential stream classification algorithms [21]. The Robust Soft Learning Vector Quantization (RSLVQ) [20] is a promising probabilistic approach, which assumes class distributions as Gaussian mixture model learned via SGA and was not yet evaluated as a stream classifier so far.

In the advent of deep learning, variations of SGD algorithms receive more and more attention. A comparison is given in [19]. They showed that momentum based gradient descent algorithms like Adadelta [24] and RMSprop [23] converge faster to better optimums as traditional approaches. Both of these algorithms are extensions of the Adagrad [8] algorithm. Note that momentum based gradient descents have been applied to a prototype based learner in [14]. In this work, we reformulate Adadelta as SGA optimizer and apply it to RSLVQ.

3 Streaming Data and Concept Drift

In the context of supervised learning a data stream is given as a sequence $S = \{s_1, \ldots, s_t, \ldots\}$ of tuples $s_i = \{\mathbf{x}_i, y_i\}$, with potentially infinite length. A tuple $s_i = \{\mathbf{x}_i, y_i\}$ contains the data $\mathbf{x}_i \in \mathbb{R}$ and the respective label $y_i = \{1, \ldots, C\}$ whereby s_t arrives at time t. A classifier predicts labels $\hat{y}_t$ of unseen data $\mathbf{x}_t \in \mathbb{R}$ employing prior model h_{t-1}, i.e. $\hat{y}_t = h_{t-1}(\mathbf{x}_t)$. The prior model and s_t is subsequently included into the new model $h_t = learn(h_{t-1}, s_t)$. This behaviour is called immediate or test-than-train evaluation. It also means, that the learning algorithm trains for an infinite time and that at time t the classifier is (re-)trained on the just arrived tuple.

3.1 Concept Drift

Concept drift is the change of joint distributions of a set of samples $\mathbf{X}$ and corresponding labels $\mathbf{y}$ between two points in time:

$$\exists \mathbf{X} : p(\mathbf{X}, \mathbf{y})_t \neq p(\mathbf{X}, \mathbf{y})_{t-1} \tag{1}$$

The term virtual drift refers to a change in distribution $p(\mathbf{X})$ for two points in time. Because, we can rewrite Eq. (1) to

$$\exists \mathbf{X} : p(\mathbf{X})_t p(\mathbf{y} \mid, \mathbf{X})_t = p(\mathbf{X})_{t-1} p(\mathbf{y} \mid, \mathbf{X})_{t-1}, \tag{2}$$

Virtual drift could also be presented at real concept drift. For a comprehensive study of various concept drift types see [10].

4 Robust Soft Learning Vector Quantization

The Robust Soft Learning Vector Quantization (RSLVQ) [20] is a probabilistic prototype based classification algorithm. Given a labeled dataset $\mathbf{X} = \{(\mathbf{x}_i, y_i) \in \mathbb{R}^d \times \{1, \ldots, C\}\}_{i=1}^n$, with data points $\mathbf{x_i}$, labels y_i, C as the number of classes, d the number of features and n as the number of samples, the RSLVQ algorithm trains a prototype model such that the error on the classification task is minimized. The RSLVQ model consists of a set of m prototypes $\Theta = \{(\boldsymbol{\theta}_j, y_j) \in \mathbb{R}^d \times \{1, \ldots, C\}\}_{j=1}^m$. Each prototype represents a multi-variate Gaussian model, i.e. $\mathcal{N}(\boldsymbol{\theta}_j, \sigma)$, approximating an assumed class dependent Gaussian mixture of $\mathbf{X}$. The goal of RSLVQ algorithms is to learn prototypes, representing the class dependent distribution, i.e. corresponding class samples $\mathbf{x}_i$ should be mapped to the correct class or Gaussian mixture based on the highest probability. The RSLVQ algorithms maximizes the maximum-likelihood ratio as objective function:

$$E(\mathbf{X}, \mathbf{y}|\Theta)_R = \sum_{i=1}^{n} \frac{p(\mathbf{x}_i, y_k|\Theta)}{p(\mathbf{x}_i|\Theta)} \tag{3}$$

where $p(\mathbf{x}_i, y_k|\Theta)$ is the probability density function that $\mathbf{x}$ is generated by the mixture model of the same class and $p(\mathbf{x}_i|\Theta)$ is the overall probability density function of $\mathbf{x}$ given Θ. Equation (3) will be optimized with SGA, i.e.:

$$\boldsymbol{\theta}_l(t+1) = \boldsymbol{\theta}_l(t) - \alpha \begin{cases} (P_y(l|\mathbf{x}) - P(l|\mathbf{x}))(\mathbf{x} - \boldsymbol{\theta}_l), & c_l = y, \\ -P(l|\mathbf{x})(\mathbf{x} - \boldsymbol{\theta}_l), & c_l \neq y. \end{cases} \tag{4}$$

where $P_y(l|\mathbf{x})$ is the probability that $\mathbf{x}$ is generated by the component l, i.e. from the mixture model of same class prototype $\boldsymbol{\theta}_y$ and $P(l|\mathbf{x})$ is the probability that $\mathbf{x}$ is generated by any component. The parameter α is the learning-rate. For a more comprehensive derivation see [20].

4.1   Momentum Based Optimization

One of the most common algorithms to optimize error functions is the gradient descent/ascent algorithm and in particular the stochastic formulation SGD and SGA [19]. In the field of Deep Learning the classic SGA has been further modified to get rid of common problems, like sensitivity to steep imbalanced valleys, i.e., areas where the cost surface curves much more steeply in one dimension than in another [22], which are common around local optima.

Momentum [18] is a method that helps accelerate SGA in the relevant direction and dampens oscillations. It does this by adding a fraction γ of the update vector of the past time step to the current update vector:

$$\begin{aligned} v_t &= \gamma v_{t-1} + \eta \nabla_\theta J(\theta) \\ \theta &= \theta - v_t \end{aligned} \tag{5}$$

The momentum term γ is usually set to 0.9 or a similar value but can be seen as a hyperparameter which can be optimized, e.g. via grid search.

Effective momentum based techniques are Adagrad, Adadelta and RMSprop [19]. While Adagrad was the first publication of these three algorithms, RMSprop and Adadelta are further developments of Adagrad, which both try to reduce its aggressive, monotonically decreasing learning rate [24]. These momentum based algorithms diverge to a local optima much faster and sometimes reach better optima than SGA [19].

Due to the fact, that momentum based algorithms make larger steps per iteration, it should adapt faster to new concepts. Thus, we implemented this idea into the Robust Soft Learning Vector Quantization (RSLVQ) and replaced the SGA by Adadelta-SGA. Instead of accumulating all past squared gradients, like in the Adagrads momentum approach, Adadelta restricts the window of accumulated past gradients to some fixed size w.

While it could inefficiently store all w squared gradients, the sum of gradients is recursively defined as a decaying average of all past squared gradients instead. The running average $E[g^2]_t$ at time step t then depends (as a fraction γ similarly to the Momentum term) only on the previous average and the current gradient:

$$E[g^2]_t = \gamma E[g^2]_{t-1} + (1 - \gamma)g_t^2 \tag{6}$$

The decay rate γ should be set to a value of around 0.9 [8].

Hence, in [8] another exponentially decaying average is introduced, this time not of squared gradients but squared parameter updates:

$$E[\Delta\theta^2]_t = \gamma E[\Delta\theta^2]_{t-1} + (1 - \gamma)\Delta\theta_t^2 \tag{7}$$

Thus, the update of the root mean squared (RMS) error of parameters is:

$$RMS[\Delta\theta]_t = \sqrt{E[\Delta\theta^2]_t + \epsilon} \tag{8}$$

Because of the reason, that $RMS[\Delta\theta]_t$ is not known, it is approximated with the RMS of parameter updates until the previous time step. Finally, the learning rate η in the previous update rule is replaced with $RMS[\Delta\theta]_{t-1}$, and we receive the following equation for updating Adadelta:

$$\Delta\theta_t = -\frac{RMS[\Delta\theta]_{t-1}}{RMS[g]_t}g_t \tag{9}$$

$$\theta_{t+1} = \theta_t + \Delta\theta_t$$

Due to the fact, that the learning rate η has been eliminated, it does not have to be optimized when using Adadelta, which keeps the number of hyperparameters low [19].

To exchange the SGA learning of the RSLVQ with the Adadelta algorithm, the learning rule of the RSLVQ is replaced by the update rule provided in Eq. (9).

The prior and posterior probabilities are calculated the same way as in the $RSLVQ_{SGA}$. Based on the posterior and prior, the gradient is calculated over the objective function of the RSLVQ:

$$g_t = \begin{cases} P_y(l|x) - P(l|x)(x - \theta_l), & c_l = y, \\ -P(l|x)(x - \theta_l), & c_l \neq y. \end{cases} \tag{10}$$

In the next step, the past squared gradients $E[g^2]$ are updated by:

$$E[g^2]_t = \gamma E[g^2]_{t-1} + (1 - \gamma)g_t^2 \tag{11}$$

Now the gradient update at time step t can be calculated:

$$\Delta\theta_t = \frac{(E[\Delta\theta^2] + \epsilon)}{\sqrt{(E[\Delta\theta^2] + \epsilon)}}g_t \tag{12}$$

In the next step of a learning iteration, the update $\Delta\theta_t$ has to be applied:

$$\theta_{t+1} = \theta_t + \Delta\theta_t \tag{13}$$

Finally, the squared parameter updates $E[\Delta\theta^2]$ are stored by:

$$E[\Delta\theta^2]_t = \gamma E[\Delta\theta^2]_{t-1} + (1 - \gamma)\Delta\theta_t^2 \tag{14}$$

5   Experiments

We compared our classifier against other state-of-the-art stream classifiers. Evaluation is done using the immediate setting, as described in Sect. 3. Since accuracy can be misleading on datasets with class imbalances, we also report Kappa statistics.

To test if there are significant differences between the performance of the algorithms, the Friedman [9] test with a 95% significance level is performed followed by the Bonferroni-Dunn [6] post-hoc test.

Ten synthetic and three real data streams are used in the experiments. The synthetic data streams include abrupt, gradual, incremental drifts and one stationary data stream. The real-world data streams have been thoroughly used in the literature to evaluate the classification performance of data stream classifiers and exhibit multi-class, temporal dependencies and imbalanced data streams with different drift characteristics. Naive Bayes could not be evaluated on the RBF stream generator [12] due to negative values. Please note, that the gradual and abrupt drifts are generated by a concept drift generator which switches the class-data generator functions as described below. For comparison purposes configuration of synthetic stream generators is taken from [12].

Table 1 presents the data streams which are used in our experiment, as well as their configuration by the drift types. For a detailed description of the used data streams see [12] and [15]. The tests of the RSLVQ are done with 2 prototypes per class.[1]

Table 1. Configuration of the data streams (A: Abrupt Drift, G: Gradual Drift, I_m: Moderate Incremental Drift, I_f: Fast Incremental Drift and N: No Drift)

Dataset	#Instances	#Features	Type	Drift	#Classes
LED_a	1,000,000	24	Synthetic	A	10
LED_g	1,000,000	24	Synthetic	G	10
SEA_a	1,000,000	3	Synthetic	A	2
SEA_g	1,000,000	3	Synthetic	G	2
AGR_a	1,000,000	9	Synthetic	A	2
AGR_g	1,000,000	9	Synthetic	G	2
RTG	1,000,000	10	Synthetic	N	2
RBF_m	1,000,000	10	Synthetic	I_m	5
RBF_f	1,000,000	10	Synthetic	I_f	5
HYPER	1,000,000	10	Synthetic	I_f	2
POKR	829,201	11	Real	-	10
GMSC	120,269	11	Real	-	2
ELEC	45,312	8	Real	-	2

---

[1] Source code on https://github.com/foxriver76/rslvq-adadelta.

Table 2. Accuracy of the algorithms on synthetic and real-world streams. Winner marked bold. Tests summarize five runs of cross-validation on one million samples or complete real-world streams per run.

Dataset	RSLVQ SGA	RSLVQ Adadelta	NB	HT	HAT	OB	OBA	SAM-KNN
LED$_A$	67.2	99.6	**100**	**100**	**100**	**100**	**100**	**100**
LED$_G$	67.1	99.8	**100**	**100**	99.9	**100**	**100**	**100**
SEA$_A$	82.9	88.7	71.6	**90.0**	83.9	86.1	86.1	89.0
SEA$_G$	82.2	88.9	67.7	**89.6**	83.2	85.6	85.6	88.7
AGR$_A$	50.1	54.3	50.8	**99.9**	79.8	53.5	53.5	54.7
AGR$_G$	52.7	53.3	55.5	**86.7**	79.3	62.4	62.3	62.5
RTG	58.8	68.5	73.1	**84.0**	60.5	67.2	67.1	65.6
RBF$_F$	59.7	59.8	-	59.8	**68.8**	91.6	90.0	91.2
RBF$_M$	53.8	57.6	-	68.1	**74.4**	94.4	94.3	95.8
HYPER	56.7	**87.3**	73.6	80.3	84.9	56.2	56.2	56.0
Synthetic average	63.1	75.8	74.0	**85.8**	81.5	79.7	79.5	80.35
ELEC	85.3	85.3	58.0	80.9	**88.3**	73.1	73.0	72.5
GMSC	71.5	93.1	14.8	**93.2**	88.1	92.3	92.3	92.8
POKR	74.7	75.1	38.9	71.2	71.3	80.2	79.1	**79.5**
Real-world average	77.2	**84.5**	37.2	81.8	82.6	81.9	81.5	81.6
Overall average	66.4	77.8	64.0	**84.9**	81.7	80.8	80.5	80.1

5.1 Results

In the following we show the experimental results. Please note, that in all tables, the values are represented in percent, which means a Kappa score of 1 equals 100% in the table.

Table 2 shows the performance of the algorithms based on their accuracy. Based on the overall average the classical Hoeffding Tree seems to perform best. Also, it stands out, that the adaptive version of the RSLVQ seems to perform far better than the classical version based on SGA. The better performance does not seem to be related to a specific drift type and can also be seen on the RTG generator, which contains no concept drift at all. The Friedman test has shown that there are significant differences between the algorithms. Because of this reason, the Bonferroni Dunn post-hoc was applied and showed statistical relevant differences between:

- Hoeffding Tree, HAT, SAM-KNN, OB and OBA show higher accuracy than Naive Bayes and RSLVQ$_{SGA}$
- RSLVQ$_{ADA}$ shows higher accuracy than Naive Bayes and RSLVQ$_{SGA}$

This leads to the insight that $RSLVQ_{ADA}$, HT, HAT, OB and OBA outperform $RSLVQ_{SGA}$ and Naive Bayes in the immediate streaming setting measured by the accuracy score.

Table 3 shows the performance of the algorithms measured by the kappa score. Based on the overall results it can be assumed, that Hoeffding Tree also performed best. $RSLVQ_{ADA}$ also performs far better than the SGA version, measured by Kappa. Application of the Friedman tests with Bonferroni-Dunn leads to the following results:

- Hoeffding Tree, HAT, OB, OBA, SAM-KNN perform better w.r.t. Kappa score than $RSLVQ_{SGA}$ and Naive Bayes on the tested data streams

Hence, both Hoeffding Tree versions as well as OB, OBA and SAM-KNN outperform $RSLVQ_{SGA}$ and Naive Bayes, measured by the Kappa score.

Both tables show, that the non-tree classifiers lack accuracy and Kappa on the AGR streams. This is probably caused by the fact, that the generator was designed to scale up decision tree learners [1]. The study shows that $RSLVQ_{ADA}$

Table 3. Kappa statistic of the algorithms on synthetic and real-world streams. Winner marked bold. Tests summarize five runs of cross-validation on one million samples or complete real-world streams per run.

Dataset	RSLVQ SGA	RSLVQ Adadelta	NB	HT	HAT	OB	OBA	SAM-KNN
LED_A	63.5	99.6	99.9	**100**	99.9	**100**	**100**	**100**
LED_G	63.4	99.8	**100**	**100**	99.9	**100**	**100**	**100**
SEA_A	59.0	72.6	29.9	**75.6**	60.7	66.6	66.6	72.8
SEA_G	62.8	76.8	30.5	**78.2**	64.9	70.1	70.1	76.3
AGR_A	0.3	−0.1	0.3	**99.9**	58.6	2.1	2.0	3.5
AGR_G	4.1	5.9	10.6	**73.3**	57.6	25.0	24.9	24.0
RTG	17.3	31.5	36.2	**66.0**	20.7	33.5	30.0	30.4
RBF_F	15.6	15.4	-	13.0	24.1	**83.2**	80.1	82.3
RBF_M	6.3	14.0	-	35.9	43.2	87.9	88.5	**91.2**
HYPER	13.5	**74.6**	47.1	60.6	69.8	12.4	12.4	12.0
Synthetic average	30.6	49.0	44.3	**70.3**	59.9	47.6	46.8	49.1
ELEC	70.0	70.0	6.3	60.8	**76.0**	43.9	43.7	43.2
GMSC	−0.4	0.0	0.5	**19.3**	1.2	0.1	0.1	0.2
POKR	55.0	56.2	6.1	47.5	49.0	**64.0**	61.7	62.6
Real-world average	41.5	42.1	4.3	**42.5**	42.1	36.0	35.2	35.3
Overall average	33.1	47.4	33.4	**63.9**	55.8	41.8	41	42.2

is overall very competitive to other stream classifiers. Also it is the best real-world data streams classifier w.r.t. accuracy, while this is not the case for Kappa score. E.g. RSLVQ$_{ADA}$ has very high performance on the GMSC stream, but has a Kappa score of 0, which indicates, that the algorithm may troubles with imbalanced data. Thus, there should be future work, which compares the algorithms on a broader range of real world data streams, especially with imbalanced labels, to validate if the high accuracy is only caused by the chosen streams. Besides, the integration of Adadelta into the RSLVQ leads to a significant increase of prediction performance over RSLVQ$_{SGA}$ overall.

6 Conclusion

In summary, the integration of Adadelta into RSLVQ leads to a statistically significant improvement in prediction performance over the RSLVQ$_{SGD}$. Further, the RSLVQ$_{ADA}$ does not perform statistically significant worse than state-of-the-art stream classifiers like HT overall. Concerning accuracy and kappa score the RSLVQ$_{ADA}$ has the best accuracy and a very competitive kappa score on real-world streams and the Hyperplane generator. In further work the performance should be verified by conducting an comprehensive study on real-world data streams. Additionally, Adadelta should handle different drift types more explicitly, which should be addressed in subsequent work.

References

1. Agrawal R, Imielinski T, Swami A (1993) Database mining: a performance perspective. IEEE Trans Knowl Data Eng 5(6):914–925
2. Augenstein C, Spangenberg N, Franczyk B (2017) Applying machine learning to big data streams: an overview of challenges. In: 2017 IEEE 4th international conference on soft computing machine intelligence (ISCMI), pp 25–29
3. Bifet A, Gavaldà R (2007) Learning from time-changing data with adaptive windowing. In: Proceedings of the seventh SIAM international conference on data mining, 26–28 April 2007, Minneapolis, Minnesota, USA, pp 443–448
4. Bifet A, Gavaldà R (2009) Adaptive learning from evolving data streams. In: Adams NM, Robardet C, Siebes A, Boulicaut J (eds) Advances in intelligent data analysis VIII, 8th international symposium on intelligent data analysis, IDA 2009, Lyon, France, 31 August–2 September 2009, Proceedings. Lecture notes in computer science, vol 5772. Springer, pp 249–260
5. Bifet A, Zhang J, Fan W, He C, Zhang J, Qian J, Holmes G, Pfahringer B (2017) Extremely fast decision tree mining for evolving data streams. In: Proceedings of the 23rd ACM SIGKDD international conference on knowledge discovery and data mining, Halifax, NS, Canada, 13–17 August 2017. ACM, pp 1733–1742
6. Demšar J (2006) Statistical comparisons of classifiers over multiple data sets. J Mach Learn Res 7:1–30
7. Domingos PM, Hulten G (2000) Mining high-speed data streams. In: Proceedings of the sixth ACM SIGKDD international conference on Knowledge discovery and data mining, Boston, MA, USA, 20–23 August 2000, pp 71–80

8. Duchi JC, Hazan E, Singer Y (2011) Adaptive subgradient methods for online learning and stochastic optimization. J Mach Learn Res 12:2121–2159
9. Friedman M (1937) The use of ranks to avoid the assumption of normality implicit in the analysis of variance. J Am Stat Assoc 32(200):675–701
10. Gama J, Zliobaite I, Bifet A, Pechenizkiy M, Bouchachia A (2014) A survey on concept drift adaptation. ACM Comput Surv 46(4):1–37
11. Gomes HM, Barddal JP, Enembreck F, Bifet A (2017) A survey on ensemble learning for data stream classification. ACM Comput Surv 50(2):23:1–23:36
12. Gomes HM, Bifet A, Read J, Barddal JP, Enembreck F, Pfharinger B, Holmes G, Abdessalem T (2017) Adaptive random forests for evolving data stream classification. Mach Learn 106(9–10):1469–1495
13. Kohonen T (1995) Learning vector quantization. Springer, Heidelberg, pp 175–189
14. LeKander M, Biehl M, de Vries H (2017) Empirical evaluation of gradient methods for matrix learning vector quantization. In: 2017 12th international workshop on self-organizing maps and learning vector quantization, clustering and data visualization (WSOM), pp 1–8
15. Losing V, Hammer B, Wersing H (2017) KNN classifier with self adjusting memory for heterogeneous concept drift. In: Proceedings - IEEE international conference on data mining, ICDM, vol 1, pp 291–300
16. Losing V, Hammer B, Wersing H (2017) Self-adjusting memory: how to deal with diverse drift types. In: Proceedings of the twenty-sixth international joint conference on artificial intelligence, IJCAI 2017, Melbourne, Australia, 19–25 August 2017, pp 4899–4903
17. Oza NC (2005) Online bagging and boosting. In: 2005 IEEE international conference on systems, man and cybernetics, vol 3, pp 2340–2345
18. Qian N (1999) On the momentum term in gradient descent learning algorithms. Neural Netw 12(1):145–151
19. Ruder S (2016) An overview of gradient descent optimization algorithms. CoRR abs/1609.04747
20. Seo S, Obermayer K (2003) Soft learning vector quantization. Neural Comput 15(7):1589–1604
21. Straat M, Abadi F, Göpfert C, Hammer B, Biehl M (2018) Statistical mechanics of on-line learning under concept drift. Entropy 20(10):775
22. Sutton RS (1986) Two problems with backpropagation and other steepest-descent learning procedures for networks. In: Proceedings of the eighth annual conference of the cognitive science society, Erlbaum, Hillsdale
23. Tieleman T, Hinton G (2012) Lecture 6.5—RMSProp: divide the gradient by a running average of its recent magnitude. In: COURSERA: neural networks for machine learning. https://www.cs.toronto.edu/~tijmen/csc321/slides/lecture_slides_lec6.pdf
24. Zeiler MD (2012) ADADELTA: an adaptive learning rate method. CoRR abs/1212.5701

Prototype-Based Classifiers
in the Presence of Concept Drift:
A Modelling Framework

Michael Biehl[1(✉)], Fthi Abadi[1], Christina Göpfert[2], and Barbara Hammer[2]

[1] Bernoulli Institute for Mathematics, Computer Science and Artificial Intelligence,
University of Groningen, P.O. Box 407, 9700 AK Groningen, The Netherlands
m.biehl@rug.nl, fthialem@gmail.com
[2] Center of Excellence - Cognitive Interaction Technology, CITEC,
Bielefeld University, Inspiration 1, 33619 Bielefeld, Germany
{cgoepfert,bhammer}@techfak.uni-bielefeld.de
http://www.cs.rug.nl/~biehl

Abstract. We present a modelling framework for the investigation of
prototype-based classifiers in non-stationary environments. Specifically,
we study Learning Vector Quantization (LVQ) systems trained from
a stream of high-dimensional, clustered data. We consider standard
winner-takes-all updates known as LVQ1. Statistical properties of the
input data change on the time scale defined by the training process. We
apply analytical methods borrowed from statistical physics which have
been used earlier for the exact description of learning in stationary environments. The suggested framework facilitates the computation of learning curves in the presence of virtual and real concept drift. Here we focus
on time-dependent class bias in the training data. First results demonstrate that, while basic LVQ algorithms are suitable for the training in
non-stationary environments, *weight decay* as an explicit mechanism of
forgetting does not improve the performance under the considered drift
processes.

Keywords: LVQ · Concept drift · Weight decay · Supervised learning

1 Introduction

The topic of learning under *concept drift* is currently attracting increasing interest in the machine learning community. Terms like *lifelong learning* or *continual
learning* have been coined in this context [1].

In the standard set-up, machine learning processes [2] are conveniently separated into two stages: In the so-called *training phase,* a hypothesis or model of
the data is inferred from a given set of example data. Thereafter, this hypothesis
can be applied to novel data in the *working phase,* e.g. for the purpose of classification or regression. Implicitly, the training data is assumed to represent the

© Springer Nature Switzerland AG 2020
A. Vellido et al. (Eds.): WSOM 2019, AISC 976, pp. 210–221, 2020.
https://doi.org/10.1007/978-3-030-19642-4_21

target task faithfully also after completing the training phase: Statistical properties of the data and the task itself should not change in the working phase.

Frequently, however, the separation of training and working phase appears artificial or unrealistic, for instance in human or other biological learning processes [3]. Similarly, in technical contexts, training data is often available in the form of non-stationary data streams, e.g. [1,4–7]. Two major types of non-stationary environments have been discussed in the literature: In *virtual drifts*, statistical properties of the available training data are time-dependent, while the actual target task remains unchanged. Scenarios in which the target itself, e.g. the classification or regression scheme, changes in time are referred to as real drift processes. Frequently both effects coincide, further complicating the detection and handling of the drift.

In general, the presence of drift requires the *forgetting* of older information while the system is adapted to recent example data. The design of efficient forgetful training schemes demands a thorough theoretical understanding of the relevant phenomena. To this end, the development of suitable modelling frameworks is instrumental. Overviews of earlier work and recent developments in the context of non-stationary learning environments can be found in e.g. [1,4–7].

Here, we study a basic model of learning in a non-stationary environment. In the proposed framework we can address both virtual and real drift processes. An example study of the latter has been presented in [8], recently, where the specific case of random displacements of cluster centers in a bi-modal input distribution was considered. Here, however, the focus is on the study of localized, but explicitly time-dependent densities of high-dimensional inputs in a stream of training examples. More specifically, we consider Learning Vector Quantization (LVQ) as a prototype-based framework for classification [9–11]. LVQ systems are most frequently trained in an online setting by presenting a sequence of single examples for iterative adaptation [10–12]. Hence, LVQ should constitute a natural tool for incremental learning in non-stationary environments [4].

Methods developed in statistical physics facilitate the mathematical description of the training dynamics in terms of typical learning curves. The statistical mechanics of on-line learning has helped to gain insights into the typical behavior of various learning systems, see e.g. [13–15] and references therein.

Clustered densities of data, similar to the one considered here, have been studied in the modelling of unsupervised learning and supervised perceptron training, see e.g. [16–18]. In particular, online LVQ in stationary situations was analysed in [12]. Simple models of concept drift have been studied before within the statistical physics theory of the perceptron: Time-varying linearly separable classification rules were considered in [19,20].

We focus on the question whether LVQ learning schemes are able to cope with drift in characteristic model situations and whether extensions like weight decay can further improve the performance of LVQ in such settings.

2   Models and Methods

First, we introduce Learning Vector Quantization for classification tasks with emphasis on the basic LVQ1 scheme. We propose a model density of data, which was previously investigated in the mathematical analysis of LVQ training in stationary environments. Finally, we extend the approach to the presence of concept drift and consider *weight decay* as an explicit mechanism of forgetting.

2.1   Learning Vector Quantization

The family of LVQ algorithms is widely used for practical classification problems [10, 11]. The popularity of LVQ is due to a number of attractive features: It is quite easy to implement, very flexible and intuitive. Multi-class problems can be handled in a natural way by introducing at least one prototype per class. The actual classification scheme is most frequently based on Euclidean metrics or other simple measures, which quantify the distance of data (inputs, feature vectors) from the class-specific prototypes obtained from the training data. Moreover, in contrast to many other methods, LVQ facilitates direct interpretation since the prototypes are defined in the same space as the data [10, 11].

Nearest Prototype Classifier
We restrict the analysis to the simple case of only one prototype per class in binary classification problems. Hence we consider two prototypes $w_S \in I\!R^N$ with the subscript $S = \pm 1$ (or $\pm$ for short) indicating the represented class of data. The system parameterizes a Nearest Prototype Classification (NPC) scheme in terms of a distance measure $d(w, \xi)$: Any given input $\xi \in I\!R^N$ is assigned to the class label $S = \pm 1$ of the closest prototype. A variety of distance measures have been used in LVQ, enhancing the flexibility of the approach even further [10, 11]. Here, we restrict the analysis to the - arguably - simplest choice: the (squared) Euclidean measure $d(w, \xi) = (w - \xi)^2$.

The LVQ1 Algorithm
A sequence of single example data $\{\xi^\mu, \sigma^\mu\}$ is presented to the LVQ system in the on-line training process [9, 12]: At a given time step $\mu = 1, 2, \ldots$, the feature vector ξ^μ is presented together with the class label $\sigma^\mu = \pm 1$.

Incremental LVQ updates are of the quite general form (see [12])

$$w_S^\mu = w_S^{\mu-1} + \Delta w_S^\mu \text{ with } \Delta w_S^\mu = \frac{\eta}{N} f_S \left[d_{+,-}^\mu, \sigma^\mu, \ldots \right] \left(\xi^\mu - w_S^{\mu-1} \right), \quad (1)$$

where the vector w_S^μ denotes the prototype after presentation of μ examples and the constant learning rate η is scaled with the input dimension N. The actual algorithm is defined through the so-called *modulation function* $f_S[\ldots]$, which typically depends on the labels of the data and prototypes and on the relevant distances of the input from the prototype vectors.

Taking over the NPC concept, the LVQ1 training algorithm [9] modifies only the the so-called *winner*, i.e. the prototype closest to the current training input. The LVQ1 update for two competing prototypes corresponds to Eq. (1) with

$$f_S[d_+^\mu, d_-^\mu, \sigma^\mu] = \Theta\left(d_{-S}^\mu - d_{+S}^\mu\right) S\sigma^\mu \quad \text{where } \Theta(x) = 1 \text{ if } x > 0 \text{ and } 0 \text{ else.} \quad (2)$$

The prefactor $S\sigma^\mu = \pm 1$ specifies the direction of the update: the *winner* is moved towards the presented feature vector if it carries the same class label, while its distance from the data point is further increased if the labels disagree.

2.2 The Dynamics of LVQ

Statistical physics based methods have been used very successfully in the analysis of various learning systems [14,15]. The methodology is complementary to other frameworks of computational learning theory and aims at the description of typical learning dynamics in simplifying model scenarios. Frequently, the approach is based on the assumption that a sequence of statistically independent, randomly generated N-dimensional input vectors is presented to the learning system. Further simplifications and the consideration of the thermodynamic limit $N \to \infty$ facilitate the mathematical representation of the learning dynamics by ordinary differential equations (ODE) and the computation of *learning curves*.

Here, we extend earlier investigations of LVQ training in the framework of a simplifying model situation [12]: High-dimensional training samples are generated independently according to a mixture of two overlapping Gaussian clusters. The input vectors are labelled according to their cluster membership and presented to the LVQ1 system, sequentially. Similar models have been investigated in the context of other learning scenarios, see for instance [16–18].

The Data

We consider random input vectors $\boldsymbol{\xi} \in I\!\!R^N$ which are generated independently according to a bi-modal distribution of the form [12]

$$P(\boldsymbol{\xi}) = \sum_{\sigma=\pm 1} p_\sigma \, P(\boldsymbol{\xi}|\sigma) \text{ with } P(\boldsymbol{\xi}|\sigma) = \frac{1}{(2\pi v_\sigma)^{\frac{N}{2}}} \exp\left[-\frac{1}{2\,v_\sigma}\left(\boldsymbol{\xi} - \lambda\boldsymbol{B}_\sigma\right)^2\right]. \quad (3)$$

The class-conditional densities $P(\boldsymbol{\xi} \mid \sigma = \pm 1)$ represent isotropic, spherical clusters with variances v_σ and means given by $\lambda\boldsymbol{B}_\sigma$. Prior weights of these Gaussian clusters are denoted by p_σ with $p_+ + p_- = 1$. For simplicity, we assume that the vectors $\boldsymbol{B}_\sigma$ are normalized, $\boldsymbol{B}_+^2 = \boldsymbol{B}_-^2 = 1$, and orthogonal with $\boldsymbol{B}_+ \cdot \boldsymbol{B}_- = 0$. The target classification for each input is given by its class-membership $\sigma = \pm 1$. The problem is not linearly separable since the clusters overlap.

Conditional averages over $P(\boldsymbol{\xi} \mid \sigma)$ will be denoted as $\langle \cdots \rangle_\sigma$, while mean values of the form $\langle \cdots \rangle = \sum_{\sigma=\pm 1} p_\sigma \langle \cdots \rangle_\sigma$ are defined for the full density (3). In a particular cluster σ, input components ξ_j are statistically independent and display the variance v_σ. We will use, e.g., the following (conditional) averages:

$$\langle \xi_j \rangle_\sigma = \lambda(\boldsymbol{B}_\sigma)_j, \quad \langle \boldsymbol{\xi}^2 \rangle_\sigma = v_\sigma N + \lambda^2, \quad \langle \boldsymbol{\xi}^2 \rangle = (p_+ v_+ + p_- v_-) N + \lambda^2. \quad (4)$$

Mathematical Analysis

We briefly recapitulate the theory of on-line learning as it has been applied to LVQ in stationary environments and refer to [12] for details.

The *thermodynamic limit* $N \to \infty$ is instrumental in the following. As one of the simplifying consequences we can neglect the terms λ^2 in Eq. (4). Moreover, the limit $N \to \infty$ facilitates the following key steps which, eventually, yield an exact mathematical description of the training dynamics in terms of ODE:

(I) Order parameters: The large number of adaptive prototype components can be characterized in terms of only very few quantities. The definition of these order parameters follows directly from the mathematical structure of the model:

$$R_{S\sigma}^\mu = \boldsymbol{w}_S^\mu \cdot \boldsymbol{B}_\sigma \quad \text{and} \quad Q_{ST}^\mu = \boldsymbol{w}_S^\mu \cdot \boldsymbol{w}_T^\mu \quad \text{for all } \sigma, S, T \in \{-1, +1\}. \tag{5}$$

The index μ represents the number of examples that have been presented to the system. Obviously, Q_{++}^μ, Q_{--}^μ and $Q_{+-}^\mu = Q_{-+}^\mu$ relate to the norms and overlaps of prototypes, the $R_{S\sigma}^\mu$ specify projections onto the cluster vectors $\{\boldsymbol{B}_+, \boldsymbol{B}_-\}$.

(II) Recursion relations: For the above introduced order parameters, recursion relations can be derived directly from the learning algorithm (1):

$$N^{-1}\left(R_{S\sigma}^\mu - R_{S\sigma}^{\mu-1}\right) = \eta\, f_S\left(\boldsymbol{B}_\sigma \cdot \boldsymbol{\xi}^\mu - R_{S\sigma}^{\mu-1}\right) \quad \text{and} \quad N^{-1}\left(Q_{ST}^\mu - Q_{ST}^{\mu-1}\right) = \ldots$$

$$\ldots \eta\left[f_S\left(\boldsymbol{w}_T^{\mu-1}\cdot\boldsymbol{\xi}^\mu - Q_{ST}^{\mu-1}\right) + f_T\left(\boldsymbol{w}_S^{\mu-1}\cdot\boldsymbol{\xi}^\mu - Q_{ST}^{\mu-1}\right)\right] + \eta^2 f_S f_T (\boldsymbol{\xi}^\mu)^2/N. \tag{6}$$

The modulation function is denoted as $f_\pm$ here, omitting its arguments. Terms of order $O(1/N)$ have been discarded; note that $(\boldsymbol{\xi}^\mu)^2 = O(N)$ according to (4).

(III) Averages over the data: Applying the central limit theorem (CLT) we can perform the average over the random sequence of independent examples. The current input $\boldsymbol{\xi}^\mu$ enters the r.h.s. of Eq. (6) only through its length and

$$h_S^\mu = \boldsymbol{w}_S^{\mu-1} \cdot \boldsymbol{\xi}^\mu \quad \text{and} \quad b_\sigma^\mu = \boldsymbol{B}_\sigma \cdot \boldsymbol{\xi}^\mu. \tag{7}$$

Since the scalar products correspond to sums of many independent random quantities, the CLT applies and the projections in Eq. (7) are correlated Gaussian quantities for large N. Hence, their joint density is fully specified by the moments

$$\langle h_S^\mu \rangle_\sigma = \lambda R_{S\sigma}^{\mu-1}, \quad \langle b_\tau^\mu \rangle_\sigma = \lambda\delta_{ST}, \quad \langle h_S^\mu h_T^\mu \rangle_\sigma - \langle h_S^\mu \rangle_\sigma \langle h_T^\mu \rangle_\sigma = v_\sigma\, Q_{ST}^{\mu-1}$$

$$\langle h_S^\mu b_\tau^\mu \rangle_\sigma - \langle h_S^\mu \rangle_\sigma \langle b_\tau^\mu \rangle_\sigma = v_\sigma\, R_{ST}^{\mu-1}, \quad \langle b_\rho^\mu b_\tau^\mu \rangle_\sigma - \langle b_\rho^\mu \rangle_\sigma \langle b_\tau^\mu \rangle_\sigma = v_\sigma\, \delta_{\rho\tau} \tag{8}$$

where $\delta_{...}$ is the Kronecker-Delta. The joint density is therefore fully specified by the order parameters of the previous time step and by the model parameters $\lambda, p_\pm, v_\pm$. This enables us to perform an average of the recursions (6) over the latest example in terms of elementary Gaussian integrations. Moreover, the result is obtained in in closed form in $\{R_{S\sigma}^{\mu-1}, Q_{ST}^{\mu-1}\}$, see [12] for details.

(IV) Self-averaging properties of the order parameters allow us to restrict the description to their mean values, see [21] for a mathematical discussion in the specific context of on-line learning. Random fluctuations vanish as $N \to \infty$ and, as a consequence, Eq. (6) correspond to the deterministic dynamics of means.

(V) Continuous time limit and learning curves: For large N, we can interpret the ratios on the left hand sides of Eq. (6) as derivatives with respect to the continuous learning time $\alpha = \mu/N$. This corresponds to the natural expectation that the number of examples required for successful training should be proportional to the number of degrees of freedom in the system. The set of coupled ODE obtained from Eq. (6) is of the generic form (see [12] for details)

$$\left[dR_{S\tau}/d\alpha\right] = \eta \left(\langle b_\tau f_S \rangle - R_{S\tau} \langle f_S \rangle\right) \tag{9}$$

$$\left[dQ_{ST}/d\alpha\right] = \eta \left(\langle h_S f_T + h_T f_S \rangle - Q_{ST} \langle f_S + f_T \rangle\right) + \eta^2 \sum_{\sigma=\pm 1} v_\sigma p_\sigma \langle f_S f_T \rangle_\sigma .$$

The (numerical) integration yields the temporal evolution of order parameters in the course of training. We consider prototypes initialized as independent random vectors of squared norm $\hat{Q}$ with no prior knowledge of the cluster structure:

$$Q_{++}(0) = Q_{--}(0) = \hat{Q}, Q_{+-}(0) = 0 \text{ and } R_{S\sigma}(0) = 0 \text{ for } S, \sigma = \pm 1. \tag{10}$$

The success of training is quantified in terms of the generalization error, i.e. the probability for misclassifying novel, random data. Under the assumption that the density (3) represents the actual target classification, we can work out the class-specific errors for data from cluster $\sigma = 1$ or $\sigma = -1$:

$$\epsilon = p_+ \epsilon^+ + p_- \epsilon^- \quad \text{with} \quad \epsilon^\sigma = \langle \Theta \left(d_{+\sigma} - d_{-\sigma}\right) \rangle_\sigma . \tag{11}$$

For the full derivation of the conditional averages as functions of order parameters we refer to [12]. Exploiting self-averaging properties (IV) again, we obtain the learning curve $\epsilon(\alpha)$, i.e. the performance after presenting (αN) examples.

Weight Decay

Next, we extend the LVQ1 update by a so-called *weight decay* term as an element of explicit *forgetting*. To this end, we consider the multiplication of all prototype components by a factor $(1 - \gamma/N)$ before the generic learning step (1):

$$\boldsymbol{w}_S^\mu = \left(1 - \gamma/N\right) \boldsymbol{w}_S^{\mu-1} + \Delta \boldsymbol{w}_S^\mu. \tag{12}$$

Since the multiplications with $(1 - \gamma/N)$ accumulate in the course of training, weight decay results in an increased influence of the most recent training data as compared to *earlier* examples. Similar modifications of perceptron training in the presence of drift were discussed in [19, 20].

Other motivations for the introduction of weight decay in machine learning range from the modelling of *forgetful memories* in attractor neural networks [22, 23] to regularization in order to reduce over-fitting [2]. As an example for the latter, weight decay in layered neural networks was analysed in [24].

The modified ODE for LVQ1 training with weight decay, cf. Eq. (12), are obtained in a straightforward manner and read

$$\left[dR_{S\tau}/d\alpha\right]_\gamma = \left[dR_{S\tau}/d\alpha\right] - \gamma R_{S\tau}; \quad \left[dQ_{ST}/d\alpha\right]_\gamma = \left[dQ_{ST}/d\alpha\right] - 2\gamma Q_{ST} \tag{13}$$

with the terms [...] on the r.h.s. formally given by Eq. (9) for $\gamma = 0$.

2.3   LVQ Dynamics Under Concept Drift

The analysis summarized in the previous section concerns learning in stationary environments with densities and targets of the form (3). Here, we discuss the effect of including concept drift within our modelling framework.

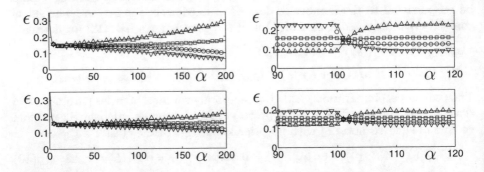

Fig. 1. LVQ1 in the presence of concept drift. Solid lines correspond to the integration of ODE with initialization as in Eq. (10). Cluster variances are $v_+ = v_- = 0.4$ and $\lambda = 1$ in the density (3). Upper graphs correspond to LVQ1 without weight decay, lower graphs display results for $\gamma = 0.05$ in (12). In addition, Monte Carlo results for $N = 100$ are shown: class-wise errors $\epsilon^\pm(\alpha)$ are displayed as upward (downward) triangles, respectively; squares mark the reference error $\epsilon_{ref}(\alpha)$ (14); circles correspond to $\epsilon_{track}(\alpha)$ (15). **Left panel:** drift with linearly increasing $p_+(\alpha)$ given by $\alpha_o = 20$, $\alpha_{end} = 200$, $p_{max} = 0.8$ in (16). **Right panel:** sudden change of class weights according to to Eq. (17) with $\alpha_o = 100$ and $p_{max} = 0.75$. Only the α-range close to α_o is shown.

Real Drift of the Cluster Centers

In the presented framework, a real drift can be modelled by processes that displace the cluster centers in N-dim. feature space while the training follows the stream of data. As a specific example, in [8] the authors study the effects of a random *diffusion* of time-dependent vectors $\boldsymbol{B}_\pm(\mu)$. The results show that simple LVQ1 is capable of tracking randomly drifting concept to a non-trivial extent.

Time Dependent Input Densities

Strictly speaking, virtual drifts affect only the statistical properties of observed example data, while the actual target classification remains the same. In our modelling framework, we can readily consider time-dependent parameters of the density (3), e.g. $\lambda(\alpha)$ and $v_\sigma(\alpha)$ by inserting them in Eq. (9). Here, we will focus on non-stationary prior weights $p_+(\alpha) = 1 - p_-(\alpha)$ for the generation of example data. In this case, a varying fraction of examples represents each of the classes in the stream of training data. Non-stationary class bias complicates the training significantly and can lead to inferior performance in practical situations [25].

(A) Drift in the training data only
Here we assume that the target classification is defined by a fixed *reference density* of data. As a simple example we consider equal priors $p_+ = p_- = 1/2$ in a symmetric reference density (3) with $v_+ = v_-$. On the contrary, the characteristics of the observed training data is assumed to be time-dependent. In particular, we study the effect of time-dependent $p_\sigma(\alpha)$ and weight decay.

Given the order parameters of the learning systems in the course of training, the corresponding *reference generalization error*

$$\epsilon_{ref}(\alpha) = \left(\epsilon^+ + \epsilon^-\right)/2 \tag{14}$$

is obtained by setting $p_+ = p_- = 1/2$ in Eq. (11), but inserting $R_{S\tau}(\alpha)$ and $Q_{ST}(\alpha)$ as obtained from the integration of the ODE (9) or (13) with time dependent $p_+(\alpha) = 1 - p_-(\alpha)$ in the training data.

(B) Drift in training and test data
In the second interpretation we assume that the time-dependence of $p_\sigma(\alpha)$ affects both the training and test data in the same way. Hence, the change of the statistical properties of the data is inevitably accompanied by a modification of the target classification: For instance, the Bayes optimal classifier and its best linear approximation will depend explicitly on the current priors $p_\sigma(\alpha)$ [12].

The learning system is supposed to track the drifting concept and we denote the corresponding generalization error, cf. Eq. (11), by

$$\epsilon_{track} = p_+(\alpha)\epsilon^+ + p_-(\alpha)\epsilon^-. \tag{15}$$

In terms of modelling the training dynamics, both scenarios, (A) and (B), require the same straightforward modification of the ODE system: the explicit introduction of α-dependent quantities $p_\sigma(\alpha)$ in Eq. (9). However, the obtained temporal evolution translates into the reference error $\epsilon_{ref}(\alpha)$ for the case of drift in the training data (A), and into $\epsilon_{track}(\alpha)$ in interpretation (B).

3   Results and Discussion

Here we present and discuss first results obtained by integrating the systems of ODE for LVQ1 with and without weight decay under different time-dependent drifts. For comparison, averaged learning curves as obtained by means of Monte Carlo simulations are also shown. All results are for constant learning rate $\eta = 1$ and the LVQ systems were initilized according to Eq. (10).

We study three example scenarios for the time-dependence $p_+(\alpha) = 1-p_-(\alpha)$:

Linear Increase of the Bias
We consider a time-dependent bias of the form $p_+(\alpha) = 1/2$ for $\alpha < \alpha_o$ and

$$p_+(\alpha) = 1/2 + \frac{(p_{max}-1/2)\,(\alpha - \alpha_o)}{(\alpha_{end} - \alpha_o)} \text{ for } \alpha \geq \alpha_o. \tag{16}$$

where the maximum class weight $p_+ = p_{max}$ is reached at learning time α_{end}. Figure 1 (left panel) shows the learning curves as obtained by numerical integration of the ODE together with Monte Carlo simulation results for ($N = 100$)-dimensional inputs and prototype vectors. As an example we set the parameters to $\alpha_o = 25, p_{max} = 0.8, \alpha_{end} = 200$. The learning curves are displayed for LVQ1 without weight decay (upper) and with $\gamma = 0.05$ (lower panel). Simulations show excellent agreement with the ODE results.

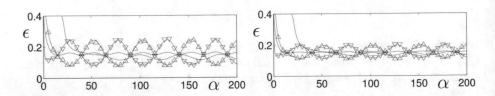

Fig. 2. LVQ1 in the presence of oscillating class weights according to Eq. (18) with parameters $T = 50$ and $p_{max} = 0.8$, without weight decay $\gamma = 0$ (left) and for $\gamma = 0.05$ (right). For clarity, Monte Carlo results are only shown for the class-conditional errors ϵ^+ (upward) and ϵ^- (downward triangles). All other settings as in Fig. 1.

The system adapts to the increasing imbalance of the training data, as reflected by a decrease (increase) of the class-wise error for the over-represented (under-represented) class, respectively. The weighted over-all error ϵ_{track} also decreases, i.e. the presence of class bias facilitates smaller total generalization error, see [12]. The performance with respect to unbiased reference data detoriorates slightly, i.e. ϵ_g grows with increasing class bias as the training data represents the target less faithfully.

The influence of the class bias and its time-dependence is reduced significantly in the presence of weight decay with $\gamma > 0$, cf. Fig. 1 (lower panel). Weight decay restricts the norm of the prototypes, i.e. the possible offset of the decision boundary from the origin. Consequently, the tracking error slightly increases, while ϵ_{ref} with respect to the reference density is decreased compared to the setting without weight decay, respectively.

Sudden Change of the Class Bias

Here we consider an instantaneous switch from high bias $p_{max} > 1/2$ to low bias:

$$p_+(\alpha) = p_{max} \text{ for } \alpha < \alpha_o \text{ and } p_+(\alpha) = 1 - p_{max} \text{ for } \alpha \geq \alpha_o. \quad (17)$$

We consider $p_{max} = 0.75$ as an example, the corresponding results from the integration of ODE and Monte Carlo simulations are shown in Fig. 1 (right panel) for training without weight decay (upper) and for $\gamma = 0$ (lower panel).

We observe similar effects as for the slow, linear time-dependence: The system reacts rapidly with respect to the class-wise errors and the tracking error ϵ_{track} maintains a relatively low value. Also, the reference error ϵ_{ref} displays robustness with respect to the sudden change of p_+. Weight decay, as can be seen in the lower right panel of Fig. 1 reduces the over-all sensitivity to the bias and its change: Class-wise errors are more balanced and the weighted ϵ_{track} slightly increases compared to the setting with $\gamma = 0$.

The weight decay does not seem to have a notable effect on the promptness of the system's adaptation to the changing bias. While it significantly regularizes the system, the expected effect of *forgetting* previous information in favor of the most recent examples cannot be observed.

Periodic Time Dependence
As a third scenario we consider an oscillatory modulation of the class weights in training:

$$p_+(\alpha) = 1/2 + (p_{max} - 1/2) \cos [2\pi \, \alpha/T] \tag{18}$$

with periodicity T on α-scale and maximum amplitude $p_{max} < 1$.

Example results are shown in Fig. 2 for $T = 50$ and $p_{max} = 0.8$. Monte Carlo results for $N = 100$ are only displayed for the class-wise errors show excellent agreement with the numerical integration of the ODE for training without weight decay (left panel) and for $\gamma = 0.05$ (right panel). The observations confirm our findings for slow and sudden changes of p_+: In the main, weight decay limits the reaction of the system to the presence of a bias and its time-dependence.

4   Summary and Outlook

In summary, we have presented a mathematical framework in which to study the influence of concept drift on prototype-based classifiers systematically. In all specific drift scenarios considered here, we observe that simple LVQ1 can track the time-varying class bias to a non-trivial extent: In the interpretation of the results in terms of real drift, the class-conditional performance and the tracking error $\epsilon_{track}(\alpha)$ clearly reflect the time-dependence of the prior weights.

In general, the reference error $\epsilon_{ref}(\alpha)$ with respect to class-balanced test data, displays only little deterioration due to the drift in the training data.

The main effect of introducing weight decay is a reduced overall sensitivity to bias in the training data: Figs. 1 and 2 display a decreased difference between the class-wise errors ϵ^+ and ϵ^- for $\gamma > 0$. Naïvely, one might have expected an improved tracking of the drift due to the imposed *forgetting*, resulting in, for instance, a more rapid reaction to the sudden change of bias in Eq. (17). However, such an improvement cannot be confirmed. Our findings are in contrast to a recent study [8], in which we observe increased performance by weight decay for a different drift scenario, i.e. the randomized displacement of cluster centers.

The precise influence of weight decay clearly depends on the geometry and relative position of the clusters. Its dominant effect, however, is the regularization of the LVQ system by reducing the norms Q_{++} and Q_{--} of the prototype vectors. Consequently, the NPC classifier is less flexible to reflect class bias which would require significant offset of the prototypes and decision boundary from the origin. This mildens the influence of the bias (and its time-dependence) and results in a more robust behavior of the employed error measures.

Alternative mechanisms of *forgetting* should be considered which do not limit the flexibility of the LVQ classifier, yet facilitate *forgetting* of older information. As one example strategy we intend to investigate the accumulation of additive noise in the training process. We will also explore the parameter space of the model density and in greater depth and study the influence of the learning rate systematically.

Acknowledgement. The authors would like to thank A. Ghosh, A. Witoelar and G.-J. de Vries for useful discussions of earlier projects on LVQ training in stationary environments.

References

1. Zliobaite I, Pechenizkiy M, Gama J (2016) An overview of concept drift applications. In: Big data analysis new algorithms for a new society. Springer
2. Hastie, T, Tibshirani, R, Friedman, J (2001) The elements of statistical learning: data mining, inference, and prediction. Springer
3. Amunts K, et al (ed) (2014) Brain-inspired computing. In: Second international workshop BrainComp 2015. LNCS, vol 10087. Springer
4. Losing V, Hammer B, Wersing H (2017) Incremental on-line learning: a review and of state of the art algorithms. Neurocomputing 275:1261–1274
5. Ditzler G, Roveri M, Alippi C, Polikar R (2015) Learning in nonstationary environment: a survey. Comput Intell Mag 10(4):12–25
6. Joshi J, Kulkarni P (2012) Incremental learning: areas and methods - a survey. Int J Data Mining Knowl Manag Process 2(5):43–51
7. Ade R, Desmukh P (2013) Methods for incremental learning - a survey. Int J Data Mining Knowl Manag Process 3(4):119–125
8. Straat M, Abadi F, Göpfert C, Hammer B, Biehl M (2018) Statistical mechanics of on-line learning under concept drift. Entropy 20(10):775
9. Kohonen, T (2001) Self-organizing maps. Springer series in information sciences, 2nd edn., vol 30. Springer
10. Nova D, Estevez PA (2014) A review of learning vector quantization classifiers. Neural Comput Appl 25(3–4):511–524
11. Biehl M, Hammer B, Villmann T (2016) Prototype-based models in machine learning. Cognit Sci 7(2):92–111 Wiley Interdisciplinary Reviews
12. Biehl M, Ghosh A, Hammer B (2007) Dynamics and generalization ability of LVQ algorithms. J Mach Learn Res 8:323–360
13. Saad D (ed) (1999) On-line learning in neural networks. Cambridge University Press, New York
14. Engel A, van den Broeck C (2001) The statistical mechanics of learning. Cambridge University Press, New York
15. Watkin TLH, Rau A, Biehl M (1993) The statistical mechanics of learning a rule. Rev Mod Phys 65(2):499–556
16. Biehl M, Freking A, Reents G (1997) Dynamics of on-line competitive learning. Europhys Lett 38:73–78
17. Barkai N, Seung HS, Sompolinsky H (1993) Scaling laws in learning of classification tasks. Phys Rev Lett 70(20):L97–L103
18. Marangi C, Biehl M, Solla SA (1995) Supervised learning from clustered input examples. Europhys Lett 30:117–122

19. Biehl M, Schwarze H (1993) Learning drifting concepts with neural networks. J Phys A Math Gen 26:2651–2665
20. Vicente R, Caticha N (1998) Statistical mechanics of on-line learning of drifting concepts: a variational approach. Mach Learn 32(2):179–201
21. Reents G, Urbanczik R (1998) Self-averaging and on-line learning. Phys Rev Lett 80(24):5445–5448
22. Mezard M, Nadal JP, Toulouse G (1986) Solvable models of working memories. J de Phys (Paris) 47(9):1457–1462
23. van Hemmen JL, Keller G, Kühn R (1987) Forgetful memories. Europhys Lett 5(7):663–668
24. Saad D, Solla SA (1997) Learning with noise and regularizers in multilayer neural networks. In: Neural information processing system (NIPS 9). MIT Press, pp 260–266
25. Wang, S, Minku LL, Yao X (2017) A systematic study of online class imbalance learning with concept drift. CoRR abs/1703.06683

Theoretical Developments in Clustering, Deep Learning and Neural Gas

Soft Subspace Topological Clustering over Evolving Data Stream

Mohammed Oualid Attaoui[1,2]([envelope]), Mustapha Lebbah[1], Nabil Keskes[2], Hanene Azzag[1], and Mohammed Ghesmoune[1]

[1] University of Paris 13, Sorbonne Paris City LIPN-UMR 7030 - CNRS, 99, av. J-B Clément, 93430 Villetaneuse, France
`attaoui@lipn.univ-paris13.fr`
[2] Higher School of Computer Science (ESI-SBA), LabRI Laboratory, Sidi Bel-Abbes, Algeria

Abstract. Subspace clustering has been successfully applied in many domains and its goal is to simultaneously detect both clusters and subspaces of the original feature space where these clusters exist. A Data stream is a massive sequences of data coming continuously. Clustering this type of data requires some restrictions in time and memory. In this paper we propose a new method named S2G-Stream based on clustering data streams and soft subspace clustering. Experiments on public datasets showed the ability of S2G-Stream to detect simultaneously the best features, subspaces and the best clustering.

1 Introduction

Data stream clustering is a technique that performs cluster analysis of data stream and is able to produce results in real time. The ability to process data in a single pass and summarize it, while using limited memory, is crucial to stream clustering. Several algorithms and methods have been proposed to deal with those constraints. A large number of algorithms use two phases. An online phase processes data stream points and produces summary statistics, an offline component phase uses the summary data to generate the clusters [1]. An alternative solution proposes to generate final clusters without using offline phase [2].

Subspace clustering is an extension of feature selection which tries to identify clusters in different subspaces of the same dataset. As feature selection, subspace clustering needs a search method and an evaluation criterion. In addition, subspace clustering must somehow restrict the scope of the evaluation criterion so as to consider different subspaces for each different cluster.

The objective of our contribution is to propose an innovative algorithm of subspace clustering evolving data stream, whose main features and advantages are described as follows: (a) The topological structure is represented by a graph wherein each node represents a cluster. (b) The use of an exponential fading function to reduce the impact of old data, outdated nodes can be removed. (c) Introduction of a double weighting system for features and subspaces in order

© Springer Nature Switzerland AG 2020
A. Vellido et al. (Eds.): WSOM 2019, AISC 976, pp. 225–230, 2020.
https://doi.org/10.1007/978-3-030-19642-4_22

to detect clusters in different dimensions. The rest of the paper is organized as follows. In Sect. 2, we explain our algorithm Subspace Micro-Batching Growing Neural Gas for Clustering Data Streams (S2G-Stream) and then evaluate its performance in Sect. 3. Finally, we concludes this paper in Sect. 4.

2 Model Proposition

In this section we introduce Soft Subspace Micro Batching Neural Gas for Clustering Data Streams. S2G-Stream based on Growing Neural Gas (GNG) model, which is an incremental self-organizing approach that belongs to the family of topological maps such as Self-Organizing Maps (SOM) [3]. We assume that the data stream consists of a sequence $\mathcal{X} = \{\mathbf{x}_1, \mathbf{x}_2, ..., \mathbf{x}_n\}$ of n (potentially infinite) elements of a data stream arriving at times $t_1, t_2, ..., t_n$, where $\mathbf{x}_i = (x_i^1, x_i^2, ..., x_i^d)$. At each time, S2G-Stream is represented by a graph $\mathcal{C}$ with K nodes, where each node represents a cluster. Each node $c \in \mathcal{C}$ has: (1) A prototype $\mathbf{w}_c = (w_c^1, w_c^2, ..., w_c^d)$ representing its position. (2) A weight π_c (3) An error variable $error(c)$ representing the distance between this node and the assigned data-points. For each pair of nodes (r, c) we have $\delta(c, r)$ the length of the shortest chain linking r and c on the graph. $\mathcal{K}^T(\delta) = \mathcal{K}(\delta/T)$ is the neighborhood function, T controls the width of $\mathcal{K}$.

S2G-Stream introduce a double weight system for features denoted by α and subspaces denoted by β. The features $\mathcal{F}$ are divided into B subspaces $\mathcal{F} = \cup_{b=1}^{B} \mathcal{F}_b$ where $\mathcal{F}_b = \{x^j, j = 1, ..., p_b\}$ is the feature set where $p_1 + ... + p_b + ... + p_B = d$. Thus, α is a matrix of $K \times B$ where α_c^b is the weight of the subspace b in the node c of $\mathcal{X}$. β is a matrix $K \times d$ where β_b is a matrix of $K \times p_b$, where $\beta_{cb}^j (j = 1, ..., p_b)$ is the weight of the jth feature in the subspace b for the node c with $\sum_{j=1}^{p_b} \beta_{cb}^j = 1$ and $\sum_{b=1}^{B} \alpha_c^b = 1, \forall c \in \mathcal{C}$. From the two types of weights, the subspaces of clusters can be revealed. Based on [4,5], we propose to minimize the new cost function defined below for data batch $\mathcal{X}^{(t+1)} = \{\mathbf{X}_1, \mathbf{X}_2, ..., \mathbf{X}_{t+1}\}$:

$$\mathcal{J}_{S2G-Stream}^{(t+1)}(\phi, \mathcal{W}, \alpha, \beta) = \sum_{c \in C} \sum_{b \in B} \sum_{\mathbf{x}_i \in \mathcal{X}^{(t+1)}} \mathcal{K}^T(\delta(c, \phi(\mathbf{x}_i))) \alpha_c^b \mathcal{D}_{\beta_{cb}} + J_{cb} + I_c$$

(1)

where $\mathcal{D}_{\beta_{cb}} = \sum_{j=1}^{p_b} \beta_{cb}^j (x_i^j - \omega_c^j)^2$. $I_c = \lambda \sum_{b=1}^{B} \alpha_c^b \log(\alpha_c^b)$ and $J_{cb} = \eta \sum_{j=1}^{p_b} \beta_{cb}^j \log(\beta_{cb}^j)$ represent respectively the weighted negative entropies associated with the subspaces weight vectors and the features weight vectors respectively. The parameters λ and η are used to adjust the relative contributions made by the features and subspaces to the clustering. The optimization of the cost function is performed alternately for each batch $\mathcal{X}^{(t+1)}$ in four steps corresponding the four parameters $\mathcal{W}, \phi, \alpha$ and β:

1. **Assignment function:** For a fixed $\mathcal{W}, \alpha$ and β, the assignment function $\phi(\mathbf{x}_i)$ is described in Eq. (2). In order to reduce the computational time,

neighborhood nodes are not considered in the assignment.

$$\phi(\mathbf{x}_i) = \arg\min_{c \in \mathcal{C}} \left(\sum_{b=1}^{B} \alpha_c^b \sum_{j=1}^{p_b} \beta_{cb}^j \left(x_i^j - \omega_c^j \right)^2 \right) \tag{2}$$

2. **Update prototypes** $\mathcal{W}$: For a fixed ϕ, α and β the prototypes $\mathbf{w}_c$ are updated for every batch of data following the equation defined below.

$$\mathbf{w}_c^{(t+1)} = \frac{\mathbf{w}_c^{(t)} n_c^{(t)} \gamma + \sum_{r \in C} \mathcal{K}^T(\delta(r,c)) \mathbf{z}_r^{(t)} m_r^{(t)}}{n_c^{(t)} \gamma + \sum_{r \in C} \mathcal{K}^T(\delta(r,c)) m_r^{(t)}} \tag{3}$$

where $\mathbf{w}_c^{(t)}$ is the previous prototype, $n_c^{(t)}$ is the number of points assigned to the cluster, $\mathbf{z}_r^{(t)}$ is the previous prototype for the cluster r (which is a neighbor of c) and $m_r^{(t)}$ is the number of points added to the cluster r in the current batch: $n_c^{(t+1)} = n_c^{(t)} + m_c^{(t)}$.

3. **Update α weights:** for a fixed ϕ, $\mathcal{W}$ and β, the parameter α is updated for every batch following the expression defined below:

$$\alpha_c^b = \frac{e^{\frac{-D_{cb}}{\lambda}}}{\sum_{s=1}^{B} e^{\frac{-D_{cs}}{\lambda}}} \tag{4}$$

with $D_{cb} = \sum_{\mathbf{x}_i \in \mathcal{X}^{(t)}} \mathcal{K}^T(\delta(\phi(\mathbf{x}_i), c)) \sum_{j=1}^{p_b} \beta_{cb}^j \left(\mathbf{x}_i^j - w_c^j \right)^2$

4. **Update β weights:** for a fixed ϕ, $\mathcal{W}$ and α, the parameter β is updated for every batch following the expression defined below:

$$\beta_{cb}^j = \frac{e^{\frac{-E_{cb}^j}{\eta}}}{\sum_{h \in B_j} e^{\frac{-E_{ch}}{\eta}}} \tag{5}$$

with $E_{cb}^j = \sum_{\mathbf{x}_i \in \mathcal{X}^{(t)}} \alpha_c^{b_j} \mathcal{K}^T(\delta(\phi(\mathbf{x}_i), c)) \left(\mathbf{x}_i^j - w_c^j \right)^2$, where b_j is the subspace where the jth feature belongs.

The complete description of the S2G-Stream algorithm can be found in Algorithm (1). Due to size limitation, we only discussed the main functions above.

Algorithm 1. S2G-Stream

Initialize the graph with two nodes, initialize α and β weights randomly;
while *there is a micro-batch to proceed* **do**
    -Get the micro-batch of data points arrived at time interval t;
    1. **Assignment Step**
      -Find the nearest and the second nearest node bmu_1 and bmu_2 and create an edge between them. If it already exists: set the age to zero;
      -Assign each point to the closest center following Equation (2);
    2. **Update Step**
      -Update the new centroid as described in Equation (3) ;
      -Update the $error$ of each node: $error(bmu_1) = error(bmu_1) + ||x_i - bmu_1||^2$;
      -Update nodes weight : $\pi_c^{(t+1)} = \pi_c^{(t)}\gamma$, where γ is the decay factor;
    3. **Add nodes**
      -Find the node q with the largest error and his neighbor f with the largest accumulated error;
      -Add the new node r between nodes q and f: $\mathbf{w}_r = 0.5(\mathbf{w}_q + \mathbf{w}_f)$;
      -Decrease the error variables of q and f by multiplying them with a constant v where: $0 < v < 1$ and assign to r the error value of q ;
      -Decrease the error of all nodes by multiplying them with a constant s, and delete isolated nodes ;
    4. **Update weights**
      -Update feature weights following Equation (4) and subspace weights following Equation (5) ;

---

3 Experimental Evaluation

We evaluate our approach on real world datasets *waveform, Image Segmentation (IS), Cardiotocography (CDT)* and *pendigits* extracted from UCI repository[1]. We evaluate clustering quality of S2G-Stream compared to 4 algorithms: Growing Neural Gas (GNG) from the Scala Smile repository[2], CluStream and DStream from R package streamMOA[3]. For the quality measures, we used Normalized Mutual Information (NMI) and Adjusted Rand index (ARAND). The results are reported in Table 1. It is noticeable that the ARAND results for S2G-Stream are higher than the results of other algorithms. The NMI values are also higher compared to the others except for GNG on *pendigits*, and for DStream on *CDT*. These results are due to the fact that S2BG-Stream detects noisy features, which allows the best features to contribute more to clustering. This is also due to the notion of fading which reduces the impact of irrelevant data.

Due to space restrictions we settled for *CTG* in the following experiment. CTG dataset describes fetal cardiotocograms and is composed of 3 subspaces. Subspace 1 contains 7 features related to the heart rate of a fetus. Subspace 2 contains 4 features describing heart rate variability. Subspace 3 is composed of 10 features defining histograms of fetal cardiography. Figure 1 represents prototypes

---

[1] http://archive.ics.uci.edu/ml/datasets.html.

[2] http://haifengl.github.io/smile/.

[3] https://github.com/mhahsler/streamMOA.

Table 1. Comparing S2G-Stream with different algorithms.

Dataset	Metrics	S2G-Stream	GNG	CluStream	DStream
waveform 20 features 2 subspaces (10,10) 5000 observations	NMI	**0.397(0.002)**	0.306(0.078)	0.393(0.065)	0.507(0.003)
	ARAND	**0.137(0.007)**	0.006(0.103)	0.010(0.001)	0.040(0.001)
IS 19 features 2 subspaces (9,10) 2310 observations	NMI	**0.550(0.05)**	0.542(0.010)	0.506(0.065)	0.435(0.07)
	ARAND	**0.418(0.04)**	0.102(0.051)	0.098(0.010)	0.134(0.002)
CTG 21 features 3 subspaces (7,4,10) 2126 observations	NMI	0.270(0.009)	0.375(0.004)	0.086(0.06)	**0.471(0.170)**
	ARAND	**0.118(0.005)**	0.030(0.011)	0.019(0.008)	0.017(0.002)
pendigits 17 features 2 subspaces (10,7) 10992 observations	NMI	0.572(0, 038)	**0.585(0.019)**	0.285(0.099)	0.554(0.15)
	ARAND	**0.408(0.060)**	0.027(0.085)	0.011(0.006)	0.016(0.011)

$\mathcal{W}$, β weights and α weights for the final batch of CTG dataset. We observe in Fig. 1b that weights of features (8,9,10,11) which are respectively ASTV , MSTV, ALTV and MLTV are higher than the weights of the other features for most clusters. We observe that this 4 features influence better the clustering process and are more important than the other features for most clusters. In Fig. 1c we observe that the α weight of the second subspace that contains this 4 features is also higher than the weights of the other two subspaces. We conclude from this experiment that heart rate variability influence better the clustering of fetal cardiotocograms.

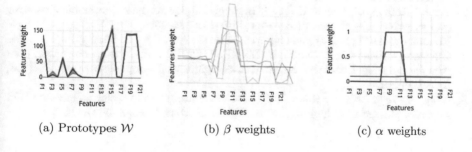

| (a) Prototypes $\mathcal{W}$ | (b) β weights | (c) α weights |

Fig. 1. Results of weights α and β, and prototypes $\mathcal{W}$ for the final batch for CTG dataset. Each color represent a node.

Figure 2 shows an example of the evolution of the graph S2G-Stream on *waveform* dataset, using sammon's nonlinear mapping, as the data flows (colored points represent labelled data points and black points represents nodes of the graph with edges in black lines). We can clearly see that S2G-Stream, begins with two randomly chosen prototypes, manages to recognize the structures of the data stream as the data flows, at the end we can see that the topology recover all the data structure. it is noticeable that our method also can detect clusters of arbitrary shapes.

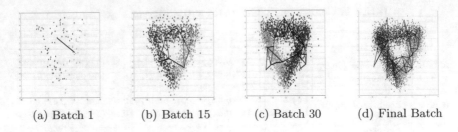

|(a) Batch 1|(b) Batch 15|(c) Batch 30|(d) Final Batch|

Fig. 2. Evolution of graph creation of S2G-Stream on waveform dataset.

4   Conclusion

In this paper, we have proposed S2G-Stream, an efficient method for subspace clustering of an evolving data stream in an online manner. We introduced a double weighting system for both features and subspaces. We also introduced the notion of fading to delete outdated nodes. Experimental evaluation and comparison with well-known clustering methods demonstrate the effectiveness and efficiency of S2G-Stream in clustering results, discovering clusters of arbitrary shape and also detecting relevant features and subspaces.

References

1. Shukla M, Kosta YP, Jayswal M (2017) A modified approach of optics algorithm for data streams. Eng Technol Appl Sci Res 7(2):1478–1481
2. Khan I, Huang JZ, Ivanov K (2016) Incremental density-based ensemble clustering over evolving data streams. Neurocomputing 191:34–43
3. Kohonen T (2012) Self-organizing maps, vol 30. Springer, Heidelberg
4. Ouattara M, Keita NN, Badran F, Mandin C (2013) Soft subpace clustering pour données multiblocs basée sur les cartes topologiques auto-organisées SOM: 2S-SOM. In: SFDS 2013
5. Chen X, Ye Y, Xu X, Huang JZ (2012) A feature group weighting method for subspace clustering of high-dimensional data. Pattern Recogn 45(1):434–446

Solving a Tool-Based Interaction Task Using Deep Reinforcement Learning with Visual Attention

Sascha Fleer[✉] and Helge Ritter

Neuroinformatics Group, EXC Cognitive Interaction Technology (CITEC),
Bielefeld University, Bielefeld, Germany
sfleer@techfak.uni-bielefeld.de

Abstract. We propose a reinforcement learning approach that combines an *asynchronous actor-critic model* with a *recurrent model of visual attention*. Instead of using the full visual information of the scene, the resulting model accumulates the foveal information of controlled glimpses and is thus able to reduce the complexity of the network. Using the designed model, an artificial agent is able to solve a challenging "mediated interaction" task. In these tasks, the desired effects cannot be created through direct interaction, but instead require the learner to discover how to exert suitable effects on the target object through involving a "tool". To learn the given mediated interaction task, the agent is "actively" searching for salient points within the environment by taking a limited number of fovea-like glimpses. It then uses the accumulated information to decide which action to take next.

Keywords: Reinforcement learning · Deep learning ·
Recurrent neural networks · Mediated interaction · Visual attention ·
REINFORCE algorithm · Tool-based interaction

1 Introduction

In the last years, a lot of exciting attempts were developed in the field of deep reinforcement learning. Numerous different models were proposed which endow an artificial agent with the capability to use the raw visual input of the scene in order to learn to grasp objects [1] or to play games [2,3]. Despite these rapid developments, there are some open problems left to solve for this kind of models that undermine and slow down the learning process. One issue is given by the fact that the whole visual scene of the environment is processed through the presented models, although only a part of it may contain the relevant information that is needed to solve the given problem. Humans are not perceiving their environment as a whole image. Instead, they only see parts of the scenery, while the location of the fixations are depending on the current task [4,5]. Directed by image-based and task-dependent saliency cues, they are able to gather the important information about the environment [6,7]. The information contents from these

© Springer Nature Switzerland AG 2020
A. Vellido et al. (Eds.): WSOM 2019, AISC 976, pp. 231–240, 2020.
https://doi.org/10.1007/978-3-030-19642-4_23

foveal "glimpses" is then combined in order to get an accumulated understanding of the visible scene.

Based on this observation from neuro- and cognitive science, some successful attempts were made to integrate this human-like way of visual perception into deep learning approaches [8]. One recently introduced example is the *recurrent model of visual attention* (RAM) [9]. The presented model is not only able to classify the MNIST training data set using a small number of fovea-like glimpses, but also to solve a simplified pong task, where the agent has to catch a ball with a small paddle. The model was then extended with the ability to localize and recognize multiple objects within one image in order to transcribe house number sequences from Google Street View images [10].

In this work, we present a modified and extended version of the RAM that is able to solve a challenging "mediated interaction task". This kind of task can only be solved when the agent recognizes to use a "mediator object" as a tool to reach its goal. To enable the model to cope with such difficult kind of tasks, we combine the RAM with the famous *asynchronous actor critic algorithm* (A3C) [11] by designing an "asynchronous recurrent attention architecture" where multiple agents explore the environment in parallel in order to solve the given problem.

Section 2 gives a short introduction to reinforcement learning and the REIN-FORCE algorithm which is used in this work. The next Sect. 3 describes the designed model in detail. Section 4 introduces the learning domain, together with some additional information on the composure of the environment. After describing the learning scenarios and the experiments, we present an evaluation of the agent's learning performance for the proposed approach (Sect. 5). Section 6 concludes with a final discussion of the achieved results, together with some suggestions to future work.

2   Reinforcement Learning

Reinforcement learning is a well-known class of machine learning algorithms for solving sequential decision making problems through maximization of a cumulative scalar reward signal [12]. We use the standard formulation of a Markov decision process defined by the tuple $(S, A, P^A, \mathcal{R}, \gamma, S_0)$, where S denotes the set of states and A the set of admissible actions. P^A is the set of transition matrices, one for each action $a \in A$ with matrix elements $P^a_{\mathbf{s},\mathbf{s}'}$ specifying the probability to end up in state $\mathbf{s}' \in S$ after taking action a when in state $\mathbf{s} \in S$. Finally, $r \in \mathcal{R} \subset \mathbb{R}$ is a scalar valued reward the agent receives after ending up in $\mathbf{s}'$, γ is the discount factor and $S_0 \subseteq S$ is the set of starting states.

Reinforcement algorithms are now aiming for the optimal policy, given as the probability $\pi(\mathbf{s}_t)$ to choose an action a while in state $\mathbf{s}_t$ at time-step t in order to maximize the accumulated discounted future reward $R_t = \sum_{k=0}^{\infty} \gamma^k r_{t+k}$. The discount factor $\gamma \in [0,1)$ balances the weighting between present rewards and rewards that lie increasingly in the future. In order to find the optimal policy, the state-value

$$V^\pi(\mathbf{s}) = \mathbb{E}^\pi [R_t | \mathbf{s}_t = \mathbf{s}], \ \forall \mathbf{s} \in S \tag{1}$$

is defined. $V^\pi(\mathbf{s})$ has the interpretation of the discounted future reward, expected from following the current policy π after taking a single freely choosable (and possibly sub-optimal) action a_t starting from state $\mathbf{s}_t$. Additionally, the action-state-value $Q^\pi(\mathbf{s}, a)$ is defined, that also takes the chosen action a under consideration and is related to the state-value via $V^\pi(\mathbf{s}) = \mathrm{argmax}_a Q^\pi(\mathbf{s}, a)$.

2.1   The REINFORCE Algorithm

Suppose a neural network has to solve a reinforcement learning task, i.e. its output is an entity that should maximize an arbitrary reward function R_t. It is then possible to update the weights according to the REINFORCE update rule [13]. The general rule for updating the corresponding weights Θ of the network is thus given by

$$\Delta_\Theta = \alpha(R_t - b_t) \cdot \zeta, \tag{2}$$

where α is the learning rate, and $\zeta = \frac{\partial \log f}{\partial \Theta}$ the *characteristic eligibility* of the function f that characterizes the policy (in the simplest case $f = \pi$, the policy itself). In order to reduce the variance, a suitable baseline b_t is subtracted from the accumulated reward R_t. As the baseline tries to approximate the total accumulated reward, it is common to use the state-value function, i.e. $b_t \approx V_t^\pi(\mathbf{s}_t)$. Doing this, $(R_t - b_t)$ can be interpreted as an estimate of the advantage function $A^\pi(\mathbf{s}_t, a_t) = Q^\pi(\mathbf{s}_t, a_t) - V^\pi(\mathbf{s}_t)$ which indicates how beneficial it is to take the action a_t for solving the given task with respect to the expected outcome. The characteristic eligibility is depending on the used kind of network output. If e.g., the network directly outputs a differentiable parametrization of the policy $\pi(\mathbf{s}_t, \Theta)$, ζ becomes

$$\zeta_\pi = \nabla_\Theta \log \pi(\mathbf{s}_t, \Theta). \tag{3}$$

Reinforcement learners that are updating their action-policy based on the accumulated reward which is then rated by some kind of *critic*, like the state-value function, are called *actor-critic models*.

Additionally it is also possible to develop learning algorithms for an output that is determined via stochastic distributions which might depend on multiple parameters, like an adaptable Gaussian with the mean μ and the standard deviation σ. Instead of directly modeling the Gaussian, the mean μ and the standard deviation σ are used as the variable outputs of the network. The characteristic eligibilities for μ and σ are then given by

$$\zeta_\mu = \frac{\partial \log \mathcal{N}(x; \mu, \sigma)}{\partial \mu} = \frac{x - \mu}{\sigma^2} \quad \text{and} \quad \zeta_\sigma = \frac{\partial \log \mathcal{N}(x; \mu, \sigma)}{\partial \sigma} = \frac{(x - \mu)^2 - \sigma^2}{\sigma^3}. \tag{4}$$

3   The RAA3C MODEL

The basic framework of our approach is the *asynchronous actor-critic model* [11]. The key idea of this model is to use multiple, independent actor-critic workers that all are exploring their own copy of the domain in parallel. For training, they independently compute their own gradient, which is then used to asynchronously update a master-network. This asynchronous update disentangles the gathered training data and enables a stable learning process. The workers then update themselves, receiving a copy of the master-networks weights at the beginning of each n-th episode.

The design of the individual workers is inspired by the *recurrent model of visual attention* [9,10]. Like the original model, they are composed of three different modules: the "glimpse network", the "location network" and the "policy network". Rather than using one simple recurrent unit [9] or 2 LSTM units [10], these modules are connected by up to 3 LSTM units [14] (see Fig. 1) which all have internal states that are built out of 256 neurons. If not stated otherwise, all layers use a *rectified linear unit* (ReLu) as their activation function. Instead of receiving the raw visual information of the whole scene, the agent is only able to process a small part of the scenery at a time in a foveal manner. While the first "glimpse" of $h_g \times w_g$ pixels is random, the agent chooses a location of the scenery, using the "location network" at each consecutive time-step in order to select a new region where to look next. As suggested in [10], the received patch is concatenated with a coarser image patch. This coarser image is covering twice as much pixels as the first patch and is then also rescaled to a size of $w_g \times h_g$. The combined foveal glimpse is processed through the "glimpse network", together with its corresponding location $\mathbf{x}_g$. The extracted features are transferred through the LSTMs in order to generate the next glimpse location $\mathbf{x}'_g$ with the help of the location network. After a predefined number of these glimpses, the agent takes the accumulated information of the state $\mathbf{s}$ to choose an available action a based on its current policy $\pi(\mathbf{s}; \Theta)$, using the "policy network".

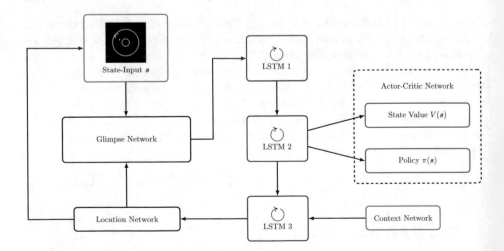

Fig. 1. An illustration of the workers within the presented model.

3.1   The Location Network

The location network uses the accumulated features of *LSTM 3* (see Fig. 1) in order to generate a new two-dimensional pair of coordinates $\mathbf{x}_g$ that define the center of the next glimpse **g**. The coordinates are chosen using a stochastic policy, modeled by a two-dimensional Gaussian. The output of the network are therefore the mean $\boldsymbol{\mu}$ and the standard deviation[1] $\boldsymbol{\sigma}$. The Gaussian is restricted to the range between -1 and 1. The used activation functions for the outputs are the *tanh* for the mean $\boldsymbol{\mu} \in [-1, 1]$ and the *sigmoid* for the standard deviation $\boldsymbol{\sigma} \in [0, 1]$. While the glimpse network is using the raw sampled coordinates $\mathbf{x}_g \in [-1, 1]$, they have to be transformed from the given coordinate system to the corresponding pixels of the input image (I_x, I_y) for the creation of a new glimpse. Additionally, a factor η is introduced that defines the pixel range that the location policy is able to cover and suppresses the output of locations at the image border.

3.2   Glimpse Network

Like the input, the glimpse network receives a vector **g** of the dimension $\dim(\mathbf{g}) = 2 \cdot w_g \cdot h_g$. It encodes the flattened information of the current foveal glimpse, taken at location $\mathbf{x}_g$. The module then combines the current glimpse vector **g** with its corresponding location $\mathbf{x}_g \in [-1, 1]$ in order to generate suitable features that are processed through the rest of the network to generate the policy and the new location $\mathbf{x}'_g$ defining where to look next. Both the glimpse vector **g** and the location $\mathbf{x}_g$ are processed through 2 linear layers with 128 neurons each. Instead of using an element-wise addition or multiplication as proposed in [9,10], the two pipelines are concatenated and then again streamed through 2 additional layers of 256 neurons in order to generate the input for *LSTM 1*.

3.3   Context Network

The context network is introduced in [10,15] to help the model to decide on the best location of the first manually created glimpse[2]. Its input is a low resolution image of the whole scenery that, in our approach, is propagated through a linear layer with 256 neurons in order to generate suitable features. These features are then taken as the initial states of *LSTM 3*, whose output is only used to generate the new glimpse locations. The other LSTMs are always initialized with zeros.

3.4   The Actor-Critic Network

The policy network is connected to *LSTM 2* as illustrated in Fig. 1. The output of the recurrent layer is the input of a linear layer with a size of 256 neurons, which

---

[1] In [9] and [10] only the mean is learned, while the standard deviation is set to a fixed value.

[2] The very first glimpse of each step is always random.

then outputs the probabilistic action policy $\pi(\mathbf{s}_t, \Theta)$ using the *softmax* activation function. The output of the same LSTM unit is also used to approximate the state-value function $\hat{V}(\mathbf{s}_t, \Theta)$.

3.5   Training

Both the action- and the location-policy can be trained using the REINFORCE algorithm (2). For the action policy, the characteristic eligibility (3) can be used. During the current episode, the transition tuples $(\mathbf{s}_t, a_t, r_t, V_t)$ are saved into a worker dependent experience buffer $\mathcal{E}$ of size N_{EB}. Here, $\mathbf{s}_t$ corresponds to the full image of the environment. After the end of an episode or if the episode buffer is full, the corresponding worker computes the weight update using the sequence of stored tuples and clears the buffer thereafter. To compute the gradient for one transition tuple, the network receives the recorded image $\mathbf{s}_t$ as its input, which is then used for a full forward-pass through the network, including the creation of the glimpses $\mathbf{g}$ and its corresponding locations $\mathbf{x}_g$, in order to get the action-policy $\pi(\mathbf{s}_t, \Theta)$ and the approximated state-value $\hat{V}(\mathbf{s}_t, \Theta)$. For the training of the stochastic location policy, the output layer for the mean $\boldsymbol{\mu}$ is trained using the REINFORCE algorithm with the characteristic eligibility ζ_μ, while the standard deviation $\boldsymbol{\sigma}$ is trained using the characteristic eligibility ζ_σ (4). It is worth mentioning that the gradient, propagating through the location network is stopped after *LSTM 3* during the update process. Instead training the network only on the locations of the last glimpse $\mathbf{g}_{1:S}$, the training can be improved by also using all included sub-sequences $\mathbf{g}_{1:s}$ with $s \leq S$ [15]. Thus, the characteristic eligibilities $\zeta_{\mu_{1:s}}$ are computed for all sub-sequences, while the sum is used for updating the weights as

$$\zeta_\mu = \sum_{s=1}^{S} \zeta_{\mu_{1:s}} = \sum_{s=1}^{S} \frac{(x_s - \mu_s)}{\sigma^2}. \tag{5}$$

for the mean and for the standard deviation respectively.

For training we use a hybrid-loss [10,15], including the objective-functions for the action- and location-policy, plus the mean-squared error between $\hat{V}_t$ and the current estimation of R_t, based on the tuples in the experience buffer $\mathcal{E}$. The full objective function then becomes

$$\mathcal{L} = \alpha \cdot \left[A_t(\mathcal{E}, \hat{V}(\mathbf{s}_{t+n}; \Theta)) \cdot [\zeta_\pi(\mathbf{s}_t; \Theta) + \zeta_\mu(\mathbf{s}_t; \Theta) + \zeta_\sigma(\mathbf{s}_t; \Theta)] \right.$$
$$\left. + \left(\sum_{i=0}^{n-1} \gamma^i r_{t+i} + \gamma^n \hat{V}(\mathbf{s}_{t+n}; \Theta) - \hat{V}(\mathbf{s}_t, \Theta) \right)^2 \right], \tag{6}$$

where $n - 1$ is the size of the current experience buffer. If the episode is terminated, the total reward R_t can be computed for each visited state. For the case of a full episode buffer, the bootstrapped state-value $\hat{V}(\mathbf{s}_{t+n}; \Theta)$ is used to estimate the total reward R_t, as displayed in (6). The advantage function A_t is approximated via *generalized advantage estimation* [16] using the rewards and state-values from the experience buffer $\mathcal{E}$.

The model is trained using 6 workers that are updating their weights every 10 episodes and an experience buffer with a maximal size of $N_{EB} = 32$. The discount factor is chosen to be $\gamma = 0.99$. The learning rate is starting at $\alpha = 25 \cdot 10^{-6}$ and decays exponentially every episode to $\alpha_0 = 10^{-6}$ with a decay-factor of 0.99. The weights of all layers are initialized using *He normal initialization* with a bias of 0.01. The model is trained, using the *Adam optimizer*.

4 Learning Domain

As our testbed for investigating "tool-based mediated interaction learning" we employ a 2D world with physics, simulated with the open source `Box2D` physics engine[3]. The world is illustrated in Fig. 2a and consists of an agent, a disc-shaped "target object" (green) and an L-shaped object "mediator object" ("tool", orange). The task of the agent is to bring the goal object into the shaded circle in the center ("goal area"). To this end, the agent can at each time step "pick" the mediator object at one of a discrete set of "picking locations" (three in the current example, marked as black dots in Fig. 2a, with the chosen picking location highlighted blue) and exert a (discretized) force/torque at the chosen picking location. This allows the agent to move the object about a fixed distance of one unit per time step within the admissible interaction range, which has a radius of 10 units. The four directions of the movement are determined by the non-static coordinate system, visualized by the dotted lines in Fig. 2a, where the mediator-object can be navigated along the coordinate axis. Additionally the agent is able to rotate it around the active picking location by $\pi/4$ per time-step. Instead of using friction, a linear and angular damping factor is implemented. Furthermore, there is an additional picking location in the center of the domain, which deals as an unbiased starting location for the agent and is further integrated to be an absorbing state that increases the stability of the applied learning algorithm. However, the agent can only reach picking locations that lie inside the circular area. Consequently, the agent can choose between 10 actions to interact with the environment.

During the learning, the agent receives information about the environment in form of raw pixels normalized between 0 and 1. The visualized image of the simulation world is downscaled from 300×300 to 84×84 pixels. In addition, the downscaled image is first processed through a greyscale filter followed by color inversion. The resulting image is shown in Fig. 2b. For all experiments the pixel conversion factor η is chosen in a way that hinders the location policy to reach the 5 rows of pixels at the image border. Learning occurs in discrete episodes, each episode being limited to 100 interaction-steps. If the agent is able to navigate the target object in the goal area[4], it receives a fixed reward

---

[3] https://box2d.org.

[4] i.e. the distance to the domains origin becomes smaller than the radius of the goal area.

of $r = 1$. To speed up the learning, the agent gets an intermediate reward of $r = 0.1$ for the first collision of the tool and the target-object "once" in every episode. Additionally, artificial noise is integrated into the system that makes the agent execute a random action with a probability of 0.1.

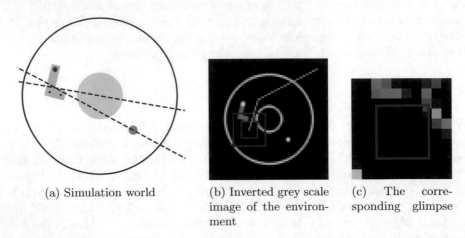

(a) Simulation world   (b) Inverted grey scale image of the environment   (c) The corresponding glimpse

Fig. 2. (a) Illustration of the presented simulation world. The blue dot (currently placed at the tool) indicates the current active picking location that the agent is using. (b) shows a preprocessed version of the domain, in which the agent selects the glimpses. (c) shows the current glimpse.

5   Experiments/Results

At the beginning of a learning episode, the tool and the target locations are sampled from uniform distributions over the admissible ranges. The agent is able to control only the tool and has now to learn how to move the target into the goal area. After successfully solving the exercise or exceeding a fixed limit of 100 interaction-steps, the task starts anew with different object positions. Using the described scenario, we tested the effectiveness of our designed model. In a first experiment, we let the agent learn to solve the given "mediated interaction task" using 3, 6, and 8 glimpses of 10×10 pixels, while the input image of the context network has the same size as the used glimpses. In a next step, the model is trained again with 6 glimpses while additionally using a 20×20 and a 40×40 image as the input for the context network[5].

For evaluating the efficiency of the learning processes, we depict the number of learning steps as a function of the average reward per episode $\langle R \rangle$ that is achieved by the agent. To compute $\langle R \rangle$, the learning performance under the current policy was evaluated over 100 episodes for each of the 40 evaluated data points. In the end, the 3 best learning runs were averaged, where the standard deviation of the mean is used as the error.

---

[5] A short movie of the learned policy for the model using 6 glimpses and a 20×20 context image can be found at https://doi.org/10.4119/unibi/2934182.

The results are presented in Fig. 3. Using a 10×10 context image, the model only receives roughly 10% of the possible reward when trained with 3 and 6 glimpses. When trained on 8 glimpses, it is able to achieve an average reward of $\langle R \rangle \approx 0.8$, corresponding to an average success rate of roughly 70%. It seems to be possible to improve the performance of the model by using images with higher resolution as an input to the context network. Using an image of twice the size, the model trained on 6 glimpses is able to achieve a learning performance that is three times higher after 8000 episodes. By again doubling the size of the image, the learning performance of the model further increases and now achieves an average reward of $\langle R \rangle \approx 0.8$ instead of $\langle R \rangle \approx 0.15$.

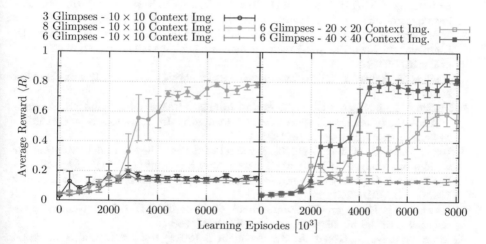

Fig. 3. In the left graph, the average reward $\langle R \rangle$ of the proposed model for 3, 6 and 8 glimpses per state s_t is plotted over the number of training episodes. The right graph shows the learning performance using 6 glimpses for different sizes of context images. The learning run using 6 glimpses with a 10×10 context image is plotted in both graphs.

6   Conclusion

This work presents an approach that enables an artificial agent to learn mediated interaction tasks using a visual input composed out of fovea-like glimpses. The designed model is a first attempt to solve more complex reinforcement learning tasks that need to cope with sparse and attention guided visual input by combining the *recurrent model of visual attention* with an *asynchronous actor-critic learner*. While the model still needs to be improved further, this work might be a good starting point for further investigation of reinforcement learning that use an attention guided visual input. A possible next step could either be to improve the model in order to learn much more complex tasks, like e.g. ATARI games, or to think about a way to combine the *recurrent model of visual attention* with a *Deep Q-Learner* [2].

Acknowledgment. This research/work was supported by the Cluster of Excellence Cognitive Interaction Technology 'CITEC' (EXC 277) at Bielefeld University, which is funded by the German Research Foundation (DFG).

References

1. Levine S, Finn C, Darrell T, Abbeel P (2016) End-to-end training of deep visuo-motor policies. J Mach Learn Res 17(1):1334–1373
2. Mnih V, Kavukcuoglu K, Silver D, Rusu AA, Veness J, Bellemare MG, Graves A, Riedmiller M, Fidjeland AK, Ostrovski G, Petersen S, Beattie C, Sadik A, Antonoglou I, King H, Kumaran D, Wierstra D, Legg S, Hassabis D (2015) Human-level control through deep reinforcement learning. Nature 518(7540):529–533
3. Jaderberg M, Mnih V, Czarnecki WM, Schaul T, Leibo JZ, Silver D, Kavukcuoglu K (2016) Reinforcement learning with unsupervised auxiliary tasks. arXiv:1611.05397
4. Hayhoe M, Ballard D (2005) Eye movements in natural behavior. Trends Cogn Sci 9(4):188–194
5. Mathe S, Sminchisescu C (2013) Action from still image dataset and inverse optimal control to learn task specific visual scanpaths. In: Advances in neural information processing systems, pp 1923–1931
6. Itti L, Koch C, Niebur E (1998) A model of saliency-based visual attention for rapid scene analysis. IEEE Trans Pattern Anal Mach Intell 20(11):1254–1259
7. Itti L, Koch C (2001) Computational modelling of visual attention. Nature Rev Neurosci 2(3):194–203
8. Hassabis D, Kumaran D, Summerfield C, Botvinick M (2017) Neuroscience-inspired artificial intelligence. Neuron 95(2):245–258
9. Mnih V, Heess N, Graves A, Kavukcuoglu K (2014) Recurrent models of visual attention. arXiv:1406.6247
10. Ba J, Mnih V, Kavukcuoglu K (2014) Multiple object recognition with visual attention. arXiv:1412.7755
11. Mnih V, Badia AP, Mirza M, Graves A, Lillicrap TP, Harley T, Silver D, Kavukcuoglu K (2016) Asynchronous methods for deep reinforcement learning. arXiv:1602.01783
12. Sutton RS, Barto AG (2018) Reinforcement learning: an introduction, 2nd edn. MIT Press, Cambridge
13. Williams RJ (1992) Simple statistical gradient-following algorithms for connectionist reinforcement learning. Mach Learn 8(3–4):229–256
14. Hochreiter S, Schmidhuber J (1997) Long short-term memory. Neural Comput 9(8):1735–1780
15. Larochelle H, Hinton GE (2010) Learning to combine foveal glimpses with a third-order Boltzmann machine. In: Lafferty JD, Williams CKI, Shawe-Taylor J, Zemel RS, Culotta A (eds) Advances in Neural Information Processing Systems 23. Curran Associates, Inc., pp 1243–1251 (2010)
16. Schulman J, Moritz P, Levine S, Jordan MI, Abbeel P (2015) High-dimensional continuous control using generalized advantage estimation. arXiv:1506.02438

Approximate Linear Dependence as a Design Method for Kernel Prototype-Based Classifiers

David N. Coelho[1(✉)] and Guilherme A. Barreto[2]

[1] Federal University of Ceará, Campus of Sobral, Fortaleza, Ceará, Brazil
`david.coelho@sobral.ufc.br`
[2] Department of Teleinformatics Engineering, Federal University of Ceará,
Center of Technology, Campus of Pici, Fortaleza, Ceará, Brazil
`gbarreto@ufc.br`

Abstract. The approximate linear dependence (ALD) method is a sparsification procedure used to build a dictionary of samples extracted from a dataset. The extracted samples are approximately linearly independent in a high-dimensional kernel reproducing Hilbert space. In this paper, we argue that the ALD method itself can be used to select relevant prototypes from a training dataset and use them to classify new samples using kernelized distances. The results obtained from intensive experimentation with several datasets indicate that the proposed approach is viable to be used as a standalone classifier.

Keywords: Prototype-based classifiers · Sparsification ·
Approximate linear dependence · Kernel classifiers · Kernel SOM

1 Introduction

Kernel-based methods have been introduced with the aim of developing nonlinear versions of linear supervised or unsupervised machine learning algorithms [7]. The underlying idea is to apply a kernel function $k(\cdot, \cdot) : \mathcal{X} \times \mathcal{X} \to \mathbb{R}$ to any pair of training vectors so that the result can be interpreted as an inner product of a mapping function $\phi(\mathbf{x})$, where $\phi : \mathcal{X} \to F$, and F is a high-dimensional reproducing kernel Hilbert space (RKHS) (a.k.a. the feature space) [16]: $k(\mathbf{x}_i, \mathbf{x}_j) = \phi(\mathbf{x}_i)^T \phi(\mathbf{x}_j)$. It should be noted that the nonlinear feature mapping $\phi(\cdot)$ is usually unknown. Thus, by means of the kernel function, inner products in the feature space are computed implicitly, i.e. without using the feature vectors directly. This appealing property of kernel methods has then been referred to as the *kernel trick*.

The process of *kernelization* has also been applied to prototype-based algorithms, such as the K-means [12], the self-organizing map (SOM) [10], the neural gas (NG) network [13] and the learning vector quantization (LVQ) [9], producing their kernelized versions: the kernel K-means [17], the kernel SOM (KSOM) [11], the kernel NG (KNG) [14] and the kernel LVQ (KLVQ) [6].

© Springer Nature Switzerland AG 2020
A. Vellido et al. (Eds.): WSOM 2019, AISC 976, pp. 241–250, 2020.
https://doi.org/10.1007/978-3-030-19642-4_24

The performances of standard and kernelized versions of prototype-based classifiers are highly dependent on the number of labeled prototypes. Although it is possible to make the set of prototypes either adaptive [1] or optimally determined by means of evolutionary algorithms [15], in the vast majority of the applications that number is set by trial and error or exhaustive grid search.

Bearing this issue in mind, in this paper we develop a simple design scheme for building kernelized prototype-based classifiers by means of the approximate linear dependence (ALD) method, which is a sparsification procedure widely used in the field of kernel adaptive filtering [5]. The proposed approach automatically selects a dictionary of samples extracted from the original dataset. The dictionary, the size of which is a function of a single scalar parameter, is then used to classify a new sample using a kernelized nearest neighbor scheme. A set of experiments with benchmarking datasets confirm the viability of the proposed approach.

The remainder of this paper is organized as follows. In Sect. 2, the fundamentals of prototype-based classification and some kernel-based methods are briefly reviewed. In Sect. 3, the approximate linear dependence method and the proposed framework is developed. The simulation results are reported and discussed in Sect. 4. The paper is concluded in Sect. 5.

2   Basics of Prototype-Based Classification

Prototype based algorithms, as SOM and LVQ, learn from samples $\{(\mathbf{x}_i, y_i) \in \mathbb{R}^p \times \{1, ..., C\} \,|\, i = 1, ..., N\}$ a mapping (projection) from a high-dimensional continuous input space $\mathcal{X}$ onto a low-dimensional discrete space $\mathcal{A}$ of Q neurons. The map $i^*(\mathbf{x}) : \mathcal{X} \to \mathcal{A}$, defined by the weight matrix $\mathbf{W} = \{\mathbf{w}_1, \mathbf{w}_2, \ldots, \mathbf{w}_Q\}, \mathbf{w}_i \in \mathbb{R}^p \subset \mathcal{X}$, assigns to each input vector $\mathbf{x}(n) \in \mathbb{R}^p \subset \mathcal{X}$ a winning prototype $i^*(n) \in \mathcal{A}$, determined by

$$i^*(n) = \arg \min_{\forall i} \|\mathbf{x}(n) - \mathbf{w}_i(n)\|^2, \tag{1}$$

where $\|\cdot\|$ denotes the Euclidean distance and n symbolizes a discrete time step associated with the iterations of the algorithm. This function can be kernelized as

$$i^*(n) = \arg \min_{\forall i} \|\phi(\mathbf{x}(n)) - \phi(\mathbf{w}_i(n))\|^2, \tag{2}$$
$$= \arg \min_{\forall i} J_i(\mathbf{x}(n)),$$

where $J_i(\mathbf{x}(n))$ can be defined as

$$
\begin{aligned}
J_i(\mathbf{x}(n)) &= \|\phi(\mathbf{x}(n)) - \phi(\mathbf{w}_i(n))\|^2, \\
&= (\phi(\mathbf{x}(n)) - \phi(\mathbf{w}_i(n)))^T (\phi(\mathbf{x}(n)) - \phi(\mathbf{w}_i(n))), \\
&= \phi(\mathbf{x}(n))^T \phi(\mathbf{x}(n)) + \phi(\mathbf{w}_i(n))^T \phi(\mathbf{w}_i(n)) - 2\phi(\mathbf{x}(n))^T \phi(\mathbf{w}_i(n)), \\
&= k(\mathbf{x}(n), \mathbf{x}(n)) + k(\mathbf{w}_i(n), \mathbf{w}_i(n)) - 2k(\mathbf{x}(n), \mathbf{w}_i(n)).
\end{aligned}
\tag{3}
$$

Prototype-based algorithms are distinguished by the update rules of their weight matrices. With the SOM algorithm, all the prototypes are updated by the following rule

$$\mathbf{w}_i(n+1) = \mathbf{w}_i(n) + \eta(n)h(i^*, i; n)[\mathbf{x}(n) - \mathbf{w}_i(n)] \tag{4}$$

where $0 < \eta(n) < 1$ is the learning rate and $h(i^*, i; n)$ is a weighting function which limits the neighborhood of the winning neuron. On the other hand, in LVQ1, just the winner prototype is updated by the following rule

$$\mathbf{w}_{i^*}(n+1) = \begin{cases} \mathbf{w}_{i^*}(n) + \eta(n)\left[\mathbf{x}(n) - \mathbf{w}_{i^*}(n)\right], y(\mathbf{x}(n)) = y(\mathbf{w}_{i^*}(n)) \\ \mathbf{w}_{i^*}(n) - \eta(n)\left[\mathbf{x}(n) - \mathbf{w}_{i^*}(n)\right], y(\mathbf{x}(n)) \neq y(\mathbf{w}_{i^*}(n)) \end{cases} \tag{5}$$

where $y(\mathbf{x}(n))$ and $y(\mathbf{w}_{i^*}(n))$ are the labels of the sample and the winning prototype respectively.

Weight updating rules can be also kernelized. In the *energy function kernel SOM* (EF-KSOM), for example, the updating rule is defined as

$$\mathbf{w}_i(n+1) = \mathbf{w}_i(n) + \eta(n) h(i^*, i, n) \nabla J_i(\mathbf{x}(n)), \tag{6}$$

where the gradient vector $\nabla J_i(\mathbf{x}(n))$ is defined as $\nabla J_i(\mathbf{x}(n)) = \frac{\partial J_i(\mathbf{x}(n))}{\partial \mathbf{w}_i(n)}$. In the next subsection, some functions that can be used as kernels are briefly described.

2.1   Kernel Functions

The linear kernel is the simplest one, where this function's output is equal to the dot product of two input vectors. This kernel, for two given vectors, $\mathbf{x} \in \mathbb{R}^p$ and $\mathbf{y} \in \mathbb{R}^p$, can be formally defined as

$$k(\mathbf{x}, \mathbf{y}) = \mathbf{x}^T \mathbf{y}. \tag{7}$$

The Gaussian kernel function has the following general form:

$$k(\mathbf{x}, \mathbf{y}) = \exp\left(-\frac{\|\mathbf{x} - \mathbf{y}\|^2}{2\gamma^2}\right), \tag{8}$$

where $\gamma > 0$ is a scale parameter (a.k.a. the width parameter, in the current context). A suitable value for the hyperparameter γ should be carefully tuned to the problem at hand [4]. If it is overestimated, the exponential behaves almost linearly and the projection to high-dimensional feature space loses its nonlinear character. If it is underestimated, the function will lack regularization and decision boundaries tend to become highly sensitive to noise in training data.

The Cauchy kernel function has the following general form:

$$k(\mathbf{x}, \mathbf{y}) = \left(1 + \frac{\|\mathbf{x} - \mathbf{y}\|^2}{\gamma^2}\right)^{-1}, \tag{9}$$

where $\gamma > 0$ is a scale parameter.

This kernel function is a long-tailed kernel, a term borrowed from Probability for denoting distributions in which too small or too large values have a large probability to occur, in contrast to the Gaussian distribution for which values far from the mean rarely occur. For this reason, the Cauchy kernel can be used to give long-range influence and sensitivity over the high-dimensional feature space [4].

The Log kernel function was introduced in [2] and its expression is given by

$$k(\mathbf{x}, \mathbf{y}) = -\log\left(1 + \frac{\|\mathbf{x} - \mathbf{y}\|^2}{\gamma^2}\right), \tag{10}$$

where log denotes the natural logarithm.

The Log kernel function belongs to a class of "not strictly positive definite" kernel functions, named *conditionally definite positive* kernel functions[1], which has been shown anyway to perform very well in many practical applications.

In the next section, the ALD method is briefly described and the proposed framework is shown.

3   The Proposed Approach

As mentioned, the training method to be proposed is based on the ALD criterion [5], which is a sparsification procedure for the construction of a dictionary consisting of a subset of the training samples $\mathcal{D}_{t-1} = \{\tilde{\mathbf{x}}_j\}_{j=1}^{m_{t-1}}$. The samples in $\mathcal{D}_{t-1}$ are *approximately* linearly independent feature vectors. The goal of the proposed approach is to take the samples of the dictionary as prototype vectors in feature space, so that they can be used in a kernelized nearest neighbor classification scheme.

At training time step t $(2 \leq t \leq N)$, with N denoting the number of training samples, after having observed $t - 1$ training samples, the dictionary $\mathcal{D}_{t-1}$ is comprised of a subset of m_{t-1} relevant training inputs $\{\tilde{\mathbf{x}}_j\}_{j=1}^{m_{t-1}}$. When a new incoming training sample $\mathbf{x}_t$ is available, one must test if it should be added or not to the dictionary. In order to do this, it is necessary to estimate a vector of coefficients $\mathbf{a} = \left(a_1, ..., a_{m_{t-1}}\right)^T$ satisfying the ALD criterion

$$\delta_t \overset{def}{=} \min_{\mathbf{a}} \left\| \sum_{j=1}^{m_{t-1}} a_j \phi\left(\tilde{\mathbf{x}}_j\right) - \phi\left(\mathbf{x}_t\right) \right\|^2 \leq \nu, \tag{11}$$

where ν is the sparsity level parameter. Developing the minimization problem in Eq. (11) and using $\kappa(\mathbf{x}, \mathbf{y}) = \langle \phi(\mathbf{x}), \phi(\mathbf{y}) \rangle$, we can write

---

[1] Let $\mathcal{X}$ be a nonempty set. A kernel $k(\cdot, \cdot)$ is called *conditionally positive definite* if and only if it is symmetric and $\sum_{j,k}^n c_j c_k k(\mathbf{x}_j, \mathbf{x}_k) \geq 0$, for $n \geq 1$, $c_1, \ldots, c_n \in \mathbb{R}$ with $\sum_{j=1}^n c_j = 0$ and $\mathbf{x}_1, \ldots, \mathbf{x}_n \in \mathcal{X}$.

$$\delta_t \stackrel{def}{=} \min_{\mathbf{a}} \left\{ \sum_{i,j=1}^{m_{t-1}} a_i a_j \kappa\left(\tilde{\mathbf{x}}_i, \tilde{\mathbf{x}}_j\right) - 2 \sum_{i,j=1}^{m_{t-1}} a_j \kappa\left(\tilde{\mathbf{x}}_i, \mathbf{x}_t\right) + \kappa\left(\mathbf{x}_t, \mathbf{x}_t\right) \right\}, \qquad (12)$$

or, using the matrix notation,

$$\delta_t = \min_{a} \left\{ \mathbf{a}^T \tilde{\mathbf{K}}_{t-1} \mathbf{a} - 2\mathbf{a}^T \tilde{\mathbf{k}}_{t-1}\left(\mathbf{x}_t\right) + k_{tt} \right\}, \qquad (13)$$

where $\left[\tilde{\mathbf{K}}_{t-1}\right]_{i,j} = \kappa\left(\tilde{\mathbf{x}}_i, \tilde{\mathbf{x}}_j\right)$, $\left(\tilde{\mathbf{k}}_{t-1}\left(\mathbf{x}_t\right)\right)_i = \kappa\left(\tilde{\mathbf{x}}_i, \mathbf{x}_t\right)$, and $k_{tt} = \kappa\left(\mathbf{x}_t, \mathbf{x}_t\right)$, with $i,j = 1, ..., m_{t-1}$. Solving (13) leads to the optimal $\mathbf{a}_t$, given by

$$\mathbf{a}_t = \tilde{\mathbf{K}}_{t-1}^{-1} \tilde{\mathbf{k}}_{t-1}\left(\mathbf{x}_t\right), \qquad (14)$$

so that the ALD condition can be rewritten as

$$\delta_t = k_{tt} - \tilde{\mathbf{k}}_{t-1}\left(\mathbf{x}_t\right)^T \mathbf{a}_t \leq \nu. \qquad (15)$$

If $\delta_t > \nu$, then the sample $\mathbf{x}_t$ must be added to the dictionary; that is, $\mathcal{D}_t = \mathcal{D}_{t-1} \cup \{\mathbf{x}_t\}$ and $m_t = m_{t-1} + 1$. However, if $\delta_t < \nu$, the sample is approximate linear dependent and must not be added to the dictionary.

For the purpose of classification, the ALD-based selection of prototype vectors for the dictionary can be carried out in two very simple ways, which are described next.

Design Method 1 - Randomly select an initial data sample. This sample will be the first element of the dictionary. Then, take the remaining samples of the training dataset, one-by-one, and apply the ALD criterion according to (14) and (15). Note that each prototype vector in $\mathcal{D}_t$ carries its class label for the sake of classification. The classifier designed by this method will be henceforth referred to by the acronym KNN-ALD-1 (*kernel nearest neighbor via ALD criterion 1*).

Design Method 2 - According to this method we have to build one dictionary per class. For a problem with C classes, it is required C dictionaries $\mathcal{D}_t^{(k)}$, $k = 1, 2, ..., C$. For this purpose, apply the Design Method 1 to the data samples of the k-th class, $k = 1, 2, ..., C$. Repeat this procedure for all classes individually. Merge the class-conditional dictionaries into a single larger dictionary: $\mathcal{D}_t = \mathcal{D}_t^{(1)} \cup \mathcal{D}_t^{(2)} \cup \cdots \cup \mathcal{D}_t^{(C)}$. The classifier designed by this method will be henceforth referred to as the KNN-ALD-2 (*kernel nearest neighbor via ALD criterion 2*).

For the classification of a new data sample, use the kernelized distance in Eq. (2) in order to find the closest prototype. The search is executed over the samples in the dictionary. Assign to that sample, the same class of the nearest prototype.

It should be noted that the only hyperparameters of the proposed approach are ν (the sparsity level) and those associated with the chosen kernel function (as the scale parameter γ). However, since the kernel parameters are common to all kernel-based methods, the only tunable parameter of the proposed approach is the sparsity level ν.

Table 1. Preliminary tests with Iris dataset and linear kernel.

Method	ν	Kernel	γ	acc_tr	acc_ts	#prototypes	#class 1	#class 2	#class 3
KNN-ALD-1	0.001	linear	.	0.956	0.905	49	15	13	21
KNN-ALD-1	**0.01**	**linear**	.	**0.867**	**0.849**	**12**	**3**	**3**	**6**
KNN-ALD-1	0.1	linear	.	0.759	0.739	5	1	2	2
KNN-ALD-2	0.001	linear	.	0.995	0.936	89	30	30	29
KNN-ALD-2	0.01	linear	.	0.942	0.912	30	10	10	10
KNN-ALD-2	**0.1**	**linear**	.	**0.907**	**0.890**	**13**	**4**	**4**	**5**

4   Results and Discussion

In this section, we report the results of comprehensive computer simulations verifying the classification performance of the proposed algorithm, with different kernels, when applied to real-world datasets. For all the datasets, the z-score normalization is used, so that all attributes have zero empirical mean and unit variance. Moreover, the algorithms were implemented from scratch using MATLAB's script language and the simulations were run on an HP notebook with 2.70 GHz Intel Core i7 processor, 16 GB of RAM memory and Windows 10 Home operating system.

4.1   Initial Tests

We start with the well known Iris dataset[2] to analyze how the number of prototypes and the classifier accuracy change as we modify the hyperparameters ν and γ, the kernel functions, and the dictionary building method. This dataset contains 3 classes with 50 samples each, where each class refers to a type of iris plant, and each sample is a vector of 4 attributes.

The results of the proposed approaches using the linear kernel are shown in Table 1. As this kernel function does not have any hyperparameter, only the dictionary building method and the sparsity level ν are analyzed here. It should be noted that larger values of ν lead to dictionaries with fewer prototypes because it becomes harder to a new sample not be considered approximate linear dependent - see Eq. (15). For small enough values of ν, all the samples from the training dataset will be used as prototypes. Comparing the results in boldface, one can observe that high accuracy rates are achieved by the KNN-ALD-2 classifier with basically the same number of prototypes of the KNN-ALD-1, in both the training and test sets. It is important to mention that this quantity of prototypes corresponds to ≈12% of the entire training set.

When using the Gaussian and Cauchy kernels, as we increase the value of the scale parameter γ, the number of selected prototypes decreases. In order to have good classification performance and few prototypes, both ν and γ should be optimized. Also, with these kernel functions, the KNN-ALD-2 classifier achieved

---

[2] https://archive.ics.uci.edu/ml/datasets/iris.

Table 2. Preliminary tests with Iris dataset and Gaussian kernel.

Method	ν	Kernel	γ	acc_tr	acc_ts	#prototypes	#class 1	#class 2	#class 3
KNN-ALD-1	0.001	Gaussian	20	0.987	0.931	88	30	28	30
KNN-ALD-1	0.01	Gaussian	20	0.914	0.899	21	7	5	9
KNN-ALD-1	0.1	Gaussian	2	0.949	0.915	24	7	7	10
KNN-ALD-2	0.001	Gaussian	20	1	0.932	99	33	32	34
KNN-ALD-2	0.01	Gaussian	20	0.954	0.909	32	9	10	13
KNN-ALD-2	0.1	Gaussian	2	0.957	0.917	30	8	10	12
KNN-ALD-2	0.1	Gaussian	5	0.936	0.911	18	5	5	8
KNN-ALD-2	0.1	Gaussian	10	0.911	0.882	12	3	4	5

the best results. These concepts are illustrated in Table 2, where results for the Gaussian kernel are shown.

Unlike the previously mentioned kernel functions, the number of prototypes increases as the value of γ also increases when using the Log kernel. Also, the values for ν should be negative, as the maximum value of this kernel is zero - see Eq. (10).

4.2 More General Tests

In the following experiments, for each evaluated dataset, we test 8 variants of the proposed algorithm, consisting of 4 different kernel functions (linear, Gaussian, Cauchy, Log) and 2 design methods (KNN-ALD-1 and KNN-ALD-2). Also, 50 independent training-testing runs are executed. For each run, three steps are performed, namely: (i) holdout (partition of the data between training and test sets), (ii) training (hyperparameters optimization and parameters update), (iii) performance testing. For the holdout step, the data is randomly divided as follows: 70% for training and the remaining 30% for testing purposes. At the end of the testing phase, several statistical figures of merit for the performance of each classifier are computed.

Finally, we perform a 5-fold cross-validation strategy in order to search the optimal values of the hyperparameters ν (from the ALD criterion) and γ (from the Gaussian, Cauchy, and Log kernel functions). The figure of merit for evaluating the performances of the algorithms while choosing the optimal hyperparameters is given by

$$J_h = \alpha - \beta.n_p \tag{16}$$

where α is the classifier's accuracy, n_p is the percentage of prototypes with respect to the number of training samples, and β is a weighting term between these two factors. By increasing β, the J_h index penalizes hyperparameters that lead the algorithm to build dictionaries with a large number of prototypes.

The motor failure dataset [3] was the first to be investigated more carefully. It consists of 294 attribute vectors, each one containing 6 harmonics of the fast Fourier Transform from a line current measurement of a three-phase induction

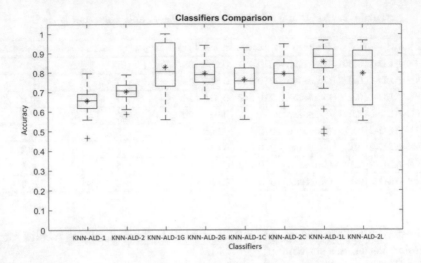

Fig. 1. Results using the motor failure dataset

motor. The samples are organized into 7 classes, where 1 is for normal operation condition (42 samples) and the other 6 correspond to short-circuit operation conditions (252 samples). In [3], the best results were reached when the problem was treated as a binary one and when the classes were balanced (adding 210 artificial samples of normal condition to the dataset). So, the same methodology was used in this paper.

The results for this methodology, using the dataset of 504 samples (252 for each class) and $\beta = 0.5$ - see Eq. (16) - are depicted in Fig. 1. First, it is possible to infer that the classification performance is improved when nonlinear kernel functions are used. If we just analyze the best performance of the proposed approach, that using the Gaussian Kernel and the KNN-ALD-1 classifier, the following results were achieved: this classifier achieved 100% of accuracy but it needed to use 62.5% of the samples from the training dataset (220 prototypes from 352 training samples).

In order to evaluate the proposed classifiers on datasets with more attributes and samples, the Pap-smear dataset[3] was chosen. It consists of 917 images of Pap-smear cells where each cell is described by 20 numerical features, and the cells fall into 7 classes. Samples from 3 classes originate from normal cells (totaling 242 samples) and samples from the other 4 classes are correspond to abnormal cells (totaling 675 samples) [8]. The classification problem for this dataset was also treated as a binary one. The results achieved by the proposed classifiers, using $\beta = 1$ are represented in Fig. 2. The worst mean and maximum accuracy rates were achieved by using the log kernel. Finally, if we just analyze, for example, the best result of the proposed approach, using the Gaussian Kernel and the KNN-ALD-1 classifier, the following results were achieved: this classifier

---

[3] http://mde-lab.aegean.gr/downloads.

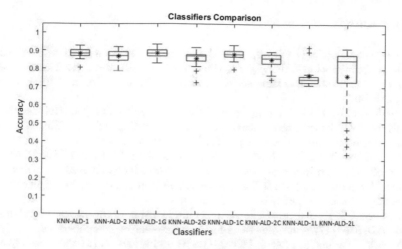

Fig. 2. Results using the Pap-smear dataset

reached 92% of accuracy using just 3.27% of samples from the training dataset (21 prototypes out of 641 training samples).

5   Conclusions and Further Work

In this paper, we presented a new methodology for the design of kernel prototype-based classifiers. The proposed approach relies on the ALD method for selecting the relevant prototypes of a training dataset and uses kernelized distances to classify new samples. This method has the advantage of having just a few hyperparameters (sparsity level ν and the kernel functions' parameter γ) to optimize, although some questions still need to be further investigated, such as the choice of the first element to enter the dictionary.

In preliminary tests with the Iris dataset, we analyzed the effects of hyperparameters' choice and some characteristics of the proposed dictionary building methods. In the general tests, the maximum results were achieved when using the Gaussian kernel function and the KNN-ALD-1 classifier. In more general tests, we have shown that this approach can be successfully applied to classification, as it reached 100% of maximum accuracy for the motor failure dataset and 92% with the Pap-smear dataset using 62.5% and 3.27% training samples of each dataset, respectively.

Currently, we are extending the proposed methods to handle large scale classification datasets by incorporating online dictionary adaptation strategies.

Acknowledgments. This study was financed by the following Brazilian research funding agencies: CAPES (Finance Code 001) and CNPq (grant 309451/2015-9).

References

1. Albuquerque RF, de Oliveira PD, Braga APdS (2018) Adaptive fuzzy learning vector quantization (AFLVQ) for time series classification. In: North American fuzzy information processing society annual conference (NAFIPS 2018). Springer, pp 385–397
2. Boughorbel S, Tarel JP, Boujemaa N (2005) Conditionally positive definite Kernels for SVM based image recognition. In: Proceedings of the IEEE international conference on multimedia & expo (ICME 2005), pp 1–4 (2005)
3. Coelho DN, Barreto G, Medeiros CM, Santos JDA (2014) Performance comparison of classifiers in the detection of short circuit incipient fault in a three-phase induction motor. In: 2014 IEEE symposium on Computational intelligence for engineering solutions (CIES). IEEE, pp 42–48
4. de Souza CR Kernel functions for machine learning applications. http://crsouza.com/2010/03/17/kernel-functions-for-machine-learning-applications/
5. Engel Y, Mannor S, Meir R (2004) The Kernel recursive least squares algorithm. IEEE Trans Signal Process 52(8):2275–2285
6. Hofmann D, Hammer B.: Sparse approximations for Kernel learning vector quantization. In: Proceedings of the ESANN 2013, pp 549–554 (2013)
7. Jäkel F, Schölkopf B, Wichmann FA (2007) A tutorial on Kernel methods for categorization. J Math Psychol 51(6):343–358
8. Jantzen J, Norup J, Dounias G, Bjerregaard B (2005) Pap-smear benchmark data for pattern classification. In: Nature inspired smart information systems (NiSIS 2005) pp 1–9
9. Kohonen T (1990) Improved versions of learning vector quantization. In: Proceedings of the 1990 international joint conference on neural networks (IJCNN 1990). IEEE, pp 545–550
10. Kohonen T (1990) The self-organizing map. Proc IEEE 78(9):1464–1480
11. Lau KW, Yin H, Hubbard S (2006) Kernel self-organising maps for classification. Neurocomputing 69(16):2033–2040
12. MacQueen JB (1967) Some methods for classification and analysis of multivariate observations. In: Proceedings of 5th Berkeley symposium on mathematical statistics and probability. University of California Press, pp 281–297 (1967)
13. Martinetz TM, Berkovich SG, Schulten KJ et al (1993) 'Neural-gas' network for vector quantization and its application to time-series prediction. IEEE Trans Neural Netw 4(4):558–569
14. Qinand AK, Suganthan PN (2004) Kernel neural gas algorithms with application to cluster analysis. In: Proceedings of the 17th international conference on pattern recognition (ICPR 2004). IEEE, pp 617–620
15. Soares Filho LA, Barreto GA (2014) On the efficient design of a prototype-based classifier using differential evolution. In: Proceedings of the 2014 IEEE symposium on differential evolution (SDE 2014). IEEE, pp 1–8
16. Yin H (2006) On the equivalence between Kernel self-organising maps and self-organising mixture density networks. Neural Netw 19(6):780–784
17. Zhang R, Rudnicky AI (2002) A large scale clustering scheme for kernel K-means. In: Proceedings of the 16th international conference on pattern recognition (ICPR 2002), vol 4. IEEE, pp 289–292

Subspace Quantization on the Grassmannian

Shannon Stiverson$^{(\boxtimes)}$, Michael Kirby, and Chris Peterson

Colorado State University, Fort Collins, CO 80523, USA
{stiverso,kirby,peterson}@math.colostate.edu

Abstract. We extend the K-means and LBG algorithms to the framework of the Grassmann manifold to perform subspace quantization. For K-means it is possible to move a subspace in the direction of another using Grassmannian geodesics. For LBG the centroid computation is now done using a flag mean algorithm for averaging points on the Grassmannian. The resulting unsupervised algorithms are applied to the MNIST digit data set and the AVIRIS Indian Pines hyperspectral data set.

Keywords: Grassmannian · LBG · K-means · Flag mean

1 Introduction

The Grassmann manifold provides a robust geometric framework for analyzing complex, high-dimensional data sets where observations are characterized by a variation in state. For example, variations in illumination confound pattern recognition systems given the sensitivity of the representation to the angle of illumination. The use of the Grassmannian greatly mitigates this problem [4,6]. Similarly, satellite imaging systems collect hyperspectral data with high-resolution both spectrally and spatially. A given substance, e.g., a field of corn, will show significant variability in the spectral signature over even small image patches. In one study, it was shown that the classes *soybean with tilling* versus *soybean with no tilling* could be separated with perfect accuracy using the Grassmannian, whereas the best vector space methods could not [7]. Also, since the Grassmannian is itself a manifold, this framework lends itself naturally to analysis using topological and geometric methods, see, e.g., [2,8,9,13]. Given the robust performance of the Grassmannian in these examples, it is desirable to explore the extension of core tools in data analysis in the geometric setting of the Grassmann manifold. Two building blocks for algorithms on the Grassmannian are a means to compare distances between points and a means to compute averages of points on the Grassmann manifold. These techniques are described in Sects. 2 and 3. We note that the self-organizing mapping algorithm of Kohonen has been adapted to this geometric framework with success [16]. The goal of this work is to extend the K-means and LBG vector quantization algorithms to Grassmann manifolds in order to provide a robust unsupervised method for

© Springer Nature Switzerland AG 2020
A. Vellido et al. (Eds.): WSOM 2019, AISC 976, pp. 251–260, 2020.
https://doi.org/10.1007/978-3-030-19642-4_25

quantizing data subspaces. In this work, K-means refers to the online method by MacQueen [18] while LBG refers to the the batch version of the algorithm developed by Linde, Buzo, and Gray [17].

The outline of this paper is as follows: In Sect. 2 we provide background on the Grassmann manifold. In Sect. 3 we describe how to average points on the Grassmannian via the *flag mean*. Section 4 outlines the Grassmann K-means algorithm, and Sect. 5 presents Grassmannian LBG. In Sects. 6.1 and 6.2 we present applications.

2   The Grassmannian

The real Grassmann manifold $Gr(p,n)$ is a manifold whose points parameterize the linear subspaces of dimension p in $\mathbb{R}^n$ [1]. One can construct $Gr(p,n)$ as a quotient manifold of the Stiefel manifold $St(p,n)$ [12]. This relationship allows for an intuitive representation of points on the Grassmannian that lends itself well to computations. Any point on $Gr(p,n)$ can be identified with a matrix $X \in St(p,n)$ whose column space spans the desired subspace $[X]$. Orthogonally invariant norms on the Grassmannian may be expressed in terms of principal angles θ_i between subspaces [12]. This is computationally appealing since principal angles can be determined from the singular values of the SVD of $X^T Y$ [5]. For example, the *chordal norm* is given by

$$d_c([X],[Y]) = \|\sin\theta\|_2. \tag{1}$$

Moreover, any set of points on the Grassmannnian, with distances measured in this way, can be isometrically embedded into Euclidean space using multi-dimensional scaling (MDS). In practice, the smallest angle pseudometric generally gives the best data separation, although the embedding into Euclidean space is no longer isometric [7].

3   Averaging Subspaces

The flag mean is an algorithm for computing averages of points on Grassmannians [11,19,20]. One can use such an algorithm to determine common attributes, within a set of points on the Grassmannian, expressed as a set of nested subspaces [19]. The flag mean algorithm, which we summarize below, is at the heart of the Grassmannian LBG procedure.

A flag is a nested sequence of subspaces. Given a finite collection of subspaces, the flag mean algorithm computes the best flag representation of the collection. Denote the flag by $\{[u_1], [u_2], ..., [u_r]\}$, where the u_i are orthogonal unit vectors with $r \leq n$. Let $\{[X_i]\}$ be a set of points in $Gr(p,n)$ and $\{X_i\}$ be their corresponding matrix representations. To construct the flag mean, iteratively solve the optimization problem

$$[u_j] = \arg\min_{[u] \in Gr(1,n)} \sum_{i=1}^{p} d_c([u],[X_i])^2 \tag{2}$$

subject to $[u] \perp [u_l]$ for all $l < j$

for $[u_1], ..., [u_r]$ [11]. Optimality is achieved when

$$\left(\sum_{i=1}^{m} X_i X_i^T \right) u = \lambda u$$

and the problem is reduced to an eigenvector computation [11].

4   Grassmann K-means Algorithm

The K-means algorithm operates on a stream of data, assigning each data point to its nearest center [18]. The chosen center is then updated in the direction of the new data point. Let x be the nth data point assigned to a center c. In Euclidean space, the center c is then updated by

$$c_{new} = c + \frac{1}{n}(x - c) \tag{3}$$

To adapt this algorithm to the Grassmann manifold, we require a way to move one subspace a specified distance towards another. This is accomplished by parameterizing the geodesic between two subspaces $[X]$ and $[Y]$ [1,15]. Given orthonormal matrix representations X and Y, respectively, the velocity matrix H that induces a geodesic between $[X]$ and $[Y]$ is given by

$$H = (I - XX^T)Y(X^T Y)^{-1}. \tag{4}$$

The singular value decomposition of the velocity matrix $H = U\Sigma V^T$ is then used to parameterize a geodesic curve between X and Y by

$$\Phi(t) = XV \cos(\Theta t) + U \sin(\Theta t) \tag{5}$$

where $\Theta = \arctan(\Sigma)$. Note $\Phi(0) = X$ and $\Phi(1) = \tilde{Y}$, with $[\tilde{Y}] = [Y]$. Using this, we can update K-means by letting $t = 1/n$. The K-means algorithm on the Grassmannian is then:

1. Construct points on $Gr(k, n)$ using raw data.
2. Select k random initial centers from the data on $Gr(k, n)$.
3. For each data point $[X]$:
   (a) Find the center $[C_i]$ nearest to $[X]$.
   (b) Update $[C_i]$ according to Eq. (5).
4. Calculate the average distortion error and check if it is smaller than the specified threshold. If not, repeat step 2.

This process can be applied several times to improve the clusters.

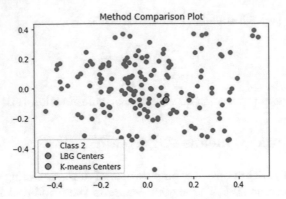

Fig. 1. MDS embedding of handwritten digit 2 and centers selected by each algorithm; the LBG center is beneath the K-means center.

5   The LBG Algorithm on the Grassmannian

The Linde-Buzo-Gray algorithm (LBG) performs vector quantization in Euclidean space by associating all points to their nearest center [17]. These centers serve as prototypes for the data in the sense that they minimize the distortion error locally. On the Grassmannian, the centroid of all points closest to a given center C_i is obtained using the flag mean $[u_1, \ldots, u_r]$. The averaging is done over elements of Voronoi sets, i.e., the collection of data points closest to a given center. The definition of a Voronoi set S_i is given by

$$S_i = \{x : d(x, c_i) \leq d(x, c_j), \quad i \neq j\}$$

The centroid is found using $c_i = M(S_i)$ where the function M represents a mapping from the members of the Voronoi set S_i to its "mean". In Euclidean space the mean is the usual centroid

$$c_i = \frac{1}{|S_i|} \sum_{x \in S_i} x \tag{6}$$

for $x \in \mathbb{R}^n$. For Grassmannians this mean is the output of Eq. (2), i.e., the flag mean. On $Gr(p, n)$, the distortion of the clusters with centroids $[C_i]$ is given by

$$D(\{S_i\}_{i=1}^k) = \sum_{i=1}^k \sum_{[X] \in S_i} d([X], [C_i])$$

The Grassmannian LBG algorithm is as follows:

1. Initialize k random centers on $Gr(p, n)$.
2. Assign each subspace $[X_i]$ to its nearest center $[C_{i*}]$.
3. Update the centers using Eq. (2).

4. Calculate the average distortion associated with the new partition using chordal distances on the Grassmannian.
5. If stopping criterion not met, then go to step 2.

The paper by Gruber and Theis contains a similar algorithm for clustering on the Grassmannian [14] based on the projection Frobenius norm, though they do not approach centroid calculations from the framework of flag subspaces.

6   Numerical Experiments

One goal of clustering methods is for each center to contain, as nearly as possible, points from only one class. We can express the *purity* of a cluster by the fraction of points belonging to the majority class for that cluster. We use this measure to establish quantitative comparisons below.

6.1   MNIST Results

We use the MNIST handwritten data set to illustrate these algorithms [10]. This data set contains 28×28 images of handwritten digits vectorized into a data point in $\mathbb{R}^{784}$. To construct points on $Gr(p, 784)$ we select p data points from the same class and form a $p \times 784$ orthonormnal matrix using the QR-decomposition. Subspaces are assigned the same label as the points used to construct them. Figure 1 illustrates the performance of both algorithms on a set of 156 data points in Gr(5,784) generated using the handwritten digit 2. One centroid was selected at random to initialize each algorithm. One LBG iteration and one K-means epoch generated essentially identical means. The visualization of the results was achieved by using MDS with the chordal matric to isometrically embed the subspaces and centroids into $\mathbb{R}^2$.

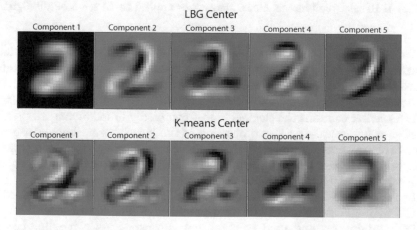

Fig. 2. Visualization of orthonormal components for each center.

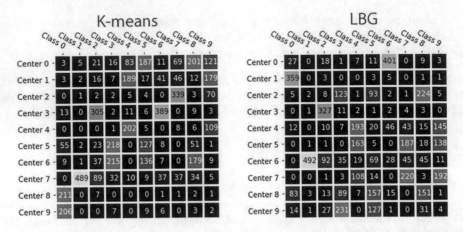

Fig. 3. Results from the Euclidean space algorithms applied to all ten MNIST digits.

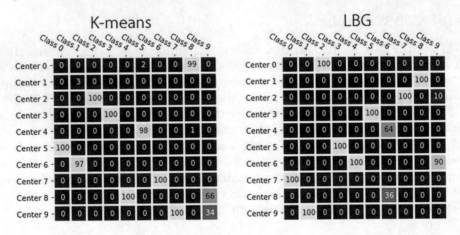

Fig. 4. Results the Grassmannian algorithms applied to all ten MNIST digits.

A closer look at the orthonormal components of each center reveals details about variations in the cluster. Figure 2 shows the five orthonormal components of each center reshaped to the original image size. In particular, because the flag mean yields an orthonormal basis ordered by energy [11], the first component of the LBG center contains the elements most commonly found among all points in the cluster, and represents first dimension of the "true" mean. Each consecutive component captures information about the most common variations from the mean, ordered from most common to least. K-means captures similar information about within-cluster variation but does not have any special ordering.

To further explore these methods, both algorithms were applied to all the MNIST digits. For each digit, 500 data points were randomly selected from the MNIST training set and used to construct subspaces. As a baseline comparison for the algorithms on the Grassmannian, the Euclidean versions of both

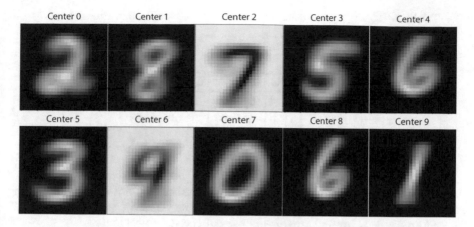

Fig. 5. Visualization of the first flag vector for each of the 10 LBG centers.

K-means and LBG were performed on the randomly selected data in Euclidean space. The Grassmannian versions of both algorithms were then tested on a data set consisting of 1000 points in Gr(5,784), with 100 points per class. All tests were performed multiple times on the data set to account for variations due to randomized starting conditions. The best result for each algorithm was chosen based on lowest cluster distortion. The average purity across the ten clusters for Euclidean K-means is 58.84% ± .22, and the average cluster purity for Euclidean LBG is 58.05% ± .23; see Fig. 3.

In contrast to the Euclidean algorithms, the Grassmannian algorithms performed the unsupervised clustering task very well; see Fig. 4. For K-means, centers 0 through 7 have purity >98%, whereas center 8 has a purity of 60.24% and center 9 has a purity of 74.63%. The average purity for the K-means algorithm is 93.19% ± .13. For the LBG trial, center 2 has a purity of 90.91%, center 6 has purity of 52.63% while all other centers are 100% pure, resulting in an average purity of 94.35% ± .14. Figure 5 shows the first component of each center chosen by LBG. Clearly, centers 4 and 8 can both be classified as the number 6, whereas center 6 appears to be a combination of digits 4 and 9. This highlights an interesting facet of subspace analysis. Because 4 and 9 are similar in overall shape, there is some amount of overlap in the subspaces spanned by these digits, making it difficult to distinguish the two. A similar effect is often seen with digits 7 and 9.

6.2   Indian Pines Results

Select classes from the Indian Pines data set [3] were used to further evaluate the performance of K-means and LBG. The classes *alfalfa* and *corn* were compared to test the algorithms on separable but unbalanced clusters. The data set contains 237 data points for *corn*, but only 46 for *alfalfa*. Points were generated in Gr(5,200) in the same manner used for the MNIST trials. Due to the class

Table 1. Results on two-class experiments on several Indian Pines classes.

	Alfalfa v. Corn				Pasture v. Trees				Pasture v. Trees			
Manifold	Gr(5,200)				Gr(5,200)				Gr(10,200)			
Method	K-means		LBG		K-means		LBG		K-means		LBG	
Class	1	4	1	4	5	6	5	6	5	6	5	6
Center 0	0	47	9	0	94	0	2	145	0	23	4	8
Center 1	9	0	0	47	2	146	94	1	4	0	0	15

size disparity and the reduction in the total number of points when generating subspaces, class 1 (*alfalfa*) contained 9 points and class 4 (*corn*) contained 47. As seen in Table 1, both algorithms clustered the data perfectly.

A second trial was performed on the classes for *pasture* and *trees*, which are unbalanced and contain overlap. Both algorithms were tested using points generated first in Gr(5,200), then in Gr(10,200). There are 483 data points for *pasture* and 730 data points for *trees*. In Gr(5,200), class 5 (*pasture*) contained 96 points and class 6 (*trees*) contained 146 points, and both methods yielded cluster purity >98%. In Gr(10,200), class 5 contained only 4 points and class 6 contained 23 points. K-means succeeded in clustering the data perfectly, but LBG had a cluster with only 66% purity. In this case, the randomly selected initial conditions were poor, which caused the algorithm to terminate in a local minimum rather than obtaining the optimal clustering. This highlights one of the pitfalls of these clustering algorithms, especially in cases where the data set is small. Results for both experiments are included in Table 1.

The final test was performed on the soybean classes *soybean with tilling* (class 10), *soybean with no tilling* (class 11), *soybeans clean* (class 12). We explore two options for clustering this data. First, we embed the points on the Grassmannian into Euclidean space by applying MDS to a matrix of pairwise smallest principal angle distances and then cluster using the standard Euclidean space algorithms. Second, we cluster directly on the manifold using the pseudometric. Constructing subspaces in Gr(5,200) similar to previous experiments resulted in 194 points in class 10, 491 points in class 11, and 118 points in class 12. A total of 10 trials were performed, and some for these experiments are displayed in Fig. 6. Once again, the Euclidean algorithms yielded mediocre results, with average cluster purity less than 60%. For the LBG algorithm, it appears to be beneficial to embed the points in Gr(5,200) back into Euclidean space using MDS before clustering. This resulted in an average cluster purity of 83.74% $\pm$ .14, verses 72.44% $\pm$ .18 when clustering was done on the manifold itself. For K-means, however, performing clustering after embedding yielded an average purity of 83.50%$\pm$.19, with all but one trial yielding at least one cluster containing only a single point. Clustering directly on the manifold raised the average purity to 85.92% $\pm$ .13 and resulted in a much more even and consistent distribution of points among centers. The best method for clustering appears to vary based on the algorithm used, and likely also changes based on the data itself.

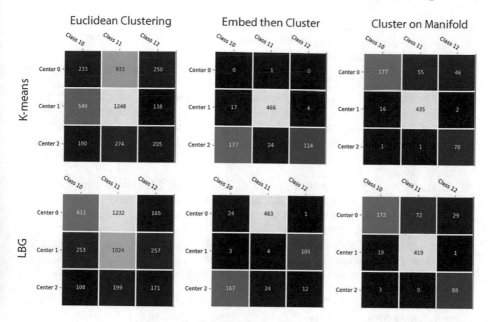

Fig. 6. A sample of the comparison trials of the K-means and LBG algorithms on classes 10–12 of the Indian Pines data set.

7 Conclusions

In this paper we extend the K-means and LBG algorithms to the framework of the real Grassmannian. We demonstrate that both approaches result in high classification purity, i.e., the cluster membership consists of either exclusively, or predominantly, data from a single label. The flag mean provides nested subspaces that capture the essence of the signature of the data in the centroid. We are able to capture the constituent patterns and their variations, which is vital for discovering new patterns of the same class but in a different variation of state. On the Indian Pines data set, we demonstrate clustering directly on the manifold and on embeddings. These algorithms for subspace quantization provide a robust means to characterize the variability in complex, high-dimensional data sets.

References

1. Absil PA, Mahony R, Sepulchre R (2007) Optimization algorithms on matrix manifolds. Princeton University Press, Princeton
2. Adams H, Chepushtanova S, Emerson T, Hanson E, Kirby M, Motta F, Neville R, Peterson C, Shipman P, Ziegelmeier L (2017) Persistent images: a stable vector representation of persistent homology. J Mach Learn 18(1):218–252
3. Baumgardner MF, Biehl LL, Landgrebe D.A (2015) 220 band AVIRIS hyperspectral image data set: June 12, 1992 Indian Pine Test Site 3, September 2015

4. Beveridge JR, Draper BA, Chang JM, Kirby M, Kley H, Peterson C (2009) Principal angles separate subject illumination spaces in YDB and CMU-PIE. IEEE Trans Pattern Anal Mach Intell 31(2):351–363
5. Björck A, Golub GH (1973) Numerical methods for computing angles between linear subspaces. Math Comput 27(123):579–594
6. Chang JM, Kirby M, Kley H, Peterson C, Beveridge J, Draper B (2007) Recognition of digital images of the human face at ultra low resolution via illumination spaces. Lecture Notes in Computer Science, vol 4844. Springer, pp 733–743
7. Chepushtanova S, Kirby M (2017) Sparse Grassmannian embeddings for hyperspectral data representation and classification. IEEE Geosci Remote Sens Lett 14(3):434–438
8. Chepushtanova S, Kirby M, Peterson C, Ziegelmeier L (2015) An application of persistent homology on Grassmann manifolds for the detection of signals in hyperspectral imagery. In: 2015 IEEE international geoscience and remote sensing symposium (IGARSS). IEEE, pp 449–452
9. Chepushtanova S, Kirby M, Peterson C, Ziegelmeier L (2016) Persistent homology on Grassmann manifolds for analysis of hyperspectral movies. In: International workshop on computational topology in image context. Lecture Notes in Computer Science, vol. 9667. Springer, pp 228–239
10. Cireşan DC, Meier U, Gambardella LM, Schmidhuber J (2010) Deep, big, simple neural nets for handwritten digit recognition. Neural Comput 22(12):3207–3220
11. Draper B, Kirby M, Marks J, Marrinan T, Peterson C (2014) A flag representation for finite collections of subspaces of mixed dimensions. Linear Algebra Appl 451:15–32
12. Edelman A, Arias TA, Smith ST (1998) The geometry of algorithms with orthogonality constraints. SIAM J Matrix Anal Appl 20(2):303–353
13. Gallivan KA, Srivastava A, Liu X, Van Dooren P (2005) Efficient algorithms for inferences on Grassmann manifolds. In: 2003 IEEE workshop on statistical signal processing. IEEE, pp 15–318
14. Gruber P, Theis FJ (2006) Grassmann clustering. In: 2006 14th European signal processing conference. IEEE, pp 1–5
15. Kirby M, Peterson, C (2017) Visualizing data sets on the Grassmannian using self-organizing mappings. In: 2017 12th International workshop on self-organizing maps and learning vector quantization, clustering and data visualization (WSOM). IEEE, pp 1–6
16. Kohonen T (1984) Self-organization and associative memory. Springer, Berlin
17. Linde Y, Buzo A, Gray R (1980) An algorithm for vector quantizer design. IEEE Trans Commun 28(1):84–95
18. MacQueen J, et al (1967) Some methods for classification and analysis of multivariate observations. In: Proceedings of the fifth Berkeley symposium on mathematical statistics and probability, vol 1, Oakland, CA, USA, pp 281–297
19. Marrinan T, Beveridge JR, Draper B, Kirby M, Peterson C (2015) Flag manifolds for the characterization of geometric structure in large data sets. In: Numerical mathematics and advanced applications-ENUMATH 2013. Springer, pp 457–465
20. Marrinan T, Draper B, Beveridge JR, Kirby M, Peterson C (2014) Finding the subspace mean or median to fit your need. In: 2014 IEEE conference on computer vision and pattern recognition (CVPR). IEEE, pp 1082–1089

Variants of Fuzzy Neural Gas

Tina Geweniger[1,2] and Thomas Villmann[2(✉)]

[1] Westsächsische Hochschule Zwickau – University of Applied Sciences,
Zwickau, Germany
tina.geweniger@fh-zwickau.de
[2] Hochschule Mittweida – University of Applied Sciences, Mittweida, Germany
villmann@hs-mittweida.de

Abstract. Neural Gas is a prototype based clustering technique taking the ranking of the prototypes regarding their distance to the data samples into account. Previously, we proposed a fuzzy version of this approach, yet restricted our method to probabilistic cluster assignments. In this paper we extend this method by combining possibilistic and probabilistic assignments. Further we provide modifications to handle non-vectorial data.

Keywords: Fuzzy Neural Gas · Possibilistic neural gas ·
Median data · Relational data

1 Introduction

Prototype based vector quantization covers a whole group of algorithms designed for clustering and data compression. The term *protoype based* implies that the data is represented by a smaller number of alike objects resulting in interpretable models. The most famous clustering method from this family is c-means [1] (or also known as k-means [2] or ISODATA [3]) where the data is approximated by c (or k) representatives. But also Self Organizing Maps (SOM) [4] and Neural Gas (NG) [5] apply this principle. The basic versions of all methods have in common, that each data sample is assigned uniquely to one prototype. Therefore, they are also called (crisp) vector quantizers. Yet in practice, very often data is overlapping and using a cluster algorithm resulting in soft labels might help to identify these regions and contribute to a better understanding of the structure of the data.

In the last decades several methods dealing with soft assignments of the data to the prototypes have been developed. For examples Fuzzy c-Means (FCM) [2], Fuzzy Self Organizing Maps (FSOM) [6], and Fuzzy Neural Gas (FNG) [7]. Yet, according to common understanding *fuzzy* almost always refers to probabilistic cluster assignments. KRISHNAPURAM AND KELLER [8] point out that there is a substantial difference between probabilistic and possibilistic clustering.

Probabilistic as well as possibilistic assignment values are inversely related to the distance to the prototypes. Yet, while probabilistic assignments to the

© Springer Nature Switzerland AG 2020
A. Vellido et al. (Eds.): WSOM 2019, AISC 976, pp. 261–270, 2020.
https://doi.org/10.1007/978-3-030-19642-4_26

prototypes always sum up to one for each data sample, for possibilistic assignments this requirement is dropped. That is, if a data sample is equidistant to two prototypes, the probabilistic assignment value will always be 0.5 for each cluster no matter how far away from the prototype the data sample is located. The possibilistic assignment value decreases with increasing distance to the prototype. Therefore, the latter kind of assignment is also called *typicality*. Cluster algorithms incorporating the concept of possibilistic assignments are suitable to detect noise and outliers.

In further studies PAL, KELLER ET AL. proposed a stable FCM version called PFCM, which takes both paradigms into account [9]. In our present contribution we adopt this method, transfer the ideas to NG, and obtain an extended NG we call Possibilistic Fuzzy Neural Gas (PFNG). Just as suggested in [9], we also introduce balancing parameters to weight the importance of possibilistic and probabilistic assignments. With certain settings explained later in this article it is possible to obtain the pure version of FNG omitting the possibilistic part and vice versa. This makes our general method usable as a stand alone approach.

Another important aspect for clustering is the type of data. To find groups of similar objects, their similarity or dissimilarity has to be known. In case of vectorial data, this similarity can be calculated easily. Most often the (squared) Euclidean distance is used. The basic versions of the above mentioned algorithms are designed to deal with this kind of vectorial data assuming that the data is embedded in Euclidean space. In case of non-vectorial data, e. g. texts, music, and gene sequences, where no practical distance measure is feasible, the original algorithms have to be adapted to handle this kind of data provided the distances are known. For our proposed algorithm PFNG we provide the necessary modifications. Thereby, we distinguish between relation data, where a Euclidean embedding can be assumed, and median data. Notice, that we only consider the case where the distances are given as dissimilarities. If only the similarities are known, these have to be converted to dissimilarities according to [10].

In the upcoming sections we first describe the different kinds of data – vectorial data, relational data, and median data. In section three a short description of a general FCM approach incorporating probabilistic and possibilistic assignments is provided. In the next section we derive a general variant of Fuzzy Neural Gas and continue with special considerations regarding vectorial, median, and relational data. Some experiments follow in section five and the last section contains a summary and some concluding remarks.

2   Interpretation of Distance for Different Types of Data

Provided we have a set $X = \{x_1, \ldots, x_N\}$ of N data objects. In order to cluster these objects some measure of dissimilarity $D_{ij} = D(x_i, x_j)$ or distance has to be related to the samples. Depending on the type of the data we distinguish between vectorial data, relational data, and median data.

Vectorial Data. If the data samples are provided as d-dimensional real-valued vectors, i. e. $\mathbf{x} \in \mathbb{R}^d$, there are different measures to calculate their distance. The most common distance measure is the Euclidean distance respectively the squared Euclidean distance (sED) simplifying calculations in the context of clustering. For two data points $\mathbf{x}_i$ and $\mathbf{x}_j$ the sED is defined as

$$D_{ij} = d^2(\mathbf{x}_i, \mathbf{x}_j) = \sum_{t=1}^{d} (x_{it} - x_{jt})^2 = \langle \mathbf{x}_i, \mathbf{x}_i \rangle_E - 2\langle \mathbf{x}_i, \mathbf{x}_j \rangle_E + \langle \mathbf{x}_j, \mathbf{x}_j \rangle_E. \quad (1)$$

Dealing with non-vectorial data, it might not be possible to use the data objects themselves to calculate the distance. Instead, the dissimilarities D_{ij} have to be provided beforehand. Usually these values are provided as a dissimilarity matrix $\mathbf{D} \in \mathbb{R}_+^{N \times N}$ whose size is directly given by the number of data samples. We further assume that this matrix is symmetric and complete, which means that all pairwise dissimilarities are known and $D_{ij} = D_{ji}$. The dissimilarity value corresponding to identical data samples is assumed to be zero resulting in a zero diagonal respectively $D_{ii} = 0$. Further, as mentioned before, it can be distinguished between median and relational data.

Relational Data. For relational data we assume that there exists a (non-linear) mapping $g(x_i) = \mathbf{x}_i \in X_E$ which projects the data samples into a Euclidean embedding space X_E. Thus, the dissimilarities D_{ij} are interpreted as the squared Euclidean distance in X_E with

$$D_{ij} = d_E^2(g(x_i), g(x_j)) = d_E^2(\mathbf{x}_i, \mathbf{x}_j). \quad (2)$$

Notice, that the mapping $g(x)$ does not have to be known.

Median Data. For median data the assumption of an underlying Euclidean metric is not valid circumventing a vectorial interpretation. Therefore, a clustering algorithm processing median data has to be able to handle the objects themselves and to rely solely on dissimilarities provided in $\mathbf{D}$.

3 Possibilistic Fuzzy c-Means

In [9] Pal and Keller proposed an improved general variant of c-means incorporating probabilistic and possibilistic cluster assignments. For a set $V = \{v_i\}_{i=1}^{N_v}$ of N_v data samples and N_p prototypes $W = \{w_j\}_{j=1}^{N_p}$ the generalized cost function to minimize is

$$J_{PFCM} = \sum_{i=1}^{N_v} \sum_{j=1}^{N_p} \beta ij D_{ij} + \sum_{j=1}^{N_p} \gamma_j \sum_{i=1}^{N_v} (1 - t_{ij})^\eta, \quad (3)$$

where $\beta ij = (au_{ij}^m + bt_{ij}^\eta)$ is an abbreviation used to improve readability of upcoming equations. The cost function J_{PFCM} itself is an extended version of the generally known fuzzy c-means cost function [2] balancing the influence of probabilities $u_{ij} \in [0,1]$ with $\sum_{j=1}^{N_p} u_{ij} = 1$ and typicalities $t_{ij} \in [0,1]$ by the variables $a \geq 0$ and $b \geq 0$, which are positive independent scalars not required to sum up to 1. The exponents $m > 1$ and $\eta > 1$ control the degree of fuzziness and typicality. The additional term containing the sum over γ_j is added as a further control parameter influencing the typicality values. Suggestions on how to set this parameter are provided in [9].

Notice, that by setting the parameters appropriately, the basic univocal c-Means [11], the probabilistic (fuzzy) c-means [2] as well as the pure possibilistic c-means [9] are obtained.

In the variant proposed in [9] only vectorial data with $\mathbf{v}_i, \mathbf{w}_j \in \mathbb{R}^d$ and sED as distance measure were considered. In [12] we generalized this approach to handle non-vectorial median and relational data. For further information especially concerning update rules for prototypes, probabilities, and typicalities refer to the mentioned literature.

4   Possibilistic Fuzzy Neural Gas

Neural Gas (NG) [5] is cluster algorithm which is more robust than c-Means and proven to converge. It incorporates local neighborhood relations by weighting the distances between data samples and prototypes by the ranks of the prototypes. More formally, in [7] the local costs applying to a fuzzy variant of NG are defined as

$$lc_\sigma(v_i, w_j) = \sum_{l=1}^{N_p} h_\sigma(w_j, w_l) \cdot D_{il} \tag{4}$$

with neighborhood function

$$h_\sigma(w_j, w_l) = c_\sigma \cdot \exp\left(-\frac{rk_j(w_l, W)^2}{2\sigma^2}\right). \tag{5}$$

The parameter $\sigma > 0$ defines the neighborhood range and c_σ assures that $\sum_{l=1}^{N_p} h_\sigma(w_j, w_l) = 1$. Using the Heaviside function $\Theta(x)$, the ranking function

$$rk_j(w_l, W) = \sum_{k=1}^{N_p} \Theta\left(d(w_l, w_j) - d(w_l, w_k)\right) \tag{6}$$

returns the rank of prototype w_l according to prototype w_j.

Notice, that all references to data samples v and prototypes w are general not restricting them to be vectors. Now we can easily transfer the idea behind the cost function given in Eq. (3) to NG by replacing the distance by the local cost term given in Eq. (4). This way we obtain

$$J_{PFNG} = \sum_{i=1}^{N_v} \sum_{j=1}^{N_p} \beta_{ij} lc_\sigma(v_i, w_j) + \sum_{j=1}^{N_p} \gamma_j \sum_{i=1}^{N_v} (1 - t_{ij})^\eta. \tag{7}$$

To minimize this cost function an alternating optimization scheme taking turns updating prototypes w_j, probabilities u_{ij}, and typicalities t_{ij} is employed. How these updates are performed depends on the kind of data and is described in the next subsections. Yet in general, the cost function is minimized by applying the method of Lagrange multipliers taking the side condition $\sum_{j=1}^{N_p} u_{ij} = 1 \; \forall \; i$ into account. If possible, the function is solved for the respective parameters explicitly, otherwise stochastic gradient descent learning (SGDL) is performed. The Lagrange function of J_{PFNG} (7) is

$$L_{PFNG} = J_{PFNG} + \left(\sum_{k=1}^{N_v} \lambda_k \left(\sum_{j=1}^{N_p} u_{ij} - 1 \right) \right). \tag{8}$$

4.1 Vectorial Data

Given a set $V = \{\mathbf{v}_1, \ldots, \mathbf{v}_{N_v}\}$ of N_v d-dimensional data vectors, e. g. $\mathbf{v}_i \in \mathbf{R}^d$, and a set $W = \{\mathbf{w}_1, \ldots, \mathbf{w}_{N_p}\}$ of N_p prototypes with $\mathbf{w}_j \in \mathbf{R}^d$, the update rules are obtained analogously to the PFCM update rules [9] by replacing the distance with the local costs (4). Update rules are derived by considering the derivatives of Eq. (8) with respect to the parameter to optimize and to solve explicitly for that parameter. Updating probabilities and possibilities the local cost term is treated as a constant and does not further influence the derivations with respect to u_{ij} and t_{ij}, respectively.

$$u_{ij} = \left(\sum_{l=1}^{N_p} \left(\frac{lc(\mathbf{v}_i, \mathbf{w}_j)}{lc(\mathbf{v}_i, \mathbf{w}_l)} \right)^{\frac{1}{(m-1)}} \right)^{-1} \qquad t_{ij} = \left(1 + \left(\frac{b}{\gamma_i} lc(\mathbf{v}_i, \mathbf{w}_j) \right)^{\frac{1}{(\eta-1)}} \right)^{-1} \tag{9}$$

If sED is chosen as distance measure the update rule for the prototypes $\mathbf{w}_j$ can also be stated explicitly as

$$\mathbf{w}_j = \frac{\sum_{i=1}^{N_v} \sum_{l=1}^{N_p} \beta_{il} h_\sigma(\mathbf{w}_j, \mathbf{w}_l) \mathbf{v}_i}{\sum_{i=1}^{N_v} \sum_{l=1}^{N_p} \beta_{il} h_\sigma(\mathbf{w}_j, \mathbf{w}_l)}. \tag{10}$$

Otherwise, we have to apply SGDL according to

$$\Delta \mathbf{w}_j \propto - \sum_{i=1}^{N_v} \sum_{l=1}^{N_p} \beta_{il} \, h_\sigma(\mathbf{w}_j, \mathbf{w}_l) \cdot \frac{\partial d(\mathbf{v}_i, \mathbf{w}_j)}{\partial \mathbf{w}_j}. \tag{11}$$

4.2 Relational Data

Considering relational data with $g(x_i) = \mathbf{v}_i$ with $\mathbf{v}_i \in V = X_E$, the prototypes are defined as convex linear combinations of the embedded data in X_E

$$\mathbf{w}_j = \sum_{i=1}^{N_v} \alpha_{ji} \mathbf{v}_i \tag{12}$$

with coefficients $\alpha_{ij} > 0$ and $\sum_{i=1}^{N_v} \alpha_{ij} = 1$.

If the dissimilarity matrix $\mathbf{D}$ is Euclidean, i. e. $D_{ij} \leq D_{ik} + D_{kj}$ for all triples (i, j, k), then the existence of such an embedding is guaranteed and the distance between embedded data samples and prototypes yields

$$d_V^2(\mathbf{v}_i, \mathbf{w}_j) = \sum_{l=1}^{N_v} \alpha_{jl} \cdot D_{il} - \frac{1}{2} \boldsymbol{\alpha}_j^T \cdot \mathbf{D} \cdot \boldsymbol{\alpha}_j \tag{13}$$

where $\boldsymbol{\alpha}_j = (\alpha_{j1} \ldots, \alpha_{jN_v})^T$ is the vector of embedding coefficients [13].

The distance of the prototypes to each other needed within the rank function (6) can be obtained in a similar manner using (1)

$$d_V^2(\mathbf{w}_j, \mathbf{w}_k) = \sum_r \sum_s \langle \mathbf{v}_r, \mathbf{v}_s \rangle \left(\alpha_{jr}\alpha_{js} - 2\alpha_{jr}\alpha_{ks} + \alpha_{kr}\alpha_{ks} \right) \tag{14}$$

where

$$\langle \mathbf{v}_r, \mathbf{v}_s \rangle = -\frac{1}{2} \left(D_{rs} - \frac{1}{N_v} \sum_{r=1}^{N_p} D_{rs} - \frac{1}{N_v} \sum_{s=1}^{N_p} D_{rs} + \frac{1}{N_v^2} \sum_{r=1}^{N_p} \sum_{s=1}^{N_p} D_{rs} \right) \tag{15}$$

is obtained by double centering the dissimilarities D_{ij} according to [14].

Update rules for prototypes, probabilities, and typicalities can again be obtained by deriving (8) with respect to the relevant parameter. The update rules for fuzzy probabilities u_{ij} and possibilistic typicalities t_{ij} remain unaffected by the data mapping and are identical with the formulas provided in Eqs. (9), except that we have to apply Eqs. (13) and (14) to calculate the local costs.

Updating the prototypes requires adapting the coefficient vectors $\boldsymbol{\alpha}_j$ in the embedding space V_E. Therefore, the derivative of Eq. (8) with respect to α_{pq} has to be considered, yielding

$$\Delta\alpha_{pq} \propto -\frac{\partial L_{PFNG}(U, T, W; \mathbf{v}_i)}{\partial \alpha_{pq}}$$

$$= \frac{\partial \sum_{i=1}^{N_v} \sum_{j=1}^{N_p} \beta_{ij} lc_\sigma(v_i, w_j)}{\partial \alpha_{pq}} \tag{16}$$

$$+ \underbrace{\frac{\partial \sum_{j=1}^{N_p} \gamma_j \sum_{i=1}^{N_v} (1 - t_{ij})^\eta}{\partial \alpha_{pq}}}_{=0} + \underbrace{\frac{\partial \left(\sum_{k=1}^{N_v} \lambda_k \left(\sum_{j=1}^{N_p} u_{ij} - 1 \right) \right)}{\partial \alpha_{pq}}}_{=\,0}.$$

Using (4) this simplifies to

$$\Delta\alpha_{pq} \propto \sum_{i=1}^{N_v} \beta_{ip} \frac{\partial \sum_{l=1}^{N_p} h_\sigma(\mathbf{w}_p, \mathbf{w}_l) \cdot d(\mathbf{v}_i, \mathbf{w}_l)}{\partial \alpha_{pq}} \tag{17}$$

Analogously to the NG theory provided in [5], the last equation can be reduced to

$$\Delta\alpha_{pq} \propto \sum_{i=1}^{N_v} \beta_{ip} \sum_{l=1}^{N_p} h_\sigma(\mathbf{w}_p, \mathbf{w}_l) \frac{\partial d(\mathbf{v}_i, \mathbf{w}_l)}{\partial \alpha_{pq}}. \tag{18}$$

In [12] the partial derivative of $d(\mathbf{v}_i, \mathbf{w}_j)$ with respect to a particular α_{pq} was derived as

$$\frac{\partial d(\mathbf{v}_i, \mathbf{w}_j)}{\partial \alpha_{pq}} = D_{iq} - \sum_{l=1}^{N_p} D_{ql} \alpha_{pl}. \tag{19}$$

Now the update rule for the coefficients in vectorial form is

$$\Delta \alpha_{pq} \propto - \sum_{i=1}^{N_v} \beta_{ip} \sum_{l=1}^{N_p} h_\sigma(\mathbf{w}_p, \mathbf{w}_l) \left(\mathbf{D}_i - \mathbf{D}_j \sum_{i=1}^{N_v} \alpha_{ji} \right) \tag{20}$$

After the update a normalization to assure $\sum_{i=1}^{N_v} \alpha_{ij} = 1$ has to be performed.

4.3   Median Data

For median data the prototypes are restricted to be data samples. Thus, care has to be taken to select the new prototypes according to the minimization requirement laid on the cost function

$$\mathbf{w}_j = \mathbf{v}_l \quad \text{with} \quad l = \operatorname*{argmin}_{l'} \left(\sum_{i=1}^{N_v} \beta_{ij} l c_{il'} + \gamma_j \sum_{i=1}^{N_v} (1 - t_{ij})^n \right). \tag{21}$$

The update rules for probabilistic assignments and typicalities again remain unchanged being equivalent with Eqs. (9) except the fact that distances respectively dissimilarities are obtained from the provided dissimilarity matrix $\mathbf{D}$. To speed up, the minimization problem can be relaxed by heuristics, e. g. secretary problem [15], or generalized expectation maximization can be performed. Often, it is sufficient to enhance the solution instead of finding the absolute minimum. Care has to be taken to avoid multiple selections of a data sample as prototype. We advice to discard data samples already in use from the set of available data points. Convergence, at least to a local minimum can be proven analogously to the convergence of M-PFCM. The proof thereof can be found in [12].

4.4   Remarks

The algorithms for all NG variants are structurally consistent following an alternating optimization scheme. In Algorithm 1 all steps are summarized. If necessary, additional information regarding differences due to the kind of the underlying data is provided.

By an appropriate choice of parameters a, b, and γ pure variants only considering fuzzy probabilities or possibilistic typicalities can be reconstructed. Setting $b = 0$ and $\gamma_i = 0$ the FNG as proposed in [7] is obtained. By setting $a = 0$ we get a possibilistic NG.

For reasons of stability we suggest to run the algorithm twice. First in probabilistic mode to roughly set the prototypes and based theron a second time for fine tuning and calculation of the typicalities. This procedure is recommended by [9] for PFCM and also proved valuable for our method.

Algorithm 1. Possibilistic Fuzzy Neural Gas (PFNG)

1: set number N_p of prototypes
2: initialize prototypes
        - vectorial: generate random vectors $\mathbf{w}_j$ within the data space
        - relational: initialize random coefficient vectors $\boldsymbol{\alpha}_j$
        - median: select random data samples as prototypes
3: **repeat**
4:     calculate fuzzy assignments u_{ij} according to Eq. (9)
5:     calculate typicalities t_{ij} according to Eq. (9)
6:     update prototypes using u_{ij} and t_{ij} from steps 4 and 5
        - vectorial: calculate $\mathbf{w}_j$ according to Eq. (10)
        - relational: update coefficient vectors $\boldsymbol{\alpha}_j$ according to Eq. (20)
        - median: select representative data samples acc. to Eq. (21)
7: **until** convergence **or** preset number of epochs elapsed **or** manual stop

---

5   Experiments

As experiments we chose the same data sets as in [12]. This way the results can be compared directly to those obtained by the respective FCM variant also incorporating probabilistic and possibilistic assignments. We use one simple two-dimensional artificial data set of Gaussian distributed data points and one real world data set from the field of psychotherapy.

5.1   Artificial Gaussian Distributions

The artificial data set contains two Gaussian distributions with 500 data points each. The first distribution is centered around $(0, 0)$ and has a variance of $\sigma = 1.0$, the other distribution has mean $(4, 0)$ and variance $\sigma = 0.5$.

    We performed a series of runs with different parameter settings and prototype initializations to examine and evaluate the performance. Due to the shortage of space we decided to present here the results for one specific setting. We set the balancing parameter for probabilities and possibilities to $a = 0.9$ and $b = 0.2$, respectively, to push the influence of probabilistic assignments. The γ-parameter was set to $\gamma = 1.0$. For the exponents m and η we chose $m = 1.2$ and $\eta = 1.5$.

    We run PFNG for median data different times starting with various initially randomly chosen prototypes. Each time the algorithm reliably converged to the same solution within a few (less than 10) iterations. Figure 1a depicts the fuzzy assignments and typicalities. The upper two pictures refer to the fuzzy assignments of the data samples to the cluster centers, the lower ones represent the typicality values. Notice, that in contrast to FCM and variants thereof [12] the prototypes are located a little off the means moved closer to the overlapping region. This is due to the cost function being based on a dynamic neighborhood instead of fixed distances. The possibilistic assignments respectively typicality values show a peculiar behaviour. At this point we do not fully understand, why data samples within the overlapping region obtain the highest typicality values.

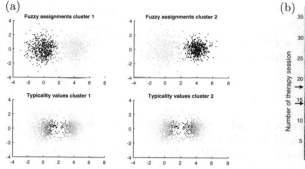
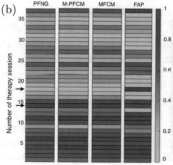

Fig. 1. (a) Gaussian data set: probabilistic and possibilistic assignments. The darker the color the higher the assignment value. Prototypes marked with large dots. (b) Psychotherapy data set: color coded probabilistic assignments to cluster 1 resp. the first therapy phase. Prototypes marked by arrows.

5.2 Clustering Transcripts of Psychotherapy Sessions

In this experiment we used transcripts of 37 psychotherapy sessions. These transcripts contain conversations between therapist and patient and belong all to one psychodynamic therapy [16]. From clinical investigations it is known that the process can be divided into two clinical stages with the culminating point (phase transition) around session 17. Several established therapy measures including among others the analysis of psycho-physiological behavior during the sessions led to this evaluation.

We used this data set in former experiments [12,17] and were able to validate the two-phase process applying our dissimilarity based median methods. Thereby we used the normalized compression distance (NCD) to obtain dissimilarity values for all dialog transcripts. NCD is s based on the Kolmogorov complexity also known as universal distance description length. This measure preserves most of the information for sufficient text comparison [5] and incorporates the file length of the compressed texts. For further technical details we refer to [17].

To apply the median variant of our proposed method MFNG we use the dissimilarity matrix and compare the results to Median FCM [17], M-PCFM [12], and Fuzzy Affinity Propagation (FAP) [18].

In Fig. 1b the fuzzy cluster assignments obtained with $a = 0.8$, $b = 0.2$, $m = 1.2$, $n = 1.5$, and $gamma = 1$ are shown. Again we were able to verify the two clusters referring to the two therapy phases. The prototypes representing sessions 14 and 18 are identical with the ones obtained in former experiments.

6   Conclusion

In the presented paper we extended Fuzzy Neural gas to a more advanced version incorporating probabilistic and possibilistic assignments by adopting the idea presented in [9] for FCM. Further we provided specific update rules to handle vectorial data as well as relational and median data. In two experimental settings we demonstrated the capability of this method and raised some questions. In further studies we have to examine the influence of the various parameters more closely and gain an in-depth understanding of the typicalities.

References

1. Bezdek JC (1980) A convergence theorem for the fuzzy isodata clustering algorithms. IEEE Trans Patt Anal Mach Intell 2(1):1–8
2. Dunn JC (1973) A fuzzy relative of the ISODATA process and its use in detecting compact well-separated clusters. J Cybern 3(3):32–57
3. Ball GH, Hall DJ (1965) Isodata an iterative method of multivariate data analysis and pattern classification
4. Kohonen T (1990) The self-organizing map. Proc IEEE 78(9):1464–1480
5. Martinetz TM, Berkovich SG, Schulten KJ (1993) 'Neural-gas' network for vector quantization and its application to time-series prediction. IEEE Trans Neural Netw 4(4):558–569
6. Karayiannis NB, Bezdek JC (1997) An integrated approach to fuzzy learning vector quantization and fuzzy c-means clustering. IEEE Trans Fuzzy Syst 5(4):622–628
7. Geweniger T, Fischer L, Kaden M, Lange M, Villmann T (2013) Clustering by fuzzy neural gas and evaluation of fuzzy clusters. Comp Int Neurosc
8. Krishnapuram R, Keller JM (1993) A possibilistic approach to clustering. IEEE Trans Fuzzy Syst 1(2):98–110
9. Pal NR, Pal K, Keller JM, Bezdek JC (2005) A possibilistic fuzzy c-means clustering algorithm. IEEE Trans Fuzzy Syst 13(4):517–530
10. Villmann T, Kaden M, Nebel D, Bohnsack A (2016) Similarities, dissimilarities and types of inner products for data analysis in the context of machine learning, pp 125–133, June 2016
11. Ball GH, Hall DJ (1967) A clustering technique for summarizing multivariate data. Behav. Sci. 12(2):153–155
12. Geweniger T, Villmann T (2017) Relational and median variants of possibilistic fuzzy C-means. In: WSOM+ 2017 12th international workshop on self-organizing maps, pp. 207–213
13. Hammer B, Hasenfuss A (2007) Relational neural gas. In: Hertzberg J, Beetz M, Englert R (eds) KI 2007: Advances in Artificial Intelligence. LNAI 4667. Springer, Heidelberg, pp 190–204
14. Pekalska E, Duin RPW (2005) The dissimilarity representation for pattern recognition: foundations and applications. In: Machine perception and artificial intelligence. World Scientific Publishing Company, December 2005
15. Freeman PR (1983) The secretary problem and its extensions: a review. Int. Stat Rev 51(2):189–206
16. Villmann T, Liebers C, Geyer M (2003) Untersuchung der psycho-physiologischen Interaktion von Patient und Therpeut im Rahmen für psychodynamische Einzeltherapien und informationstheoretische Auswertung. Psychotherapeutische Reflexionen gesellschaftlichen Wandels, pp 305–319
17. Geweniger T, Zühlke D, Hammer B, Villmann T (2010) Median fuzzy c-means for clustering dissimilarity data. Neurocomputing 73(7–9):1109–1116
18. Geweniger T, Zühlke D, Hammer B, Villmann T (2009) Fuzzy variant of affinity propagation in comparison to median fuzzy c-means. In: Advanced in self-organizing maps - proceedings of WSOM, LNCS 5629. Springer, pp 72–79

Autoencoders Covering Space
as a Life-Long Classifier

Rudolf Szadkowski[✉], Jan Drchal, and Jan Faigl

Department of Computer Science, Faculty of Electrical Engineering,
Czech Technical University in Prague, Technická 2, 166 27 Prague 6, Czech Republic
{szadkrud,drchajan,faiglj}@fel.cvut.cz
https://comrob.fel.cvut.cz/

Abstract. A life-long classifier that learns incrementally has many challenges such as concept drift, when the class changes in time, and catastrophic forgetting when the earlier learned knowledge is lost. Many successful connectionist solutions are based on an idea that new data are learned only in a part of a network that is relevant to the new data. We leverage this idea and propose a novel method for learning an ensemble of specialized autoencoders. We interpret autoencoders as manifolds that can be trained to contain or exclude given points from the input space. This manifold manipulation allows us to implement a classifier that can suppress catastrophic forgetting and adapt to concept drift. The proposed algorithm is evaluated on an incremental version of the XOR problem and on an incremental version of the MNIST classification where we achieved 0.9 accuracy which is a significant improvement over the previously published results.

Keywords: Incremental learning · Life-long learning · Auto-encoder · Catastrophic forgetting · Concept drift

1 Introduction

Incremental learning is important for domains where the incoming data must be continually integrated into a classifier. Such classifier is expected to train and predict the incoming data as long as it is in operation; hence, a life-long classifier. In a life-long time scale, we expect that the target classes can change in time, where such change is called a concept drift. Moreover, there is also a problem of catastrophic forgetting: as the classifier is continually trained it can forget some knowledge it learned earlier. Both the concept drift and catastrophic forgetting are the main challenges of incremental learning [1].

In neural networks, concept representations are distributed throughout the network weights [2]. In such distributed representations, during each training iteration, a slight change of a single parameter can lead to changes in multiple concepts at once. During life-long learning, these changes can accumulate, and concepts might get forgotten [1]. The existing approaches to the concept drift in

© Springer Nature Switzerland AG 2020
A. Vellido et al. (Eds.): WSOM 2019, AISC 976, pp. 271–281, 2020.
https://doi.org/10.1007/978-3-030-19642-4_27

neural networks are based on preferring updates only a part of the network that is relevant to the new data increments [3,4]. Probably, the most straightforward implementation of this idea is to use an ensemble of predictors [5].

Our proposed method is based on an ensemble of classifying autoencoders. An autoencoder is a neural network trained to approximate an identity transformation of the input. Training of the autoencoder is unsupervised because the loss, also called the reconstruction error, is defined as a difference between the input and its transformed image. This reconstruction error of the autoencoder has an alternative in anomaly detection. An anomalous input is detected by the autoencoder when the reconstruction error exceeds a given threshold [6,7]. A related application of the same method is the novel class detection [8,9] where the task is to detect unknown classes. In both applications, the anomaly and novel class detection, the autoencoder divides the input space into two regions according to the reconstruction error. Here, we call the region with low reconstruction error θ-wrap, where θ represents the maximum reconstruction error threshold. Hence, each θ-wrap can be understood as a manifold which covers most of the samples on which the corresponding autoencoder was trained.

In this paper, we take advantage of θ-wrap interpretation of the autoencoder and present an incremental supervised training algorithm that suppresses the catastrophic forgetting and adapts the classifier to concept drift. The classifier is implemented as an autoencoder ensemble, where each autoencoder is trained to cover its respective class by a θ-wrap. We present two methods to train θ-wrap that are referred COVER and UNCOVER. The methods train the θ-wrap to cover or uncover the given set of points while suppressing the catastrophic forgetting. The UNCOVER method is used for the case of the concept drift when a class starts intersecting with θ-wrap unrelated to that class. The conflicting θ-wrap is then trained to uncover the intersection with the UNCOVER method. Both UNCOVER and COVER methods are the building blocks for the incremental training algorithm for the whole autoencoder ensemble. For each given sample batch, the algorithm uncovers given samples by θ-wraps to which the samples do not belong, and then it covers the samples by their respective θ-wrap.

The proposed algorithm is designed under relaxing assumptions where classes are "crisp" manifolds that are sampled without noise and UNCOVER and COVER methods always work perfectly. We examine the robustness of the proposed algorithm in two dimensions, incrementally learning the XOR dataset in which we simulate the concept drift. We also test how the ensemble of classifying autoencoders withstand incremental training on the MNIST dataset.

2   Related Work

Autoencoder ensembles have been employed for detecting outliers or anomalies [10,11] where all autoencoders are trained to classify the same task, but each autoencoder is trained to its specific task [12]. As a new task appears, the algorithm builds a new autoencoder on this task in order to capture and store its representative information. This storage helps the proposed algorithm to suppress the catastrophic forgetting that is enforced structurally, where the concepts are stored in their subnetworks, and thus the new concept does not influence the previous ones. This structural solution to catastrophic forgetting is also used in our proposed algorithm. We additionally introduce a resampling algorithm that suppresses catastrophic forgetting within each subnetwork. The structural separations of concepts are not the only solution to catastrophic forgetting. Recently proposed Elastic Weight Consolidation (EWC) algorithm [3] selectively freezes the neural network weights that are important to the previously learned task/class. The EWC is compared to other approaches tackling the catastrophic forgetting in [13], where the authors report the EWC provides the best results. However, neither the algorithm [12] nor [3] can successfully deal with the concept drift problem where classes change in time. In particular, when a class is suddenly relabeled. In such a case, a classifier needs to implement a forgetting mechanism. We present such a mechanism in the following section.

3   Analysis

Let $X = (0,1)^N$ be the N-dimensional input space and let $C_i \subset X$ be the i-th class, where $i \in \{1 \dots M\}$. We assume that classes C_i form mutually disjoint manifolds. The manifold C_i is unknown, but at the time t, we get a finite pointset of class-samples $S_i^t \subset C_i$. We can only work with the given class-samples S_i^t at the time t and, moreover, any class C_i can change at the time $(C_i^t \neq C_i^{t'})$ due to the concept drift. We aim to find such a classifier $F^t : X \to \{1 \dots M\}$ at t that

$$\forall i \in \{1 \dots M\}, \forall \boldsymbol{x} \in C_i^t : F^t(\boldsymbol{x}) = i. \tag{1}$$

We propose to use M trainable manifolds $P_i(\theta) \subset X$, which we call θ-wraps, to mimic their respective classes. Ideally, each class C_i is wrapped by its respective θ-wrap $P_i(\theta)$, while all wraps are mutually disjoint:

$$\forall i \in \{1 \dots M\} : C_i \subset P_i(\theta), \tag{2}$$
$$\forall i,j \in \{1 \dots M\}, i \neq j : P_i(\theta) \cap P_j(\theta) = \emptyset. \tag{3}$$

We propose the following straightforward algorithm to reach this desired state. At each iteration t, the given class-sample set $S_i^t \subset C_i^t$ is firstly uncovered by all θ-wraps $P_j^t(\theta); j \neq i$ and then covered by $P_i^t(\theta)$ θ-wrap (see Algorithm 2). After the training, we say P covers S and P uncovers S that means $S \subset P$ and $P \cap S = \emptyset$, respectively. The classification can be then realized by the function $F(\boldsymbol{x}) = \arg\max_i [\![\boldsymbol{x} \in P_i(\theta)]\!]$. The proposed implementation of θ-wrap $P_i(\theta)$ and its training algorithm follows.

4   Method

We propose to implement θ-wraps with autoencoders, where θ-wrap $P_i(\theta)$ is defined by its underlying autoencoder $g_i : X \to X$. An autoencoder is usually

---

Algorithm 1. Train autoencoder with positive and negative samples.

Variables g: autoencoder; S_+, S_-: positive and negative samples; θ: threshold;
E: max epoch; $\mathcal{J}$: cost function;
Result g: updated autoencoder;

1: **function** TRAIN(g, S_+, S_-, θ)
2:　　$g^0 \leftarrow g$
3:　　**for** $e = 1$ **to** E **do**
4:　　　　$S_-^e \leftarrow \{s | \mathcal{L}(s, g^{e-1}(s)) \leq \theta; s \in S_-\}$　　　▷ Filters S_- points with low $\mathcal{L}$.
5:　　　　$S_+^e \leftarrow \{s | \mathcal{L}(s, g^{e-1}(s)) \geq \theta; s \in S_+\}$　　　▷ Filters S_+ points with high $\mathcal{L}$.
6:　　　　$\epsilon^e \leftarrow \mathcal{J}(g^{e-1}, S_+^e, S_-^e)$
7:　　　　$g^e \leftarrow$ gradient-descent(g^{e-1}, ϵ^e)
8:　　　　**if** $\forall s \in S_+ : \mathcal{L}(s, g^e(s)) < \theta$ **and** $\forall s \in S_- : \mathcal{L}(s, g^e(s)) > \theta$ **then**
9:　　　　　　**break**
10:　　　**end if**
11:　　**end for**
12:　　$g \leftarrow g^e$
13: **end function**

---

trained to reconstruct some target subset $A \subset X$, i.e., $\forall a \in A : g_i^*(a) = a$. The reconstruction error is measured by the Euclidean distance $\mathcal{L}(x, g(x)) = ||x - g(x)||$. θ-wrap corresponding to the i-th autoencoder is defined as $P_i(\theta) = \{x | \mathcal{L}(x, g_i(x)) < \theta, \forall x \in X\}$, i.e., it is a manifold where all distances between points x and images $g_i(x)$ are smaller than some threshold $\theta \in (0, \mathcal{L}_{\max})$ where $\mathcal{L}_{\max}$ is the maximal reachable distance from the middle of the $(0, 1)^N$ space

$$\mathcal{L}_{\max} = \min_x \sup_y \mathcal{L}(x, y) = \sqrt{N}/2; x, y \in (0, 1)^N. \tag{4}$$

θ-wrap $P_i(\theta)$ is trained by training the related autoencoder g_i using the function TRAIN(g_i, S_+, S_-) depicted in Algorithm 1, where the finite sets $S_+, S_- \subset X$ are called the positive and negative samples, respectively. Ideally, the positive samples should reside inside θ-wrap $P_i(\theta)$ while the negative samples should be outside:

$$g_i^t \leftarrow \text{TRAIN}(g_i^{t-1}, S_+^t, S_-^t, \theta), \tag{5}$$

where for samples S_+^t, S_-^t, and i-th θ-wrap $P_i^t(\theta)$ it holds that

$$(S_+^t \subset P_i^t(\theta)) \wedge (S_-^t \cap P_i^t(\theta)) = \emptyset, \tag{6}$$

and where t is the iteration index. The cost function $\mathcal{J}(g, S_+, S_-) \in \mathbb{R}^+$ in Algorithm 1 gives the cost of the autoencoder g w.r.t. the positive and negative sets S_+, S_-

$$\mathcal{J}(g, S_+, S_-) = \frac{1}{|S_+|} \sum_{s \in S_+} \mathcal{L}(s, g(s))^2 + \frac{1}{|S_-|} \sum_{s \in S_-} (\mathcal{L}_{\max} - \mathcal{L}(s, g(s)))^2. \tag{7}$$

The class is then predicted by

$$F(\boldsymbol{x}) = \arg\min_i \mathcal{L}(\boldsymbol{x}, g_i(\boldsymbol{x})). \qquad (8)$$

Let $\text{SAMP}(g_i)$ be a function that gives a set of random wrap-samples: it creates a set $\hat{X}_i$ of uniform random samples of X, and keeps only those that are in θ-wrap, $\text{SAMP}(g_i) = \{\boldsymbol{s}|\mathcal{L}(\boldsymbol{s}, g_i(\boldsymbol{s})) < \theta, \boldsymbol{s} \in \hat{X}_i\}$, where $|\hat{X}_i| = S_{\max}$. Ideally, to satisfy the properties (2) and (3), it would be sufficient to set $\forall i : g_i^* \leftarrow \text{TRAIN}(g_i^t, S_i, \hat{P}_i^t, \theta)$, where $S_i, \hat{P}_i^t$ are large enough finite point-sets sampled from C_i and $\cup_{j \neq i} \text{SAMP}(g_j)$, respectively. However, in our case, the class-sample set $S_i^t \subset C_i$ can contain just a few samples (one being the worst case scenario). If the class-sample set S_i^t does not represent its class C_i well (e.g., it contains just one sample) the catastrophic forgetting may take place. In the context of θ-wraps, the catastrophic forgetting is an event when after training θ-wrap $P_i^{t-1}(\theta)$, new θ-wrap $P_i^t(\theta)$ ceases to cover a part of C_i that was covered by its predecessor, i.e., $(C_i \cap P_i^{t-1}(\theta)) \not\subset P_i^t(\theta)$. θ-wrap $P_i^{t-1}(\theta)$ is not preserved by Algorithm 1 because the gradient descent method optimizes the cost that is calculated only from the given samples S_-^t, S_+^t at the time t. We propose to address the problem by adding wrap-samples $\text{SAMP}(g^{t-1})$ to the positive samples and define the method $g_i^t \leftarrow \text{COVER}(g_i^{t-1}, S_i^t, \theta)$ as

$$\text{COVER}(g_i^{t-1}, S_i^t, \theta) := \text{TRAIN}(g_i^{t-1}, S_i^t \cup \text{SAMP}(g_i^{t-1}), \bigcup_{j \neq i} \text{SAMP}(g_j^{t-1}), \theta). \qquad (9)$$

As discussed above, the algorithm needs to implement a forgetting mechanism to deal with the concept drift. An abrupt concept drift of class C_i is a discrete event when a class abruptly changes $C_i^t \neq C_i^{t-1}$ [13]. The class change that hurts the performance of the classifier happens when C_i^t intersects with θ-wrap of other class $P_j^{t-1}(\theta), j \neq i$. To keep the properties (2) and (3), the θ-wrap $P_j^{t-1}(\theta)$ must first uncover the intersection $P_j^{t-1}(\theta) \cap C_i^t; j \neq i$ before the θ-wrap $P_i^{t-1}(\theta)$ covers

---

Algorithm 2. Update autoencoders with the labeled samples D.

> **Variables** $\{g_i\}$: collection of M autoencoders, $i \in \{1 \ldots M\}$;
> $D = \{(\boldsymbol{s}, k)\}$: set of labeled samples where k indicates class of sample $\boldsymbol{s}$;
> θ: threshold; $\mathcal{L}$: metric function;
> **Result** $\{g_i\}$: updated autoencoders;

1: **function** UPDATE($\{g_i\}$, D, θ)
2:     **for** $i = 1$ **to** M **do**
3:         $S_i \leftarrow \{\boldsymbol{s}|k = i; (\boldsymbol{s}, k) \in D\}$
4:         **for** $j = 1$ **to** M **where** $j \neq i$ **and** $\exists \boldsymbol{s} \in S_i : \mathcal{L}(\boldsymbol{s}, g_j(\boldsymbol{s})) < \theta$ **do**
5:             $g_j \leftarrow \text{UNCOVER}(g_j, S_i, \theta)$ ▷ See (10).
6:         **end for**
7:         $g_i \leftarrow \text{COVER}(g_i, S_i, \theta)$ ▷ See (9).
8:     **end for**
9: **end function**

this intersection. Let class-sample set $\hat{S}_i^t$ be a subset of such conflicting intersection $P_j^{t-1}(\theta) \cap C_i^t; j \neq i$, then, before we cover $\hat{S}_i^t$ with $P_i^t(\theta)$ using (9) we must train $P_j^t(\theta)$ to uncover $\hat{S}_i^t$. We define the method $g_j^t \leftarrow \text{UNCOVER}(g_j^{t-1}, \hat{S}_i^t, \theta)$, that trains g_j to uncover $\hat{S}_i^t$, as

$$\text{UNCOVER}(g_j^{t-1}, \hat{S}_i^t, \theta) := \text{TRAIN}(g_j^{t-1}, \text{SAMP}(g_j^{t-1}) - \mathcal{B}(\hat{S}_i^t, \varepsilon),$$
$$\bigcup_{k \neq j} \text{SAMP}(g_k^{t-1}) \cup \hat{S}_i^t, \theta), \qquad (10)$$

where $\mathcal{B}(\hat{S}_i^t, \varepsilon) = \bigcup_{s \in \hat{S}_i^t} \{x | \mathcal{L}(s, x) < \varepsilon; x \in X\}$ is ε-neighbourhood of the whole set $\hat{S}_i^t$. With cost function (7), sample set covering (9), and uncovering (10) Algorithms 1 and 2 are completely defined.

5   Results

The proposed approach has been empirically evaluated on two datasets. First, in Sect. 5.1 we study how the algorithm handles both the catastrophic forgetting and concept drift in the simple XOR problem. Then, in Sect. 5.2 we deploy the method in MNIST dataset [14], where the proposed approach demonstrates surprising benefits of the ensemble of almost-independent autoencoders. The utilized autoencoder is composed of four hidden ReLU layers with the sizes 800, 400, 400, and 800. The input and output layers have N units each where the output layer is composed of sigmoid units. The autoencoder is overcomplete for experiments in Sect. 5.1, but cost (7) prevents the autoencoder to become an identity function. ADAM with default parameters [15] is utilized as the `gradient-descent` method in Algorithm 1. The subtraction parameter from (10) is set to $\varepsilon = 0.01$. All hyperparameters were selected empirically.

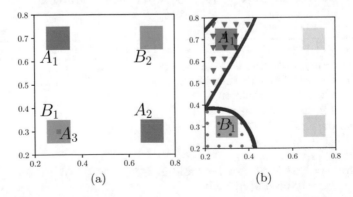

(a)   (b)

Fig. 1. (a) Squares used in the classification scenarios. Green squares are part of C_1 and orange squares are part of C_2, where $A_3 \subset B_1$. (b) Initially, the classifier is learned only on A_1 and B_1 squares. A_1 is covered by $P_1(\theta)$ (blue) and B_1 is covered by $P_2(\theta)$ (red). θ-wraps are visualized by taking samples from θ-wrap.

5.1   Incremental Training of Binary Classification

The principle of the proposed method listed in Algorithm 2 is demonstrated in three classification scenarios where two autoencoders are trained to classify classes $C_1, C_2 \subset (0,1)^2$. The input space dimension is $N = 2$, training hyperparameters are set to $\theta = \mathcal{L}_{max}/2$, and the maximum number of epochs $E = 1000$. The two classes C_1, C_2 are composed of squares A_1, A_2, A_3, B_1, and B_2[1] depicted in Fig. 1a. Each scenario starts with $C_1 = A_1$ covered by $P_1(\theta)$ and $C_2 = B_1$ covered by $P_2(\theta)$ (see Fig. 1b). The `cat-forget` scenario demonstrates the catastrophic forgetting where the classes are extended to $C_1 = A_1 \cup A_2$ and $C_2 = B_1 \cup B_2$. The class-sample batches D^t of size 100 are sampled only from $A_2 \cup B_2$ which can result into "forgetting" the boxes A_1 and B_1. We set the number of the maximum wrap-samples $S_{max} = 0$ to demonstrate the catastrophic forgetting. The `no-forget` scenario has the same setting except $S_{max} = 5000$. The difference between the scenarios can be seen in Figs. 2a and b. Finally, in the `concept-drift` scenario, the new classes are $C_1 = A_1 \cup A_3$ and $C_2 = B_1 - A_3$ to study how the method handles the concept drift. The class-sample batches D^t are of the size one, and they are sampled only from A_3, and thus P_1 tries to cover each class-sample while P_2 tries to uncover it, see Fig. 2c. Results of the statistical evaluation are listed in Table 1. Accuracies were averaged from five 10-iteration long runs and evaluated on classes $C_1 = A_1 \cup A_2$ and $C_2 = B_1 \cup B_2$, except for `concept-drift` which was evaluated on $C_1 = A_3$ and $C_2 = B_1 - A_3$.

Table 1. Classifier evaluation on XOR test data (new samples taken from respective classes). Accuracy averages taken from 5 runs, with standard deviation to show the algorithm stability.

Scenario	cat-forget	no-forget	concept-drift
Accuracy	0.62 ± 0.11	0.94 ± 0.09	0.78 ± 0.09

5.2   Catastrophic Forgetting Evaluation on MNIST

In the second scenario, we perform the evaluation introduced in [13] which measures the classifier performance on the incremental learning of the MNIST dataset. The dataset D is divided into D_{even} and D_{odd} containing even and odd number characters, respectively. The algorithm trains on D_{even} first and then on D_{odd} (with no access to D_{even}). During the training, the overall accuracy on the test set is measured. Each class-sample of MNIST dataset corresponds to a point in $[0,1]^{784}$ space which is scaled to $[0.15, 0.85]^{784}$ space to avoid plateaus during training sigmoid output units. The training hyperparameters are set to $\theta = \mathcal{L}_{max}/10$, and the maximal epoch number $E = 1000$. The maximum wrap-samples size is $S_{max} = 0$ because the sampling method is not suitable for such a

---

[1] The centers of 0.1×0.1 large squares A_1, A_2, B_1, and $B2$ are (0.3,0.7), (0.7,0.3), (0.3,0.3), and (0.7,0.7), respectively. A_3 is 0.02×0.02 large centered at (0.3,0.3).

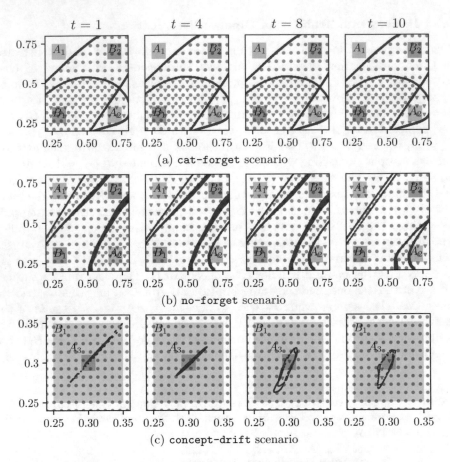

Fig. 2. Cover evolution in catastrophic forgetting, normal and concept drift scenarios. Initially (for $t = 0$), $P_1(\theta)$ (blue) and $P_2(\theta)$ (red) cover A_1 and B_1, respectively, see Fig. 1b. In (a) and (b), the goal is to cover A_2 and B_2 without uncovering A_1 and B_1. We train on batches containing 100 class-samples of A_2, B_2. (a) $S_{\max}$ is zero which leads to uncovered A_1 and B_1 being part of the both θ-wraps. (b) $S_{\max} = 5000$ and both θ-wraps managed to keep their respective squares A_1, B_1 covered. (c) A concept drift happened when the new classes became $C_1 = A_1 \cup A_3$ and $C_2 = B_1 - A_3$. We can see the detail of A_3 (green rectangle) which we want to uncover by $P_2(\theta)$ and cover by $P_1(\theta)$. We train on batches containing one class-sample taken from A_3. During ten iterations, $P_2(\theta)$ (red) formed a cavity which is covered by $P_1(\theta)$ at $t = 10$. Both θ-wraps are visualized with thick border and interior filled with sparse markers.

high-dimensional input, and thus there is no mechanism preventing catastrophic forgetting except for the fact that the autoencoders train almost independently on each other (see Algorithm 2). There are $M = 10$ autoencoders training in their respective classes. Each iteration, we sample a random batch of the size 100 from the current dataset D. For the first $T_1 = 300$ iterations, only the odd numbers D_{odd} are trained, and the next $T_2 = 300$ iterations, only the even numbers

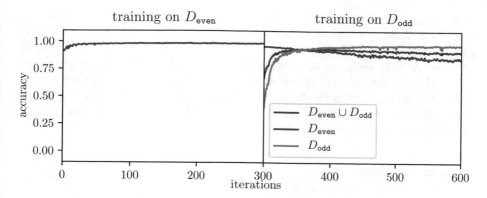

Fig. 3. Evolution of the accuracy during training on the MNIST dataset with even D_{even} and odd D_{odd} numbers. For each iteration, 100 random class-samples is taken from either D_{even} or D_{odd} and trained. On average, the UPDATE function processes 59.87 epochs at each iteration of Algorithm 2.

D_{even} are trained. Here, we differ from [13] where the training parameters are $T_1 = T_2 = 2500$, and thus the algorithms are exposed to catastrophic forgetting for more iterations. However, it depends whether one iteration is equivalent to one parametric update because our algorithm performed 66.82 updates per iteration on average (calculated from six independent runs). Therefore about 40 092 updates are performed in the total, and thus the proposed algorithm withstands the catastrophic forgetting for even more updates than, e.g., EWC [3] with one update per iteration and thus updated just 5000 times. An example of the accuracy evolution is shown in Fig. 3.

The achieved accuracy on the MNIST testing dataset for six experimental runs of 600 iterations is 0.90 ± 0.02. For a rough comparison with existing approaches, the best result reported in [13] on even-odd numbers learning task is the accuracy 0.64, which was achieved using the EWC algorithm [3].

5.3   Discussion

The idealized property (3) does not perfectly hold as can be seen in Fig. 2, where θ-wraps intersects for NO-FORGET and CONCEPT-DRIFT scenarios. The intersection is caused by TRAIN method (Algorithm 1) which does not always ensure the property (6), where the positive samples should reside inside while the negative samples outside of the trained θ-wrap. θ-wrap then loses some positive samples or keeps covered some negative samples which can lead to intersections with other θ-wraps. However, these intersections seem to become smaller with increasing iterations, hinting the robustness of the proposed algorithm.

The bottleneck of the proposed algorithm is the SAMP method, which does not scale well with the increasing dimensionality[2]. For our future work, we aim

---

[2] It is apparent from Sect. 5.2, as sampling a point that lies in θ-wrap is roughly equivalent to getting an actual image of digit by randomly sampling 28×28 pixels.

to explore possibilities on how to implement SAMP method more efficiently. However, high-dimensionality might be the reason why the proposed classifier has surprisingly good results. Most images in MNIST have high contrast, and thus the samples are clustered close to the corners of the 784-dimensional unit hypercube. These clusters are covered by autoencoders with θ-wraps, i.e., manifolds of the Euclidean space. Euclidean distances between corners of the high-dimensional hypercube can be quite large (the maximum is $\sqrt{784} = 28$), θ-wraps are probably less likely to reach across such distances and cover other classes.

6 Conclusion

In this paper, we provide an analysis of the properties of classifying autoencoders in the context of the concept drift and catastrophic forgetting, where the autoencoders are represented as trainable manifolds called θ-wraps. With the θ-wrap representation and classes defined as subspaces of the input space, we describe the training, catastrophic forgetting, and concept drift using set operations. This allows us to design an algorithm that trains the ensemble of autoencoders to incrementally cover their respective classes with their respective θ-wraps. Even though the algorithm is designed under relaxing assumptions, the results support its feasibility and relatively robustness for two-dimensional input space, and the proposed approach produces competitive results on incremental learning of the MNIST dataset.

Acknowledgments. This work was supported by the Czech Science Foundation (GAČR) under research project No. 18-18858S.

References

1. Gepperth A, Hammer B (2016) Incremental learning algorithms and applications. In: European symposium on artificial neural networks (ESANN), pp 357–368
2. Hinton GE, McClelland JL, Rumelhart DE (1986) Distributed representations. In: Parallel distributed processing: explorations in the microstructure of cognition, vol. 1: foundations, pp 77–109. MIT Press, Cambridge
3. Kirkpatrick J et al (2017) Overcoming catastrophic forgetting in neural networks. Proc Nat Acad Sci 114(13):3521–3526
4. Li Z, Hoiem D (2016) Learning without forgetting, *CoRR*, vol. abs/1606.09282
5. Krawczyk B, Minku LL, Gama J, Stefanowski J, Woźniak M (2017) Ensemble learning for data stream analysis: a survey. Inf Fusion 37:132–156
6. Chandola V, Banerjee A, Kumar V (2009) Anomaly detection: a survey. ACM Comput Surv 41(3):15:1–15:58
7. Borghesi A, Bartolini A, Lombardi M, Milano M, Benini L (2018) Anomaly detection using autoencoders in high performance computing systems, *CoRR*, vol. abs/1811.05269
8. Marchi E, Vesperini F, Squartini S, Schuller B (2017) Deep recurrent neuralnetwork-based autoencoders for acoustic novelty detection. Comput Intell Neurosci 2017

9. Mustafa AM, Ayoade G, Al-Naami K, Khan L, Hamlen KW, Thuraisingham B, Araujo F (2017) Unsupervised deep embedding for novel class detection over data stream. In: IEEE international conference on big data, pp 1830–1839
10. Chen J, Sathe S, Aggarwal C, Turaga D (2017) Outlier detection with autoencoder ensembles. In: SIAM international conference on data mining, pp 90–98
11. Mirsky Y, Doitshman T, Elovici Y, Shabtai A (2018) Kitsune: an ensemble of autoencoders for online network intrusion detection, *CoRR*, vol. abs/1802.09089
12. Triki AR, Aljundi R, Blaschko MB, Tuytelaars T (2017) Encoder based lifelong learning, *CoRR*, vol. abs/1704.01920
13. Pfülb B, Gepperth A, Abdullah S, Kilian A (2018) Catastrophic forgetting: still a problem for DNNs. In: International conference on artificial neural networks (ICANN), pp 487–497
14. LeCun Y, Cortes C (2010) MNIST handwritten digit database. http://yann.lecun.com/exdb/mnist/. Accessed 29 Jan 2019
15. Kingma DP, Ba J (2014) Adam: a method for stochastic optimization, *CoRR*, vol. abs/1412.6980

Life Science Applications

Progressive Clustering and Characterization of Increasingly Higher Dimensional Datasets with Living Self-organizing Maps

Camden Jansen[1,2] and Ali Mortazavi[1,2(✉)]

[1] Developmental and Cell Biology,
University of California Irvine, Irvine, CA 92697, USA
ali.mortazavi@uci.edu
[2] Center for Complex Biological Systems, University of California Irvine,
Irvine, CA 92697, USA

Abstract. Long-lived consortiums in genomics generate massive highly-dimensional datasets over the course of many months or years with substantial blocks of data added over time. Algorithms designed to characterize and cluster this data are designed to run once on a dataset in its entirety, and thus, any analysis of these collections must be entirely re-done from scratch every time a new block of data is added. We describe a novel progressive clustering approach using a variation of the self-organizing map (SOM) algorithm, which we call the Living SOM. Our software package is capable of clustering highly-dimensional data with all of the power of regular SOMs with the added benefit of incorporating additional datasets as they become available while maintaining the initial structure as much as possible. This allows us to evaluate the impact of the new datasets on previous analyses with the potential to keep classifications intact if appropriate. We demonstrate the power of this technique on a collection of gene expression experiments done in an embryonic time course of development for mouse from the ENCODE consortium.

Keywords: Progressive clustering · Genomics · Self-organizing maps

1 Introduction

Self-Organizing Maps (SOMs) [1] and further metaclustering [2] have been shown to effectively cluster highly-dimensional data for characterization [3, 4]. However, like other unsupervised learning algorithms, they were designed to be run on a set of data in its entirety and must be re-trained every time new data is available. In the field of genomics, it is common for large consortiums such as the Encyclopedia of DNA Elements (ENCODE) to generate huge collections of highly-dimensional data such as new RNA-seq experiments on the same genes in new tissues or ChIP-seq experiments on the same genome regions with new transcription factors over the course of many

A. Vellido et al. (Eds.): WSOM 2019, AISC 976, pp. 285–293, 2020.
https://doi.org/10.1007/978-3-030-19642-4_28

years with big blocks being released over time. The nature of this data might be interpreted as additional dimensions added to a fixed number of points in the existing dataset rather than new data points added in a similarly sized N-dimensional space. Unfortunately, unsupervised learning algorithms do not typically support dynamic datasets that change in dimensionality, as all of the down-stream classification has to be re-done after each release.

After each data release, it would be ideal to be able to use the previous analysis to help train a new clustering. However, simply using the previous SOM unit locations as the initialization point is problematic. For example, adding a dimension can potentially disassociate clustered data points. This can cause the units sitting in those clusters to settle halfway between their original associated genes, becoming stuck in local minima, and generating a sub-optimal clustering (Fig. 1).

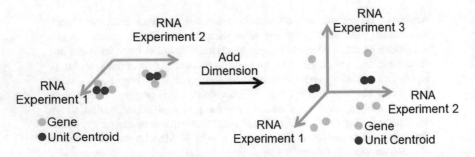

Fig. 1. An illustration of a potential effect of adding a dimension to an existing analysis. Adding a dimension can possibly disassociate clustered data points, which can cause the units sitting in those clusters to settle halfway between their original associated genes and become stuck.

Here, we present a novel method that we call the Living SOM (LSOM) that allows dimensions to be inserted, one at a time, into an already trained SOM while maintaining the original topology as much as possible. Its purpose is not only to speed up learning over complete re-training, but also to allow for the possible preservation of the down-stream classification. This method for data insertion is fast and highly reactive to SOM units becoming stuck in local minima during the addition of dimensions. We present a full comparison between this algorithm and the standard Kohonen SOM in terms of clustering reproducibility at both the SOM unit and metacluster level using a highly-dimensional genomic dataset from the ENCODE consortium. Finally, we show that drops in reproducibility after certain data insertions reveal structural novelty in that data and could be used to detect either biological novelty or erroneous data.

The results of this paper are organized as follows: Sect. 2 introduces the LSOM algorithm. Section 3 contains comparisons between Kohonen SOMs and the LSOM, both in regular usage and after simulating a data release. Section 4 details an exhaustive analysis of the importance of data insertion order. Finally, Sect. 5 contains a discussion of the results.

2   The Living SOM

There have been many modifications to the standard Kohonen SOM to cluster various modalities of data over the years. For example, modifications have been developed for streaming data where data points are added at inconsistent time steps to the overall pool, such as the Ubiquitous SOM [5]. In that algorithm, a more-aggressive organizing step is added to the standard SOM to solve a similar issue to the one described in Fig. 1, in which SOM units would get stuck between several data points as the streaming data would move to another part of data space. We began development of the LSOM from this algorithm because many of the issues with adding dimensions are similar to those from streaming data, and we use similar metrics for triggering the organizing step and computing the learning rate and radius. As such, dimensions are added one at a time in a similar manner as observations being added by streaming data.

The main metric for triggering the organizing step is when the "average drift," which is a weighted average between the average quantization error and the average neuron utility, exceeds a limit for a number of time steps. This limit is set to the average drift at the end of the organizing step or at the beginning of training and is kept between insertions. This step lasts for 1 epoch, or one pass through each data point, allowing them to influence the new position of the units. Afterwards, the LSOM returns to the beginning of regular learning. The organizing step is rare in practice, with 0–1 occurring in the LSOMs built in Sect. 3.

There are two sets of learning parameters in the LSOM that are active during the different states. During the ordering state, the learning rate and radius are set to an aggressive level dependent only on ordering time [5]. Conversely, in the default learning state, the learning parameters are dependent on learning time (to force convergence) and the current drift level compared to the drift limit (Eqs. 1 and 2 below). See the Algorithm 1 and the Parameters and Formulae section at the end for details.

Algorithm 1: Living SOM Algorithm (per insertion)

1: Input previous set of SOM units. $U_{old} \in R^{(k-1) \times (n \times m)}$, where n and m are the SOM rows and columns respectively and k is the new total number of dimensions.

2: Input previous training matrix. $M_{old} \in R^{(k-1) \times o}$, where o is the number of data points.

3: Input new vector of observations. v.

4: Input previous drift limit, d_0.

5: Input previous number of faults, f.

5: Create new training matrix $M \in R^{k \times o}$ by combining M_{old} and v by row.

6: Create new set of SOM units, $U \in R^{k \times (n \times m)}$, by adding a 0 to each unit in U_{old}.

7: Set variables, $g = true$, $i = 1$

8: Randomly reorder the rows in M.

9: **while** g **do**

10:            **if**$(f >= o)$ **do**

11:                        Perform standard SOM algorithm on organizing parameters for 1 epoch **(Organizing Step)**

12:                        set $i = 1$

13:                        Calculate drift, d

14:                        Set drift limit, $d_0 = d$

15:            Find closest unit u in U to $M[\,,i]$

16:            Calculate drift, d

17:            Calculate learning radius and learning rate

18:            **if**(current drift $> d_0$) **then**

19:                        f++

20:            Perform update step on unit u

21:            **if**(learning radius < 1) **then**

22:                        $g=false$

23:            i++

---

3 Clustering Comparisons to Kohonen SOMs

The LSOM was designed with datasets from genomic consortiums such as ENCODE in mind. Thus, to test the performance of LSOMs compared to Kohonen SOMs, we selected a set of gene expression data from a developmental time course done in mouse by the ENCODE consortium (Fig. 2). In this context, the data points correspond to the genes and the dimensions correspond to the experimental tissue-timepoint combinations. This time course was chosen due to its high quality and the high variety of the biological samples. Also, all of these experiments were done by a single lab, so batch effects should be less prevalent. The gene expression values for the first replicate of each experiment were downloaded from the ENCODE portal [6] and built into a large training matrix containing 69,691 gene expression measurements per experiment.

Mouse ENCODE embryonic time course RNA-seq datasets

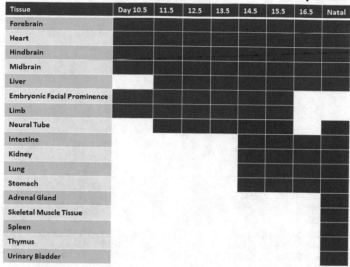

Tissue	Day 10.5	11.5	12.5	13.5	14.5	15.5	16.5	Natal
Forebrain								
Heart								
Hindbrain								
Midbrain								
Liver								
Embryonic Facial Prominence								
Limb								
Neural Tube								
Intestine								
Kidney								
Lung								
Stomach								
Adrenal Gland								
Skeletal Muscle Tissue								
Spleen								
Thymus								
Urinary Bladder								

Fig. 2. 78 datasets chosen to test the LSOM's clustering reproducibility because of their high quality and these samples are part of a time course of related samples. The 6 Day 10.5 datasets and 6 of the 8 Heart datasets (skipping Day 11.5 and Day 13.5) were also used to test the effect of data insertion order.

To ensure that the LSOM generates comparable clusterings to Kohonen SOMs, we trained a control 40×60 Kohonen SOM on all 78 data sets over 100 trials (full individual runs) with 1000 epochs. Then, we built 100 Kohonen SOMs and 100 Living SOMs on the same data. Afterwards, metaclusters were called on each of these SOMs [2]. Finally, we calculated the Jaccard indexes [7] between the Control SOM and each of the experiment SOMs (Fig. 3A). Comparison of the distributions of these indexes did not find any significant difference between the two types (Fig. 3B). Thus, the changes made to the standard algorithm did not affect its ability to cluster highly-dimensional genomic data at the unit or metacluster level.

Next, we analyzed whether the scaffold of the LSOM would maintain its structure, and thus, have a higher reproducibility, during a simulated data release. For this analysis, we built 100 LSOMs on the datasets, each with a random one removed, which is then added into the LSOMs. Again, metaclusters were called on each of these SOMs. Finally, we calculated the Jaccard indexes for the clusterings done before and after the data insertion (Fig. 4). These indexes were significantly higher than those from Kohonen SOMs at both the unit and metaclusters level. This provides evidence that the LSOM is leveraging the prior training and that the LSOM scaffold is maintaining its structure after a data insertion as intended.

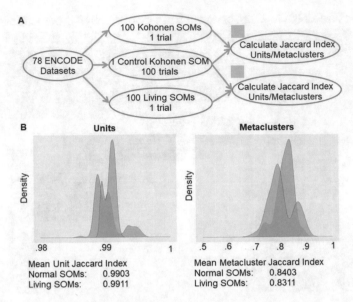

Fig. 3. (A) In order to determine whether LSOMs create similar clusterings to normal SOMs, we trained 100 Kohonen and LSOMs on the same set of data in random orders and calculated the Jaccard Index, or reproducibility, of these clusterings at the unit and metacluster scales. (B) The Jaccard indexes were very similar, indicating that LSOMs generate similar clusterings to normal SOMs.

4   Reproducibility Is Affected by Data Insertion Order

In the previous section, LSOMs were built by inserting genomic data one at a time in a random order. In the Kohonen SOM, the order of the columns in the training matrix does not matter, and we therefore wished to determine what effect, if any, data insertion has on the reproducibility of the LSOM. We built 2 sub-collections - one with six of the heart data sets and a second subset with six of the Day 10.5 data sets and built control Kohonen SOMs for each (Fig. 5A). We then built 720 LSOMs for each sub-collection, exhaustively testing every possible data insertion order and we calculated Jaccard indexes between the LSOMs and the control SOM. The distributions of these indexes show that the data insertion order does not have a significant effect on reproducibility most of the time, but we found a few low-scoring clusterings that we inspected further (Fig. 5B).

Displaying the data insertion order of the 5 clusterings of the Day 10.5 data with the worst Jaccard indexes reveals that adding the heart data set last has the potential to create a detectable decrease in the reproducibility of the LSOM (Fig. 5C). This is interesting as the heart dataset is not the most distant data when analyzing the PCA of the training matrix (Fig. 5D) with hindbrain's sample accounting for 55.8% of the variance to heart's 32.7%. However, it may signify that at those points where heart differs from the other datasets, the dataset splits a substantial number of otherwise clustered points. This suggests that the information in the heart dataset adds more novelty to the analysis.

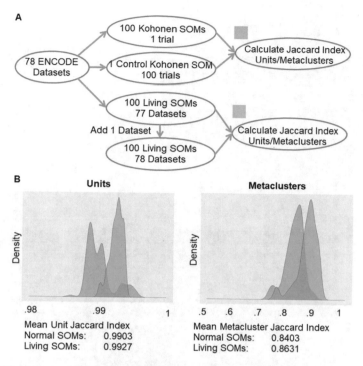

Fig. 4. (A) In order to simulate a data release, we also trained 100 living SOMs on 77 of the data sets. Then, we inserted 1 data set and compared the reproducibility at the unit and metacluster level to re-training the SOM from scratch. (B) Adding one dimension to the Living SOM was not only significantly faster than re-training a normal SOM, but the following clustering was very similar to the previous analysis, more so than re-training from scratch.

5 Discussion

In this work, we have presented a novel method, the Living SOM, for clustering datasets that grow over time without requiring a complete re-clustering on each release. LSOMs do this by using a previous SOM's units as the initialization with a gentle learning rate based on the current "drift," a weighted average between average error and neuron usage. If the average drift goes over a predetermined limit, it indicates that the LSOM has settled into local minima, and the LSOM will switch into a more aggressive re-organization mode for 1 epoch and set a new limit. Datasets are added one at a time until the data release is fully inserted.

This algorithm produces similar clusterings to the classical SOM trained on the same collection of highly-dimensional genomic datasets. Additionally, when simulating the subsequent addition of new datasets, the LSOM leverages the previous analysis to maintain the structure of the scaffold, thus generating a significantly higher reproducibility to the previous iteration compared to clustering de novo. The metaclusters in particular see a very large improvement. Finally, by showing that the order of data

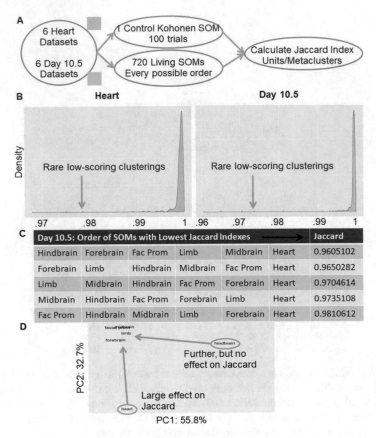

Fig. 5. (A) In order to test the effect of data insertion order on clustering, we chose (1) six datasets from heart and (2) six datasets from embryonic Day 10.5, built SOMs using every possible insertion order, and compared those to a regular SOM. (B) While the majority of LSOM runs resulted in good Jaccard indexes, there were a few orderings that were lower-scoring. (C) Visual inspection of the bottom 5 SOMs using Day 10.5 data revealed that the heart 10.5 dataset, when added last, caused this effect. (D) In a PCA of the six Day 10.5 datasets, hindbrain, not heart is the most Euclidian-distant, but the heart is the most biologically-distinct.

insertion can affect the reproducibility, we also show that the LSOM is capable of chang. To combat this issue, LSOMs could ideally be run with the most different datasets first (as calculated by hierarchical clustering), and thus, the reproducibility would never drop below acceptable values.

It may be possible to use this property of the LSOM as an advantage. By virtue of computing the reproducibility of the LSOM as we add datasets, it is possible to measure this drop. Datasets that result in a substantial drop could be inspected to assess whether they are improperly labeled or extremely error-prone data as the clustering is done. An alternative view is that monitoring the reproducibility also provides us with a metric for measuring how much "novelty" a new dataset adds to existing analyses.

Parameters and Formulae

Formulae: Most of the formulae in this work come from [5], except for the following which have been edited in this work (Table 1).

Table 1. Parameters used in analysis

Parameter	Symbol	Value	Parameter	Symbol	Value
Rows		40	Radius factor initial	σ_i	0.8
Columns		60	Radius factor final	σ_i	0.2
Learning rate initial	η_i	0.2	Beta factor	β	0.7
Learning rate final	η_f	0.08			

Learning State - Learning Rate η, Radius Factor σ

$$\eta(t) = \begin{cases} \left(\frac{\eta_f}{d(t_f)} d(t) \right)^{\left(1-\frac{t}{t_f}\right)}, & d(t) < d(t_f) \\ \eta_f, & \text{otherwise} \end{cases} \tag{1}$$

$$\sigma(t) = \begin{cases} \left(\frac{\sigma_f}{d(t_f)} d(t) \right)^{\left(1-\frac{t}{t_f}\right)}, & d(t) < d(t_f) \\ \sigma_f, & \text{otherwise} \end{cases} \tag{2}$$

Acknowledgments. We would like to thank the Wold lab at Caltech for providing the data for this work as well as Dana Wyman in the Mortazavi lab at UCI for feedback. Funding for this work was provided by NHGRI UM1 HG009443 to AM.

References

1. Kohonen T (2001) Self-organizing maps, 3rd edn. Springer, Heidelberg
2. Alhoniemi E (2000) Clustering of the self-organizing map. IEEE Trans Neural Netw 11 (3):586–600
3. Mortazavi A et al (2013) Integrating and mining the chromatin landscape of cell-type specificity using self-organizing maps. Genome Res 23:2136–2148
4. Tamayo P et al (1999) Interpreting patterns of gene expression with self-organizing maps: methods and application to hematopoietic differentiation. PNAS 96(6):2907–2912
5. Silva B, Marques N (2015) The ubiquitous self-organizing map for non-stationary data streams. J Big Data 2:27
6. Link to ENCODE datasets. https://bit.ly/2FGKWnx. Accessed 17 Jan 2019
7. Jaccard P (1912) The distribution of the Flora of the Alpine Zone. New Phytol. 11 (1912):37–50

A Voting Ensemble Method to Assist the Diagnosis of Prostate Cancer Using Multiparametric MRI

Patrick Riley[1], Ivan Olier[1], Marc Rea[2], Paulo Lisboa[1],
and Sandra Ortega-Martorell[1(✉)]

[1] Department of Applied Mathematics, Liverpool John Moores University,
Byrom Street, Liverpool L3 3AF, UK
P.J.Riley@2014.ljmu.ac.uk, {I.A.OlierCaparroso,
P.J.Lisboa,S.OrtegaMartorell}@ljmu.ac.uk
[2] Department of Medical Imaging, Clatterbridge Cancer Centre,
Bebington, Wirral CH63 4JY, UK
M.Rea@nhs.net

Abstract. Prostate cancer is the second most commonly occurring cancer in men. Diagnosis through Magnetic Resonance Imaging (MRI) is limited, yet current practice holds a relatively low specificity. This paper extends a previous SPIE ProstateX challenge study in three ways (1) to include healthy tissue analysis, creating a solution suitable for clinical practice, which has been requested and validated by collaborating clinicians; (2) by using a voting ensemble method to assist prostate cancer diagnosis through a supervised SVM approach; and (3) using the unsupervised GTM to provide interpretability to understand the supervised SVM classification results. Pairwise classifiers of clinically significant lesion, non-significant lesion, and healthy tissue, were developed. Results showed that when combining multiparametric MRI and patient level metadata, classification of significant lesions against healthy tissue attained an AUC of 0.869 (10-fold cross-validation).

Keywords: mpMRI · GTM · Data mining · SVM · Classification

1 Introduction

Prostate cancer is the second most commonly occurring cancer in men worldwide [1]. Diagnosis through Magnetic Resonance Imaging (MRI) is the second line of diagnosis, usually after a trans-rectal ultrasound biopsy which is conducted after a positive blood screening test. However, this test has a relatively low specificity, leading to over-diagnosis and therefore overtreatment.

Although MRI diagnosis can overcome this, it requires specialist knowledge to review the prostate MRI, which is time consuming. Furthermore, the abundance of available MRI data can provide difficulty with where to begin. This study uses both Generative Topographic Mapping (GTM) and Support Vector Machines (SVMs) as assisting tools, in visualization and classification respectively, for the diagnosis of prostate cancer on multiparametric MRI (mpMRI) scans, through a voting ensemble technique.

© Springer Nature Switzerland AG 2020
A. Vellido et al. (Eds.): WSOM 2019, AISC 976, pp. 294–303, 2020.
https://doi.org/10.1007/978-3-030-19642-4_29

Upon request from collaborating clinicians, the SPIE ProstateX challenge dataset [2] was extended in this study from its original use, whereby the contralateral of the lesion location may be taken as healthy prostate tissue. This extends from the two classes available in the data set – the clinically significant lesions and non-significant lesions. The latter are denoted "non-significant" as prostate cancer treatment is not always needed for a lesion of a lower Gleason score [3].

The use of the contralateral to attain healthy tissue denotes a novel point for this study. Through pairwise tests of the three classes, the voting ensemble technique gives rise to a diagnosis aiding tool which is in demand from clinicians and radiologists.

Recent research from the ProstateX challenge tackles the two class problem – clinically significant against non-significant – with various machine learning methods such as deep learning [4], convolutional neural networks [5] and SVMs [6]; the efficiency of the latter inspiring its use in this study. The use of the GTM in this work provides insight into the structure of the data set, to allow for an explanation of the results of the ensemble voting method.

The Data & Classification Methods sections detail how the ProstateX challenge data set has been pre-processed and the setup of the SVM. The Model Evaluation section shows the ensemble voting systems results, with an Application to prostate lesion findings following. Using the GTM, an unsupervised explanation into the hidden data structure is presented, providing interpretability to the supervised SVM classifier.

2 Data

The ProstateX challenge training data used for this study is a collection of 330 lesion findings over 204 patients, each with mpMRI taken around the prostate. Metadata for each lesion finding, and the findings within each scan parameter were held in separate files. For each patient, one or more prostate lesions with their location were identified through its scanner coordinate position. Each lesion within the metadata holds a level denoting its clinical significance; where it is clinically significant if the Gleason score of the lesion is 7 or higher, or non-significant where the Gleason score is 6 or below. The numeric Gleason score of a given lesion is not available in the data set.

This work extends on the original ProstateX challenge data based on a request from collaborating clinicians. Through using the contralateral of the lesion location, healthy tissue can be extracted. The total number of significant cases in the dataset is 76, while the non-significant cases amount to 254. As these are imbalanced classes, we decided to subsample the class of non-significant cases to only 75 cases. In line with this, a total of 54 healthy tissue samples were extracted. Three of the various mpMRI scan parameters are analyzed in this study: T2-weighted, K^{trans} and Apparent Diffusion Coefficient (ADC). Only transverse images were used. These modalities are all shown to be related to lesion clinical significance [7].

T2-weighted imaging is a form of spin-echo pulse sequencing, showing fatty tissue and fluid brightly. ADC is a measure of the magnitude of diffusion (of water molecules) within tissue and is calculated from diffusion weighted imaging. K^{trans}, a type of perfusion imaging, represents a measure of capillary permeability, calculated from dynamic contrast-enhanced imaging.

3   Classification Methods

For creating the classifiers we follow the steps outlined below:

(a) *Pre-processing:* The ProstateX dataset is a large data set, with each patient holding multiple scan parameters each – furthermore, the naming of the files was not consistent. Various rules were required to extract the relevant data. Metadata was collected from both the separate files, and the DICOM header data of the relevant image files; including the patient age which is used within the fitted models.

b) *Patch Extraction and vectorization of the data:* For each lesion, a centered 5 mm × 5 mm patch is extracted at a resolution of 1 px/mm for each of the three MRI scan parameters. The contralateral is defined as the opposite side of the MRI image to which the lesion is located. For example, if the lesion is located 25 mm to the left of the center of the MRI image, then the contralateral is located 25 mm to the right of the center of the MRI image, as validated by collaborating clinicians. They were used as the 'healthy' tissue data and is suitable for analysis as informed by collaborating clinicians. For the selection of those healthy samples, we identified the patient with only one lesion finding (as indicated in the metadata), which is of a certain distance away from the center of the image. In these cases, also a 5 mm × 5 mm patch was extracted at a resolution of 1 px/mm for each of the three MRI scan parameters (as in the lesion patch extraction), following the method of a successful submission to the SPIE ProstateX challenge [6]. Images were flattened into 75 dimensional vectors. All processing was carried out with 32-bit floating point pixel values, preserving large dynamic range and subtle contrast differences. These vectors can be further augmented with patient level metadata, as described.

(c) *Augmenting the vectors with patient level metadata*: For the model presented in this study, the image data vector is further augmented with patient level metadata. The first is the zonal location of the region of interest (lesion or healthy tissue) within the prostate, denoted as a dummy variable. The second is the patient age.

(d) *Data standardization*: Each input dimension is scaled – for K^{trans} images, this is through the log transform. For all other input dimensions, this is through subtraction of the mean and division of the standard deviation. This ensures that the distribution of each dimension is approximately normal, again similar to [6].

(e) *Classification using Support Vector Machines*: For this classification study upon three class labels through pairwise testing, Support Vector Machines were utilized, as first described in [8]. SVM was successfully applied to this data in [6]. The linear kernel is not suitable for this data set – it was tested but discarded due to poor performance. The kernel selected for this work is the radial basis function (RBF-SVM), as defined in Eq. 1. Through defining the kernel function there is no need to perform $\emptyset(.)$ explicitly.

$$K\left(x^i, x^j\right) = \emptyset\left(x^i\right)^T \emptyset\left(x^j\right) = \exp\left(-\gamma\left\|x^i - x^j\right\|^2\right), \gamma > 0 \tag{1}$$

(f) *Initial cross-validation on all data for hyperparameter selection:* Two hyperparameters require tuning for an RBF-SVM: the soft-margin constant, C and the kernel parameter, γ. A cross validation is used to evaluate each data combination and pairwise test, to find the best set of parameters. A grid search was carried out on $C \in \{0.1, 0.5, 1, 2, 5, 10, 20, 30, 50\}$ and $\gamma \in \{1.0, 0.2, 0.1, 0.05, 0.01, 0.001, 0.0001, 0.00001\}$. For the results presented in this paper, the 10-fold cross validation results across all the data are presented in Table 1.

4 Model Evaluation: Using a Voting Ensemble Method for Aiding Prostate Cancer Diagnosis

In this study, a voting ensemble method across the three pairwise classifiers are proposed to give a final classification of each finding. This is to evaluate the system as a classifier. The majority vote across all classifiers is taken to be the final classification label. An undecided category was added for the cases where all three classifiers predict something different and a majority vote cannot be reached. In this case, as a tool for aiding diagnosis, a clinician would be alerted to this case for an expert to classify it themselves.

The voting ensemble method is evaluated through a 10-fold cross validation. Within each fold, 90% of the data is used for training and 10% of the data is partitioned according to the frequencies of the data, for testing the voting ensemble methodology. The results of each fold are used to calculate the classification across the data set used for this study. Each of the pairwise classifier is trained using the RBF-SVM with the parameters found through the cross-validation grid search (as explained in Sect. 3). The results of applying the voting ensemble method (after cross validation) to the pairwise classifiers created using all image patch data and patient metadata is presented in Table 2.

For the significant cases, 72.4% of them were correctly classified as significant, with the majority of the misclassifications (23.6%) going to the non-significant lesions class. Only three cases (3.9%) were incorrectly classified as healthy tissue, and no cases were classified as undecided. In turn, the healthy tissue was correctly classified a 68.5% of the time, while again the majority of misclassification (25.9%) went to the non-significant class, and only 2 cases fell into the significant class.

The non-significant cases were correctly classified only 48.0% of the time, however, considering that this class is the intermediate between having a healthy prostate and a significant tissue, this is to some extent expected. Still, this is the less critical issue from the clinical viewpoint, as it is known that treatment is not always needed for non-significant prostate cancers [3] and techniques such as active surveillance may be employed within the patients care plan. Potentially, those non-significant cases predicted as significant (20%) will undergo more in-depth analysis, which may result in their final correct diagnosis as non-significant; and those predicted as healthy (32%) may hold a very low Gleason score and therefore closely resemble healthy prostate tissue. The latter group must probably would not require any treatment, therefore the risk to the patients is not considered high.

5   Classification Results: Application to Prostate Lesion Findings

An application of the ensemble voting system applied to given lesion findings and healthy tissue is presented in Fig. 1. For cases A and B, a significant and non-significant lesion finding respectively, are classified through the three classifiers. For case C, the ensemble voting system is shown to classify healthy prostate tissue. In all three of these cases, the majority vote matches the original label and is therefore correctly classified. It is intended that this aids prostate lesion diagnosis.

However, for case D, which is known to be healthy prostate tissue, there is no majority vote; all three pairwise classifiers have predicted differently. In practice, this would be brought to the attention of a clinician, for their technical knowledge to make the final decision. Nevertheless, in all cases, lesion or not, the final decision would rest with them as this tool is designed to aid diagnosis.

Table 1. 10-fold cross validation results for hyperparameter grid search.

Classifier	AUC	Standard deviation	C	γ
Significant vs. Non-significant	0.722	0.262	50	0.001
Significant vs. Healthy	0.869	0.099	0.5	0.01
Non-significant vs. Healthy	0.713	0.134	2	0.01

Table 2. Results of the voting ensemble method applied to the dataset containing all image patch data and patient metadata.

Original label	No. cases	Final predicted label			
		Significant	Non-significant	Healthy	Undecided
Significant	76	55	18	3	0
Non-significant	75	15	36	24	0
Healthy	54	2	14	37	1

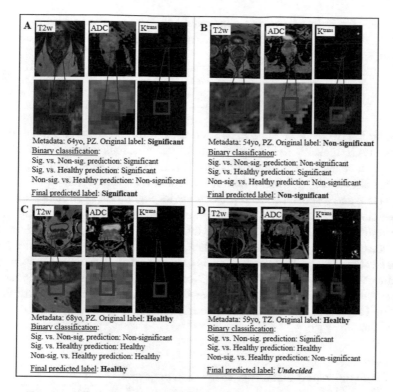

Fig. 1. Application of the voting ensemble method to lesion findings. A: classification of a clinically significant lesion. B: classification of a non-significant lesion. C: classification of healthy tissue. D: disagreement throughout the ensemble voting system. (yo – years old)

6 Discussion: Explanation of the Classification Results Using Generative Topographic Mapping

Through visualizing the data in the latent space, it is possible to explain the supervised SVM classification results using the unsupervised GTM [8]. GTM is a nonlinear latent variable model of the manifold learning family, with sound foundations in probability theory. It performs simultaneous clustering and visualization of the observed data through a nonlinear and topology-preserving mapping from a visualization latent space in (with being usually 1 or 2 for visualization purposes) onto a manifold embedded in a multi-dimensional space, where the observed data reside. The mapping that generates the manifold is carried out through a generalized additive regression function:

$$y = W\varphi(u) \tag{2}$$

where $y \in D$, $u \in$, W is the matrix that generates the mapping, and φ is a vector with the images of S basis functions φs. To achieve computational tractability, the prior distribution of u in latent space is constrained to form a uniform discrete grid of M centers, analogous to the layout of the Self-Organizing Map (SOM, [9]) units, in the form of a sum of delta functions:

$$p(\boldsymbol{u}) = \frac{1}{K} \sum_{k=1}^{K} \delta(\boldsymbol{u} - \boldsymbol{u}_k) \tag{3}$$

Similar to how the RBF kernel has been used in this application through the SVM classifier, GTM typically uses a set of radial basis functions to map the results of the unsupervised analysis. By doing this, the interpretability problem of the SVM is somewhat alleviated. The similar kernel tricks allow for interpretation.

Figures 2 and 3 show the mean projections for Significant against Healthy, and for all three classes respectively. Clustering can be seen within Fig. 2 for both the Significant and the Healthy classes within the latent space. However, with the introduction of the Non-significant class, there is no clustering and it 'invades' both the Significant and the Healthy classes. To an extent this is to be expected. Where the Gleason score of a Significant lesion is 7 or higher, some Non-significant lesions at the higher end of their possible Gleason scores, such as 5 or 6, may overlap with the Significant class. Similarly, those with a lower Gleason score, such as 1 or 2, may overlap with the Healthy class.

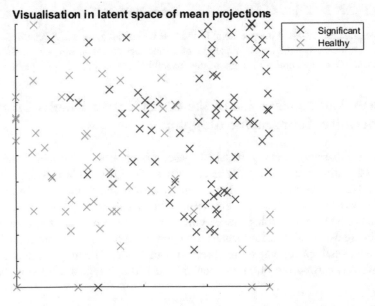

Fig. 2. Mean projections in the latent space using GTM: Significant against Healthy. Clustering is present.

Visualisation in latent space of mean projections

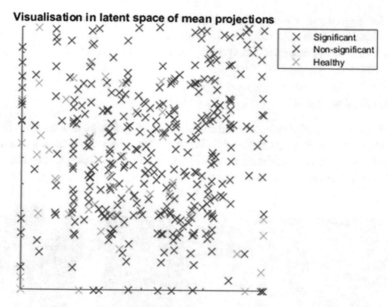

Fig. 3. Mean projections in the latent space using GTM: all three classes. The clustering becomes distorted with the introduction of the non-significant (blue) class.

Figure 4 shows the boxplot of the distributions of the mean projections of the points. All classes overlap, which is to be expected as shown in Fig. 3. The interquartile ranges of the Significant class and the Healthy class, showing separation (as shown in Fig. 2). The interquartile range of the Nonsignificant class overlaps with both the Significant class and the Healthy class.

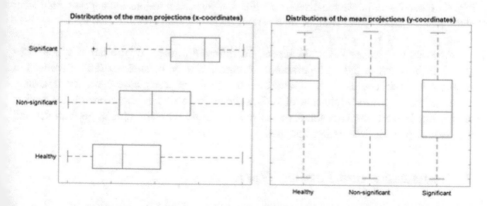

Fig. 4. Boxplots of the distributions of the mean projections for each class label.

Through the addition of the healthy class, the initial 2-class classifier problem of the ProstateX challenge can be visualized, with much overlap of the Non-significant class within the Significant class. Submissions to the ProstateX challenge were for the 2-class classifier problem using AUC and hence comparisons are not presented alongside this study; enhanced interpretability using the GTM with the binary classifiers is provided.

6.1   Borderline Cases

Figure 5 shows examples of borderline cases of Non-Significant lesions from the latent space, where case **A** is an example of a Non-significant lesion which is located within the Significant clustering, and case **B** is an example of a Non-significant lesion which is located within the Healthy clustering.

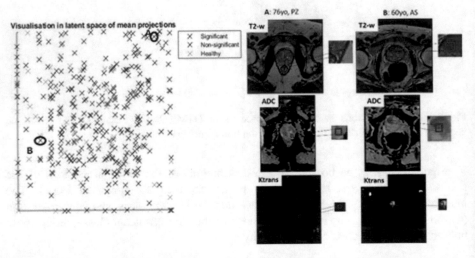

Fig. 5. Examples of borderline cases - both are Non-significant lesions but appear within other clusters in the latent space, with the multiparametric profile displayed for each case.

Reasons why cases **A** and **B** appear on different corners of the latent space visualization, in terms of the multiparametric images, may be because of the differences in brightness of the images. Although both Non-significant cases, case **A** for example may appear similar to the brightness of a Significant case and is therefore closer to that clustering. Indeed, the Gleason score is not available with the challenge data set and this would provide interesting insights.

7   Conclusions and Further Work

This study found that using an ensemble voting method across three classifiers, a majority vote system can assist clinicians in the diagnosis stage of a patient's cancer care. The use of healthy tissue has been included through demand and validation from clinicians. Through using a supervised SVM, the results can be explained, with enhanced interpretation, through an unsupervised visualization method, the GTM.

Further work will look at increasing the classifiers scope, leading to segmentation maps of the area to further aid diagnosis, particularly for non-invasive treatment techniques. Furthermore, utilizing the decision of a clinician with their classification of an undecided case through active machine learning techniques will be studied.

Acknowledgments. This work has been funded by the Liverpool John Moores University PhD Scholarship Fund. The authors would like to thank Andy Kitchen for his assistance in validating data extraction methods.

References

1. Bray F, Ferlay J, Soerjomataram I, Siegel RL, Torre L, Jemal A Prostate cancer statistics. World Cancer Research Fund. https://www.wcrf.org/dietandcancer/cancer-trends/prostate-cancer-statistics
2. Litjens G, Debats O, Barentsz J, Karssemeijer N, Huisman H (2014) Computer-aided detection of prostate cancer in MRI. IEEE Trans Med Imaging 33:1083–1092. https://doi.org/10.1109/TMI.2014.2303821
3. Gallagher J Prostate cancer treatment "not always needed" - BBC News. https://www.bbc.co.uk/news/health-37362572
4. Liu S, Zheng H, Feng Y, Li W (2017) Prostate cancer diagnosis using deep learning with 3D multiparametric MRI. Presented at the 12 March 2017
5. Mehrtash A, Sedghi A, Ghafoorian M, Taghipour M, Tempany CM, Wells WM, Kapur T, Mousavi P, Abolmaesumi P, Fedorov A (2017) Classification of clinical significance of MRI prostate findings using 3D convolutional neural networks. In: Armato III SG, Petrick NA (eds.) SPIE medical imaging 2017, computer-aided diagnosis, Orlando
6. Kitchen A, Seah J (2017) Support vector machines for prostate lesion classification. In: Armato III SG, Petrick NA (eds.) SPIE medical imaging 2017, computer-aided diagnosis. International Society for Optics and Photonics, Orlando, p 1013427
7. Langer DL, van der Kwast TH, Evans AJ, Plotkin A, Trachtenberg J, Wilson BC, Haider MA (2010) Prostate tissue composition and MR measurements: investigating the relationships between ADC, T2, Ktrans, ve, and corresponding histologic features. Radiology 255:485–494. https://doi.org/10.1148/radiol.10091343
8. Bishop CM, Svensén M, Williams CKI (1998) GTM: The generative topographic mapping. Neural Comput 10:215–234. https://doi.org/10.1162/089976698300017953
9. Kohonen T (2001) Self-organizing maps. Springer, Berlin, Heidelberg

Classifying and Grouping Mammography Images into Communities Using Fisher Information Networks to Assist the Diagnosis of Breast Cancer

Meenal Srivastava, Ivan Olier, Patrick Riley, Paulo Lisboa,
and Sandra Ortega-Martorell$^{(\boxtimes)}$

Department of Applied Mathematics,
Liverpool John Moores University, Byrom Street, Liverpool, UK
meenal.srivastava@gmail.com, {I.A.OlierCaparroso,
P.J.Lisboa,S.OrtegaMartorell}@ljmu.ac.uk,
P.J.Riley@2014.ljmu.ac.uk

Abstract. The aim of this paper is to build a computer based clinical decision support tool using a semi-supervised framework, the Fisher Information Network (FIN), for visualization of a set of mammographic images. The FIN organizes the images into a similarity network from which, for any new image, reference images that are closely related can be identified. This enables clinicians to review not just the reference images but also ancillary information e.g. about response to therapy. The Fisher information metric defines a Riemannian space where distances reflect similarity with respect to a given probability distribution. This metric is informed about generative properties of data, and hence assesses the importance of directions in space of parameters. It automatically performs feature relevance detection. This approach focusses on the interpretability of the model from the standpoint of the clinical user. Model predictions were validated using the prevalence of classes in each of the clusters identified by the FIN.

Keywords: Medical image analysis ·
Community detection in medical images · Fisher distance ·
Fisher information metric · Information geometry · Riemannian space

1 Introduction

Breast cancer is the most frequent cancer among women, impacting 2.1 million women each year, and also causes the greatest number of cancer-related deaths among women [1]. This type of cancer usually takes time to develop and symptoms become evident very late. Currently, there is no effective way to cure later stage breast cancer, therefore early and accurate detection of tumor plays a vital role in improving the prognosis, as it allows for better treatment planning.

Besides physical examination, commonly used modalities for breast screening are mammography, ultrasonography, Magnetic Resonance Imaging (MRI) and core-needle

© Springer Nature Switzerland AG 2020
A. Vellido et al. (Eds.): WSOM 2019, AISC 976, pp. 304–313, 2020.
https://doi.org/10.1007/978-3-030-19642-4_30

biopsy. Among these techniques, mammography is considered the best, cheapest way to detect the tumor [2]. However sensitivity of mammography can vary considerably due to factors like radiologist's experience, human error and image quality [3]. Visual clues during early stages are subtle and varied in appearance making diagnosis difficult. Many times, abnormalities are hidden by breast tissue structure. According to statistical reports, patients with dense breasts have high chances of receiving false negatives for lesions [4]. Appearance of normal tissue is also highly variable and complex in mammograms, making tumor identification more difficult. Micro-calcifications are frequently missed. In the current practice, pathological confirmation of malignancy and tumor grade characterization is done with a biopsy, which is an invasive, painful procedure, that also carries risk of tumor cell migration [5].

Most of existing work done in breast cancer classification is categorizing into normal and tumor class. However, tumor can be either benign or malignant, with the latter being the one requiring treatment. Hence, the two classes need to be differentiated as well. In this paper we attempt to classify mammography images into these three classes: malignant, benign or normal. Our aim is to propose a methodology that can assist clinicians in managing their breast cancer patients, by visualizing mammographic images in a different, novel way, using Fisher Information Network (FIN) [6].

FIN provides a global view of data due to the use of Fisher metric and displays a meaningful structure that implicitly informs about underlying class probabilities. The FIN framework [7] can be used both for visualization of data and for constructing interpretable retrieval-based classifier since connection weights contain accurate information about similarity between data points. This approach focusses on both accurate prediction as well as interpretability of output.

In our proposed methodology we construct a FIN using probability density estimates calculated for three classes- normal, benign, malignant. The aim is to detect underlying patterns and structure in breast cancer images. We test our approach using an existing, publicly available database [8]. By producing FIN's visualization of similarity networks we expect to elucidate the underlying data structure, community membership and class prediction. In addition, by dividing the network into communities and stratifying the data according to classification labels, we expect to be able to provide instances similar to new query image, which can then be analyzed in order to better understand new instances.

The paper is organized as follows: Sect. 2 describes the dataset used for constructing and testing the FIN. Section 3 describes the methodology to construct the network In Sect. 4, we present the results and discuss its significance and implications. Finally, Sect. 5 presents our conclusions.

2 Material

Images from the Mammographic Image Analysis Society (Mini MIAS) database [8], [available at http://peipa.essex.ac.uk/info/mias.html] are used in this paper. The dataset is arranged in pairs of films, where each pair represents the left (even filename numbers) and right (odd filename numbers) mammogram of a single patient. The dataset is composed of 326 mammograms of right and left breast, from 161 patients, where 51 observations were labeled as malignant, 66 as benign and 209 as normal/healthy. In this work we randomly under-sampled the normal observations to balance the number of cases per class in the dataset. Hence, the dataset used in this paper contains 66 benign, 51 malignant and 62 normal observations, for a total of 179.

The dataset includes radiologist's "truth" markings for locations of abnormalities present in images. For these images, coordinates for center as well as approximate radius of circle enclosing the abnormality, is available. In case of calcifications, center locations and radii are for clusters rather than individual calcifications. The dataset also contains information related to background tissue of patients, and margin/shape of abnormalities. Table 1 shows the details of the dataset used in this paper regarding tissue, abnormalities and class.

Table 1. Mini MIAS dataset used in project

Tumour shape	Tumour type		Breast type		
	Benign	Malignant	Dense	Fatty	Glandular
Architectural Distortions	9	10	7	6	6
Asymmetric	6	9	7	4	4
Calcification	13	12	14	5	6
Circumscribed	20	4	3	13	8
Miscellaneous	7	8	2	8	5
Spiculated	11	8	7	5	7

Classes	Breast type			Total per class
	Dense	Fatty	Glandular	
Benign	23	23	20	66
Malignant	17	18	16	51
Normal	11	31	20	62

Figure 1 is an image from the dataset illustrating the challenges faced in automatic evaluation/classification of these medical images. Artefacts like patient's names, labels, etc. are present in the images. Pectoral muscles having same intensity as tumors are visible in mediolateral oblique (MLO) view and needs to be removed. The intensity of normal region in dense breast is similar to that of tumor region in fatty and glandular breast (see Fig. 2b, c and f). Additionally, tumors can have varied shape, size and characteristics making classification difficult. Figure 2 shows examples of the different tissue types that can be found in this dataset: (a) fatty breast with normal tissue (b) fatty breast with mass tissue (c) dense breast with normal tissue (d) dense breast with mass tissue (e) glandular breast with normal tissue (f) glandular breast with tumour tissue.

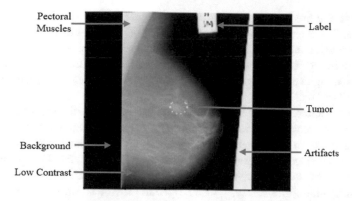

Fig. 1. Problems in mammogram image in Mini-MIAS dataset.

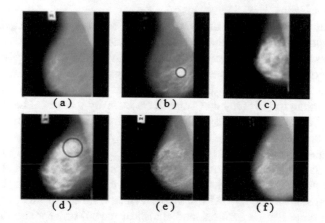

Fig. 2. Sample images from Mini MIAS dataset showing different tissue types.

3   Methodology

As discussed in the introduction section, the aim of the paper is to detect communities based on similarities detected in mammography images of cancer patients. For this, we proposed to follow the steps below (see Fig. 3):

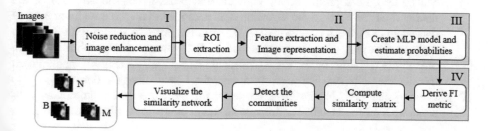

Fig. 3. Pipeline followed for creating the Fisher Information Network

(I)   Image Pre-processing: This is the first step where noise in images is removed using 3×3 size median filter followed by the removal of image artefacts using image binarization and intensity thresholding. The breast profile is extracted by removing pectoral muscles using slant line approximation. Images are then morphologically enhanced using top hat filter [9]. Structuring element used was "box" of "radius" 5.

(II) Image Representation: Before representing images with features, ROI of size 50×50, containing normal, benign tumor or malignant tumor regions, is extracted from enhanced images. For images with tumor, the center of abnormalities defined in dataset was taken as the ROI center. For the normal samples, coordinates of ROI center were manually defined. ROI extraction reduced the training time significantly. It also helped to provide an initial representative area for benign, malignant, and normal instances for our model to learn differences from. The image representation involves two stages, feature extraction and feature selection, explained below:

    a. Feature Extraction: In this paper, texture features are extracted to represent images. Texture contains information about structural arrangement of surfaces. It specifies the roughness or coarseness of an object surface and can be described as a pattern with some element of regularity. Texture features have been proven useful in differentiating mass and normal breasts tissues, and according to [10] they can outperform intensity features. The reason for this can be that tumor area exhibits low texture compared to normal parenchyma. Texture features can be extracted by many methods, for example statistical methods such as gray level co-occurrence matrix (GLCM) in [11], model based methods such as Markov random field (MRF) model [12], and transformation based methods such as wavelet decomposition [13]. However, most studies use GLCM to extract texture features. In this paper, statistical first order texture features, and statistical second order texture features from GLCM, together with gray level run length matrix (GLRLM) are extracted.

    b. Feature Selection: This step is performed to remove noisy, redundant and irrelevant features and retain the optimal set of effective and discriminating features. Specifically, features were ranked using Classification and Regression Tree (CART) [14]. As interpretability was the focus of the project, CART was preferred over other methods. CART also takes into account discriminative power of the variable with respect to target variable.

(III) Calculate probabilities: Classification for the three classes is performed using Multi-Layer Perceptron (MLP) with 10 fold cross-validation on 70% of the dataset used for training the algorithm and then testing on the remaining 30% of dataset to evaluate its performance. The hidden layer is set up with 3 hidden units and weight decay regularization of 0.1. Once the algorithm is tuned, probability densities of the three classes are calculated, which will then be used to calculate pairwise distances, producing the Fisher distance matrix.

(IV) Creating the FIN: Next step is to detect the structure in the dataset by constructing the FIN. Details for this can be found below (also illustrated in Fig. 3):

a. Fisher Information metric [6]: This metric is derived from the Fisher distance matrix, and defines a Riemannian space where distances reflect similarity with respect to a given probability distribution. This metric is informed about generative properties of data, and hence can assess the importance of directions in space of parameters. FI combines multiple variables together to assess patterns and evaluate stability in system. It is obtained by differentiating the logarithm of the conditional probability $p(x|\theta)$ with respect to x and summing over all possible classifications:

$$FI(x) = E_{p(c|x)}\{(\nabla_x \log p(c|x))^T (\nabla_x \log p(c|x))\} = \\ -E_{p(c|x)}\{\nabla_x^2 \log p(c|x)\} . \tag{1}$$

Where $E_{p(c|x)}$ denotes the expectation over the density function $p(c|x)$ and is the gradient with respect to x.

Given the FI metric, the infinitesimal distance between a pairs of neighboring points in the data space is given by the quadratic differential form:

$$d(x, x + \Delta x)^2 = \Delta x^T FI(x) \Delta x . \tag{2}$$

An important property of this metric is that it automatically scales each dimension of the data space according to its degree of relevance with respect to class membership, expanding directions along which $p(c|x)$ changes rapidly and compressing those where the variation is little. The result is a Riemannian space where the posterior class membership probability changes evenly in all directions.

b. Similarity matrix and community detection: After calculating the distance matrix with a Gaussian radial kernel, the similarity matrix is then computed from them, resulting in the adjacency matrix which defines network structure. The communities are then detected by maximizing modularity using Newman's spectral optimization [15], resulting in clusters that best represent the graph structure. In medical database, patients do not interact as in social networks. Hence in medical networks, the presence of an edge indicates similarity between patients (observations), and the weights determine the strength of links. Central nodes in each community can also be found and used as representatives of the clusters giving a set of characteristic points to associate with each of the communities.

c. Visualization of similarity network: This is done using classical Multi-Dimensional Scaling (MDS) [16]. MDS uses a distance matrix to produce representation of points in lower dimensional Euclidean space such that the distance between them approximate as closely as possible the dissimilarities between corresponding instances in the original matrix. Applying MDS is also key as the mapping of the matrix onto the Euclidean space will allow many commonly used methods from signal processing to be applied. All this, while also retaining the distance structure generated by the FI matrix.

4   Results and Discussion

Preprocessing images with median filter removed the noise and morphological enhancement increased the contrast between bright and dark areas so that the classifier performance was optimized. The balanced (taking into account the number of cases per class) accuracy of the initial MLP model was 78.98% on the separated test set, with a standard deviation of 6.89. This was considered acceptable as compared to the literature. Uppal and Naseem achieved an accuracy of 96.97% on MIAS dataset using fusion of cosine transform for classification into tumor and non-tumor [17]. MLP classifier in this paper distinguished fully between normal and tumor class. This initial MLP was created to generate the probability densities (i.e. probability of class membership) to generate the Fisher Information (FI) metric.

Figure 4 shows the three-dimensional representation of the FIN for the three classes involved, i.e. normal, malignant and benign. As expected, the normal class is well separated from both tumor classes, which are in turn also reasonably well-separated from each other, with some misclassifications.

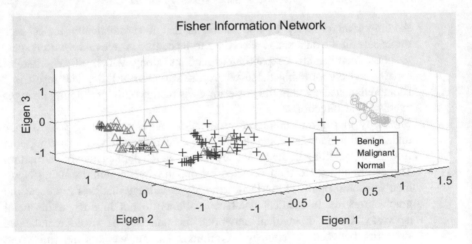

Fig. 4. FIN representation of Mini MIAS dataset.

Figure 5 contains two- and three-dimensional representations of the dataset with communities identified by the network for the three classes. In this representation, edges are displayed only between members of same community to highlight cluster membership. The Fisher network is able to separate the data in three clearly distinguishable groups of communities, with normal cases being fully separated from the tumor cases.

Figure 6 further analyses the purity of the communities detected by the FIN. It separates the normal cases entirely in three communities (1, 5 and 8) of 35, 23 and 4 subjects, respectively. These three communities not only do not contain observations from tumor samples, but also are represented very far from the rest of them representing the tumors. Communities 2, 4, 7 and 9 are mainly representing the benign

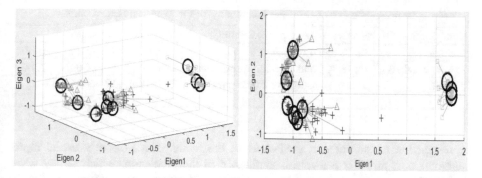

Fig. 5. Three- and two-dimensional representations of the FIN with communities identified by each network marked.

cases, while the malignant tumors are mainly represented by communities 3 and 6. There is a benign case which is not assigned any of these nine communities and exists as a singleton in the center of the network (referred to as community 10 in the paper). See Table 2 for more details.

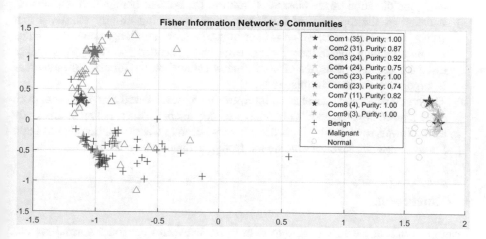

Fig. 6. FIN representation with size and purity for each community marked

The accuracies for the tumor classes were 76.5% and 87.9% for the malignant and benign classes, respectively. A total of 8 benign cases were falsely identified as malignant, and 12 malignant cases were falsely predicted as benign. None of the tumors were classified as normal, and none of the normal were classified as tumors, either benign or malignant, for a striking 100% accuracy in the separation of tumor from healthy tissue.

Another interesting aspect of the FIN representation in Fig. 6 is the shape formed by all the scattered tumor cases projected, going from community 2 (mainly benign) to community 3 (mainly malignant). Possibly, all the area in between is showing a

Table 2. Number of cases per community of the main represented class.

Communities	Number of cases per community	Main class represented
1, 5 and 8	35, 23 and 4 normal cases, resp.	Normal: 62 (out of 62) – 100%
2, 4, 7, 9 and 10	27, 18, 9, 3 and 1 benign cases, resp.	Benign: 58 (out of 66) – 87.9%
3 and 6	22 and 17 malignant cases, resp.	Malignant: (39 out of 51) – 76.5%

representation of the different gradations of the tumor. This is a very useful representation as it provides a level of confidence in the prediction, which is key when analyzing new observations. Furthermore, when projecting a new case in the map, it would be possible to look at neighboring cases and learn their characteristics, treatments, outcomes and prognosis, and with that information improve the diagnosis and prognosis of the analyzed patient.

The advantages of having used the FIN in this work can be summarized as follows:

(1) The FIN automatically filters relevant variables based on their contribution with respect to the classification problem, measured by their influence on the posterior class probabilities. This allowed us to create a model that amplifies distances along the direction of the classes of interest, i.e. normal, benign and malignant.

(2) It preserves the topology of the input space, producing affinity measures that reflect the data structure, which is later infused into the similarity network. It helps in understanding the hidden mechanisms that generated the data. This makes distances accurate measures of dissimilarity even when the number of covariates is large, as is the case in this study.

(3) The framework can be used as an interpretable retrieval-based classifier, and, even more importantly, the results obtained are interpretable. This was key in this study as it allowed us not only to produce highly accurate classifiers, but also to learn for each case the level of confidence in the predicted outcome.

5   Conclusion

It can be seen that the FIN is able to distinguish between tumor and normal regions with 100% accuracy and separates benign and malignant tumor classes reasonably with some misclassifications. Nodes at the ends of the network are clear in terms of membership while membership of center nodes is relatively unclear and needs to be investigated further. Fisher information metric, used to construct the network, is informed about the generative properties of the data, and thus assesses the importance of directions in the space of the parameters. As the metric is Riemannian, expected variation of probability density caused by a distortion in parameters is different depending on the location of the space in which it is measured. Thus, Fisher information metric provides an elegant, clearly defined and statistically rigorous solution which is visualized as communities in a network, besides being classified into benign, malignant and normal classes.

This paper enhances image morphologically, with median filter for noise removal. Statistical second order texture features were extracted with GLCM and GLRLM along with first order texture features. Other methods for image enhancement, noise removal and extracting features can also be tested in the future.

References

1. World Health Organization (2018) Breast cancer. WHO website
2. Homer MJ (1997) Mammographic interpretation: a practical approach, 2nd edn. McGraw-Hill, New York, p 376 Health Professions Division
3. Rangayyan RM, Ayres FJ, Leo Desautels JE (2007) A review of computer-aided diagnosis of breast cancer: toward the detection of subtle signs. J Franklin Inst 344(3–4):312–348
4. Oelze ML (2012) Quantitative ultrasound techniques and improvements to diagnostic ultrasonic imaging. In: IEEE international ultrasonics symposium, IUS
5. Tadayyon H, Sadeghi-Naini A, Wirtzfeld L, Wright FC, Czarnota G (2014) Quantitative ultrasound characterization of locally advanced breast cancer by estimation of its scatterer properties. Med Phys 41(1):012903
6. Ruiz H, Ortega-Martorell S, Jarman IH, Martín JD, Lisboa PJG (2012) Constructing similarity networks using the Fisher information metric. In: European symposium on artificial neural networks, computational intelligence and machine learning (ESANN), Bruges, Belgium, pp 191–196
7. Ruiz H, Jarman IH, Martín JD, Lisboa PJG (2011) The role of Fisher information in primary data space for neighbourhood mapping. In: European symposium on artificial neural networks, computational intelligence and machine learning (ESANN), Bruges, Belgium, pp 381–386
8. Suckling J, Parker J, Dance D (1994) The mammographic image analysis society digital mammogram database. In: Exerpta medica international congress series
9. Zhao D, Shridhar M, Daut DG (1992) Morphology on detection of calcifications in mammograms. In: Proceedings of the 1992 IEEE international conference on acoustics, speech, and signal processing, ICASSP-92, vol 3. IEEE, pp 129–132
10. Yao J, Chen J, Chow C (2009) Breast tumor analysis in dynamic contrast enhanced MRI using texture features and wavelet transform. IEEE J Sel Top Signal Process 3(1):94–100
11. Haralick RM, Shanmugam K, Dinstein I (1973) Textural features for image classification. IEEE Trans Syst Man Cybern SMC-3(6):610–621
12. Cross GR, Jain AK (1983) Markov random field texture models. IEEE Trans Pattern Anal Mach Intell 5(1):25–39
13. Laine A, Fan J (1993) Texture classification by wavelet packet signatures. IEEE Trans Pattern Anal Mach Intell 15(11):1186–1191
14. Breiman L, Friedman JH, Olshen RA, Stone CJ (1984) Classification and regression trees. Routledge, Abingdon
15. Newman MEJ (2004) Detecting community structure in networks. Eur Phys J B 38:321–330
16. Young G, Householder AS (1938) Discussion of a set of points in terms of their mutual distances. Psychometrika 3(1):19–22
17. Uppal MTN Classification of mammograms for breast cancer detection using fusion of discrete cosine transform and discrete wavelet transform features. Biomed Res 27(2):322–327

Network Community Cluster-Based Analysis for the Identification of Potential Leukemia Drug Targets

Adrián Bazaga[1,2](✉) and Alfredo Vellido[3]

[1] Department of Genetics, University of Cambridge, Cambridge CB2 3EH, UK
ar989@cam.ac.uk
[2] STORM Therapeutics Ltd., Cambridge CB22 3AT, UK
[3] Computer Science Department,
Intelligent Data Science and Artificial Intelligence (IDEAI) Research Center,
Universitat Politècnica de Catalunya, Barcelona, Spain
avellido@cs.upc.edu

Abstract. Leukemia is a hematologic cancer which develops in blood tissue and causes rapid generation of immature and abnormal-shaped white blood cells. It is one of the most prominent causes of death in both men and women for which there is currently not an effective treatment. For this reason, several therapeutical strategies to determine potentially relevant genetic factors are currently under development, as targeted therapies promise to be both more effective and less toxic than current chemotherapy. In this paper, we present a network community cluster-based analysis for the identification of potential gene drug targets for acute lymphoblastic leukemia and acute myeloid leukemia.

Keywords: Network analysis · Community clusters · Drug discovery · Therapeutic targets · Leukemia

1 Introduction

Leukemia is known to be a group of cancers that usually begin in the bone marrow and result in a high number of abnormal white blood cells. Although its prevalence is low, the chance of surviving is one of the lowest among cancer diseases. Moreover, the population in developed countries is aging incrementally, and taking into account that older people have a higher risk of developing leukemia, a steady increase of cases is to be expected.

Leukemia involves complex genetic factors, and identifying which ones are relevant to treat the disease can make a difference between life or death. The treatments for and reactions of each type of leukemia may vary considerably [1]. The probability of survival can be increased by methods that allow to identify types of leukemia accurately, as well as by the use of computational methods for the discovery of relevant targets in the genome that might be of interest

© Springer Nature Switzerland AG 2020
A. Vellido et al. (Eds.): WSOM 2019, AISC 976, pp. 314–323, 2020.
https://doi.org/10.1007/978-3-030-19642-4_31

for drug development [2]. In this paper, we will focus on the task of discovering promising target genes for the treatment of acute lymphoblastic leukemia (ALL) [3] and acute myeloid leukemia (AML) [4] types, which represent almost half of the totality of leukemia cases.

The main objective of this paper can be stated as follows: given the gene-interaction network related to ALL and AML, where some genes are known to be targets of certain drugs approved by the U.S. Food and Drug Administration (FDA), as they are highly significant for the disease at hand, and some others are not (or not known to be), we aim to assess if the topological structure of the sub-networks related to the genes belonging to the two different classes, namely target and non-target, have any specificity in terms of statistical properties; we also want to analyze if, by using network community cluster detection techniques, it is possible to find potential drug targets.

The rest of the paper is structured as follows: Sect. 2 introduces the basic techniques employed in our analysis, as well as the data used in the study. Then, Sect. 3 describes and discusses in detail the experimental findings, while Sect. 4 concludes the paper and points to issues deserving further research.

2 Materials and Methods

2.1 Graph-Theoretical Centralities

Degree Centrality. This paper focuses on graph community cluster analysis [5]. For this, some graph centrality measures must be defined first. The degree centrality is defined as the number of edges going into or out of a node.

Betweenness Centrality. This measure quantifies the number of times a node is in the way along the shortest path between two other nodes. For a node v, it is defined as

$$B(v) = \sum_{s \neq v \neq t \in V} \frac{\sigma_{st}(v)}{\sigma_{st}}, \tag{1}$$

where σ_{st} is the total number of shortest paths from node s to node t and $\sigma_{st}(v)$ is the number of those shortest paths that pass through v.

Closeness Centrality. The closeness centrality of a node is defined as the average length of the shortest path between the node and all other graph nodes. A variant that accounts for the possibility of having a not connected graph is known as *harmonic centrality*, which is the one used in this paper. For a node v, it is defined as

$$HC(v) = \sum_{y \neq v} \frac{1}{d(y, v)}, \tag{2}$$

where $d(y, v)$ is the distance between nodes y and v, imposing that $\frac{1}{d(y, v)} = 0$ if there is no path from y to v.

PageRank Centrality. The PageRank centrality [6], which is a variant of the Eigenvector centrality [7], was originally defined for a scenario where a user surfs the web by clicking links. The PageRank value of a website is an estimation of the probability that the user is on a web page at a given moment. Generalizing from webpages to network nodes, three elements determine the PageRank of a node: the number of incoming edges, the number of outgoing edges of the linking nodes, and the PageRank of the linking nodes.

2.2 Random Graph Null Hypothesis: The Erdös-Rényi model

The Erdös-Rényi graph [8] is a random network model where edges are connected independently between each pair of nodes with a probability p that follows a Bernoulli distribution, thus they have no community (cluster) structures. This model has been widely used as a null hypothesis to find patterns in the topology and community structure of real networks [9]. For this reason, in this work we use it to study the particular properties of gene-interaction networks.

2.3 Community Finding Algorithms

Walktrap Algorithm. The Walktrap algorithm [10] is a community detection algorithm that uses a distance metric based on performing random walks and uses a hierarchical agglomerative clustering method. Formally, if two nodes i and j belong to the same community, the probability to reach a third node k belonging to the same community by means of a random walk, should differ minimally for i and j. Then, the distance between two nodes i and j is constructed by summing these differences over all nodes, with a correction for each node degree.

Infomap Algorithm. The Infomap algorithm represents the community-cluster structure of a graph by means of a two-level nomenclature based on Huffman coding [11]. It defines the problem of finding the optimal clustering of a graph as finding a description of minimum information using random walks on the graph. Moreover, the algorithm objective function is to maximize the so-called Minimum Description Length [12].

2.4 Data Gathering and Building the Gene-Interaction Network

The data in our experiments is divided over different sources. First, we searched for which FDA-approved drugs are currently used to treat ALL and AML types of leukemia. This was obtained from the U.S. National Cancer Institute [13].

On the other hand, from the Drug-Gene Interaction Database [14], the genes that are targeted by a given drug, which total 197, were obtained. Then, in order to obtain negative samples (non-target genes), we queried human gene identifiers from HumanMine, a biological database developed by the University of Cambridge [15,16]. From a pool of 62,906 genes, 197 were randomly sampled, constrained to only those genes with at least one known interaction with another gene, and that are not known to be a target of any disease. This was done to obtained a balanced dataset for analysis.

After that, we built the gene-interaction network of both the 197 target and non-target genes. To do so, for each gene we queried the BioGRID database [17] for the genes interacting directly with each of the genes in our dataset. Consequently, for each gene we have its direct interacting neighborhood, where some interacting gene neighbors may be shared among different genes, thus leading to a connected graph. This resulted in a network comprised of 12,761 nodes and 72,634 edges (see Table 1 for summary information).

Table 1. Network composition

Network	Nodes	Edges
Target network	11966	50512
Non-target network	11966	22122
Full	12761	72634

3 Experimental Results and Discussion

3.1 Description of the Network Structure

The illustration of the full network is provided in Fig. 1, with the node sizes drawn according to their PageRank [6,7] value. These values are linearly related to the dimension of the vertex, where a greater PageRank value corresponds to a greater dimension of the node. In Fig. 1, blue nodes are the non-target genes and the red ones are the target genes, that are given by approved FDA drugs for ALL and AML. Also, orange nodes are genes that interact with either target genes, non-target genes, or both, as is the case when they have common neighbors. These orange nodes are not known to be targets or non-target genes, and thus we refer to them as unknown genes from now on.

In order to further characterize our network, we calculate two widely known graph metrics: the transitivity (also known as clustering coefficient), and the diameter. We measure these metrics for the full network, the target genes sub-network and the non-target genes sub-network. The transitivity, T, of a graph, G, is based on the relative number of triangles in the graph, compared to the total number of connected triples of nodes. Formally, transitivity T of a graph G is calculated as

$$T(G) = \frac{3 \times \text{number of triangles in the graph}}{\text{number of connected triples in the graph}}, \tag{3}$$

where the factor of three in the numerator takes into account the fact that each triangle contributes to three different connected triples in the graph, one centered at each node of the triangle.

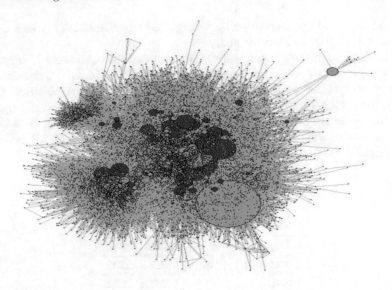

Fig. 1. Visualization of the gene-gene interaction network where node sizes are relative to their PageRank value

The diameter of a graph is the length of the longest shortest path between any two graph vertices, u and v, where $d(u, v)$ is a graph distance, that is, the largest number of vertices which must be traversed in order to travel from one vertex to another, without taking into account loops or backtracking paths. Thus, if the shortest path between the two farthest nodes in the graph has a length of 3, the diameter of the graph will have a value of 3.

Table 2. Transitivity and diameter metrics of the full graph and target/non-target subgraphs

Network	Transitivity	Diameter
Target network	0.1365	10
Non-target network	0.1436	10
Full	0.1346	10

In Table 2, we show the transitivity and diameter metrics of the full network, as well as for the target and non-target sub-networks. As can be seen, the diameter in the three scenarios is the same, thus the longest shortest path is the same. Also, the transitivity of the non-target sub-network is higher than the one of the full and target networks, thus indicating that this sub-network has a higher density of connections between nodes. Consequently, in the non-target sub-network, the number of shared interacting genes between the non-target genes is higher than the corresponding one in the other networks.

In order to study the topology of the graph, we assessed its clustering coefficient as a measure of characterization of the graph topology, with respect to how it would behave in a random scenario, using the Erdös–Rényi model as our null hypothesis. The null hypothesis we set is such that the clustering coefficient of our network is not significantly different to that of a random model, given a 0.95 of confidence ($\alpha = 0.05$). Consequently, the hypothesis we want to verify is that the clustering coefficient of our network, and hence, its topology, is due to its specific nature and not due to a random behaviour. In order to reject or accept the null hypothesis, we compute the p-value by dividing the number of times the random model has a higher value than our network, divided by the total number of times we carry out the experiment. For this study, we build 30 random graph as experimental set and compute the clustering coefficient of each one.

Moreover, given the size of the graph, and in order to speed up the computation, we carried out an optimization to calculate the metrics exactly but without scanning all the graph, bounding the values of the clustering coefficient in the null hypothesis, C_{NH}, below C_{NH}^{min} and above C_{NH}^{max}, exploring only a subset of M nodes of the network. The value of the lower bound C_{NH}^{min} is calculated as $\frac{1}{N} \cdot \sum_{i=1}^{M} C_i$, and the value of the upper bound C_{NH}^{max} as $\frac{1}{N} \cdot \sum_{i=1}^{M} C_i + 1 - \frac{M}{N}$.

Then, with the previous formulae, after calculating the clustering coefficient C_i for only the first M nodes in the network produced by the null hypothesis, we can compare it with the bounds, and assume that if $C_{NH}^{min} \geq C$, then $C_{NH} \geq C$, also if $C_{NH}^{max} < C$ then $C_{NH} < C$.

Furthermore, by allowing for a certain degree of error in evaluating C (the clustering coefficient in our network) and C_{NH}, we can further optimize the calculation by means of a Monte Carlo procedure. To do so, we order the graph by doing a uniform permutation of the vertices, and then calculate the clustering coefficient only for the first M vertices. The value of M we use is based on the fact that even when $M \ll N$, we can get a good estimation of the clustering coefficient [18], such as $100M/N = 10\%$, and with that premise we can solve as $M = \lceil \frac{0.1 * N}{100} \rceil$.

Table 3. Statistical marginal significance (p-value) after the Monte Carlo procedure on the clustering coefficient. The selected confidence value is 0.95 ($\alpha = 0.05$).

Network	Erdös–Rényi (random model)
Target network	0.03
Non-target network	0

As shown in Table 3, the p-value with respect to the random network null hypothesis is lower than the significance level $\alpha = 0.05$, leading to significant evidence for the rejection of the null hypothesis. This means that, as our network clustering coefficient is a particular characteristic of it, our network topology can be seen as having specific, non-random, characteristics.

3.2 Community Analysis

Since we are dealing with a graph with different node categories, it is worth investigating if a community detection algorithm is able to detect these underlying clusters. In order to build the communities, we used the previously described Infomap [11] and Walktrap [10] algorithms, taken from the R package *igraph*. Notice that the genes that are not known to be targets or non-targets have been excluded from the community analysis, as the main idea in this part of the analysis is to verify if target (or non-target) genes are similar enough, and, due to the interactions between them and their local topology in the network, they can be grouped in pure communities or clusters. Here, we define a pure community or cluster as a group formed of only one type of genes, either targets or non-targets. Hence, an impure community would be formed of a mixture of targets and non-targets.

After running the community finding algorithms, we measured the goodness of the communities found by the two different chosen algorithms. For this, we rely on different quality measures: Triangle Partition Ratio (TPR), expansion, conductance and modularity. TPR is the fraction of nodes within a cluster that belongs to a triad; thus a higher value translates into a clustering with a higher quality. Expansion is the number of nodes leaving the cluster; thus, a lower value means a clustering of better quality. Conductance is the fraction of total edge volume that points outside the cluster; thus, a lower value is better. Modularity is the difference between the number of edges in the cluster and the expected number of edges of a random graph with the same degree distribution; thus, a higher value is better.

Table 4. Quality measures of community structure.

Algorithm	TPR	Expansion	Conductance	Modularity	Communities
Walktrap	**0.763**	**1.216**	**0.188**	**0.6**	318
Infomap	0.68	6.906	0.394	0.56	370

The quality metrics for each of the algorithms are summarized in Table 4. From these results, it is clear that the Walktrap algorithm yields the best values for TPR, expansion, modularity and conductance. Consequently, by relying exclusively on the values of these heuristic quality metrics to judge the goodness of the community finding algorithm performance, we can say that the Walktrap algorithm produces the best segmentation of the graph into different communities. We then proceeded to analyze in more detail such communities.

In Fig. 2, we show the percentage of genes per category (target and non-target) of each community given by the result of the Walktrap community detection algorithm. From this result, we can say that there is a densely connected cluster of target genes and a densely connected cluster of non-target genes. The result given by the community finding algorithm is very interesting as, since

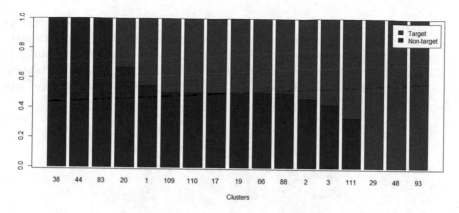

Fig. 2. Communities composition based on the Walktrap algorithm.

it is able to find pure communities of targets, it means that the target genes have a particularly characteristic topology in the network, thus leading to the conclusion that it is possible to unveil new targets by means of their structure on the network (e.g., by taking into account their graph-theoretical centrality). Then, unknown genes that have characteristics similar to the target genes may be potential targets.

3.3   Analysis of the Network Based on Centrality Measures

After ensuring that the topology of the two types of genes in our network is characteristic, we studied the network using the graph-theoretical centrality measures described in Sect. 2. Their values for the full network and the target and non-target sub-networks are summarized in Table 5.

Table 5. Average values of the centrality measures in the three networks

Network	Degree	Betweenness	Closeness
Target	0.0022	0.00021	0.29
Non-target	0.0021	0.00023	0.28
Full	0.0020	0.0002	0.29

Then, given the previous insights, we analyzed the genes with the highest values for each of the graph-theoretical based centralities, as those are the ones showing clearer predominant centrality values in the whole network.

After that, we took the genes that were shortlisted in the three sets. Thus we took the genes that are among the top with respect to their centrality values in all the centrality measures at the same time.

In Fig. 3 the set of genes that have the highest values in all the centralities is illustrated. As can be seen, this set is comprised of only target and unknown genes at the same time, pointing out that these unknown genes may be potential candidates as drug targets for AML and ALL. These genes are HSP90AA1, TRIM25, ELAVL1, APP, MCM2, CUL3, HSPA8, XPO1, EGLN3, UBC and NXF1.

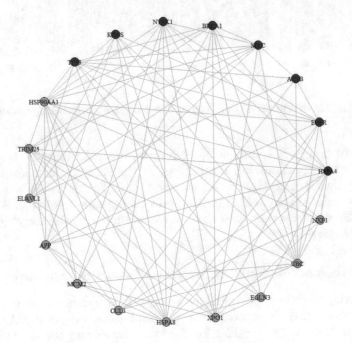

Fig. 3. Set of top shortlisted genes with the highest values in all centrality measures: degree, closeness and betweenness.

4   Conclusions

This brief paper has reported network analysis and community cluster detection for the identification of potential drug targets for ALL and AML types of leukemia. We have described in detail how we built a gene-interaction network for the genes targeted by currently FDA approved drugs for ALL and AML, and for non-target genes obtained from HumanMine, a publicly available biological database.

The non-randomness (and therefore the topological specificity) of the network has been asserted. Furthermore, by using the Walktrap and Infomap community detection algorithms, it has been shown that both target and non-target genes can be segmented in pure groups. Finally, by analyzing several graph-theoretical centrality measures, and taking the genes that hold the highest values of these measures in the network, we were able to identify a set of potential drug target candidates for ALL and AML.

This study could be extended to the complete leukemia spectrum, including chronic lymphocytic leukemia, chronic myelogenous leukemia, hairy cell leukemia, mast cell leukemia and meningeal leukemia. In addition, it would be interesting to analyze in more detail the specificities of those genes that were found to be potential targets, by carrying out a gene set enrichment analysis to check if, statistically, they have significant pathways affected and their relation to ALL and AML leukemias.

References

1. Pui CH, Evans WE (2006) Treatment of acute lymphoblastic leukemia. New Engl J Med 354(2):166–178
2. Arakawa H et al (1998) Identification and characterization of the ARP1 gene, a target for the human acute leukemia ALL1 gene. PNAS 95(8):4573–4578
3. Pui CH (2011) Acute lymphoblastic leukemia. Springer, Heidelberg
4. Lowenberg B, Downing JR, Burnett A (1999) Acute myeloid leukemia. New Engl J Med 341(14):1051–1062
5. Fortunato S (2010) Community detection in graphs. Phys Rep 486(3–5):75–174
6. Franceschet M (2011) PageRank. Commun ACM 54(6):92
7. Bonacich P, Lloyd P (2001) Eigenvector-like measures of centrality for asymmetric relations. Soc Netw 23(3):191–201
8. Erdös P, Rényi A (1960) On the evolution of random graphs. In: Publication of the mathematical institute of the hungarian academy of sciences, pp 17–61
9. Yates PD, Mukhopadhyay ND (2013) An inferential framework for biological network hypothesis tests. BMC Bioinform 14(1):94
10. Pons P, Latapy M (2005) Computing communities in large networks using random walks. In: Computer and information sciences, ISCIS 2005. Springer, Heidelberg, pp 284–293
11. Rosvall M, Bergstrom CT (2008) Maps of random walks on complex networks reveal community structure. PNAS 105(4):1118–1123
12. Rissanen J (1978) Modeling by shortest data description. Automatica 14(5):465–471
13. National Cancer Institute (2019) Drugs approved for leukemia. https://www.cancer.gov/about-cancer/treatment/drugs/leukemia
14. McDonnell Genome Institute WUSoM (2019) The drug-gene interaction database. http://www.dgidb.org/
15. Smith RN et al (2012) InterMine: a flexible data warehouse system for the integration and analysis of heterogeneous biological data. Bioinformatics 28(23):3163–3165
16. Kalderimis A et al (2014) InterMine: extensive web services for modern biology. Nucleic Acids Res 42(W1):W468–W472
17. Chatr-Aryamontri A et al (2017) The biogrid interaction database: 2017 update. Nucleic Acids Res 45(D1):D369–D379
18. Watts DJ, Strogatz SH (1998) Collective dynamics of 'small-world' networks. Nature 393(6684):440

Searching for the Origins of Life – Detecting RNA Life Signatures Using Learning Vector Quantization

Thomas Villmann[1]([✉]), Marika Kaden[1], Szymon Wasik[2], Mateusz Kudla[2], Kaja Gutowska[2,3], Andrea Villmann[1,4], and Jacek Blazewicz[2,3]

[1] Saxony Institute for Computational Intelligence and Machine Learning, University of Applied Sciences Mittweida, Mittweida, Germany
`thomas.villmann@hs-mittweida.de`
[2] Institute of Computing Science, Poznan University of Technology, Poznan, Poland
[3] Institute of Bioorganic Chemistry, Polish Academy of Sciences, Warsaw, Poland
[4] Schulzentrum Döbeln-Mittweida, Döbeln, Germany
`https://www.institute.hs-mittweida.de/webs/sicim.html`

Abstract. The most plausible hypothesis for explaining the origins of life on earth is the RNA world hypothesis supported by a growing number of research results from various scientific areas. Frequently, the existence of a hypothetical species on earth is supposed, with a base RNA sequence probably dissimilar from any known genomes today. It is hard to distinguish hypothetical sequences obtained by computer simulations from biological sequences and, hence, to decide which characteristics provide biological functionality. In the present consideration biological sequences obtained from RNA-viruses are compared with computationally generated sequences (artificial life probes). The task is to discriminate the samples regarding their origin, biological or artificial. We used the learning vector quantization (LVQ) model as the respective classifier. LVQ is a dissimilarity based classifier, which has only weak requirements regarding the underlying dissimilarity measure. This gives the opportunity to investigate several dissimilarity measures regarding their discriminating behavior for this task. Particularly, we consider information theoretic dissimilarities like the normalized compression distance (NCD) and divergences based on bag-of-word (BoW) vectors generated on the base of nucleotide-codons. Additionally, the geodesic path distance is applied taking an unary coding of sequences for a representation in the underlying Grassmann-manifold. Both, BoW and GPD allow continuous updates of prototypes in the feature space and in the Grassmannmanifold, respectively, whereas NCD restricts the application of LVQ methods to median variants.

1 Introduction

The origin of life is one of the most interesting questions for mankind. The today accepted hypothesis for this is the RNA world hypothesis [1–3]. This hypothesis

© Springer Nature Switzerland AG 2020
A. Vellido et al. (Eds.): WSOM 2019, AISC 976, pp. 324–333, 2020.
https://doi.org/10.1007/978-3-030-19642-4_32

supposes that the life originated from the RNA sequences which had an ability of self-replication [4]. Many considerations validate the evidence of this claim, for an overview we refer to [5,6].

Beginning with publications by H. QUASTLER and M. EIGEN the bioinformatics perspective becomes more and more important for analysis of RNA and DNA sequences [7–9]. One question in this context is how artificially generated RNA sequences differ from RNA sequences of biological origin and, related to that, which mathematical tools are suitable to detect those differences. A first attempt in that direction was presented in [5]. Artificially generated sequences according to theoretical standards were compared to biological sequences using the edit distance and the normalized compression distance (NCD). The comparison was processed investigating rank statistics for within class and between class dissimilarities. The results suggest that the sequences can be distinguished weakly (approx. 63% accuracy). Yet, a respective classification model was not applied in this investigation. Thus the question arises whether a classification scheme could achieve better results.

In our publication we use variants of KOHONEN'S learning vector quantization (LVQ, [10]) as an appropriate robust prototype based classification model depending on the dissimilarity measure between sequences. Particularly, we applied other dissimilarity measures than the NCD and edit distance for comparison, which are discussed to be suitable in the context of symbolic DNA/RNA sequence analysis. These measures comprise bag-of-words approaches, Grassmann-manifold embedding and other together with appropriate data dissimilarities. Depending on the mathematical properties of considered dissimilarities like differentiability, symmetry etc., gradient methods can be taken for optimization, like stochastic gradient descent learning (SGDL) for generalized LVQ (GLVQ, [11]. Otherwise, median or relational variants of GLVQ have to be applied [12].

The paper is structured as follows. First, we shortly describe the data and the aim of this investigation. Thereafter, we briefly review approaches and dissimilarity measures to compare symbolic RNA sequences. Subsequently, respective LVQ variants are considered. The result section, finally provides the achieved classification performances.

2   Biological and Artificial RNA Sequence Data - The Search for Signatures of Life

As mentioned in the introduction, the RNA world hypothesis is the starting point for our consideration. According to [5], the only known biological entities that have genomes composed of RNA are RNA viruses. RNA viruses are seen as the closest relatives to the primordial RNA. However, RNA viruses are not alive itself because they require a system of a living cell to reproduce [13,14]. Yet, it is assumed that the RNA sequences of these viruses kept remnants of earlier replication strategies [15].

The RNA virus sequences used in this consideration are taken from the NCBI Viral Genomes database [16]. Each sequence was cut randomly into 70- to 100-nucleotide long fragments. This choice was made according to one of the most significant open problems in this area: How the life was able to overcome the so-called Eigen's paradox, stating that it was not possible to evolve genetic sequences long enough to code mechanisms required by more complex organisms [17]. According to this, the theoretical, maximal length of the genetic sequence that could exist without any error correction mechanism was estimated to be about 100 nucleotides long. Accordingly, in longer sequences, the mutation ratio is so high that it destroys the information stored in the sequence over a longer time period. The lower bound was chosen as 70, because tRNA molecules that serve as the physical link between the mRNA and the amino acid sequence of proteins are typically 76 to 90 nucleotides in length [18]. Finally, a training set S_b and a test set T_b of biologically functional sequences are obtained. The detailed preparation is described in [5].

The artificial reference data sets S_a and T_a for training and test, respectively, contain sequences of different lengths uniformly distributed in the interval $[70, 100]$. The frequency statistics for the nucleotides in these set follows exactly this one of the biological data set S_b ($p_A = 28,7\%$, $p_C = 21,3\%$, $p_G = 22,9\%$, $p_U = 27,1\%$). Both data sets are taken from [5]. The number of data in each of the four subsets is 500.

3   Methods and Similarity Measures for RNA/DNA Sequence Comparison

Several measures are favored to compare symbolic sequences. Among them, those measures became attractive during the last years, which do not assume a sequence alignment for preprocessing [19–22]. Frequently, respective dissimilarity measures are based on information theoretic approaches.

Normalized Compression Distance (NCD). One of the most prominent and successful measures is the normalized compression distance

$$NCD\left(s_1, s_2\right) = \frac{\max\left(K\left(s_1|s_2\right), K\left(s_2|s_1\right)\right)}{\max\left(K\left(s_1\right), K\left(s_2\right)\right)} \tag{1}$$

as approximation of the universal description length [23,24]. There, $K\left(s\right)$ is the Kolmogorov complexity of the sequence s [25], which is usually approximated by the size of the compressed sequence using a compression tool relied on the LZ77 algorithm and Huffman coding [26,27].

Bag of Words (BoW). Another popular method to extract information of RNA/DNA sequences is the method bag-of-words (BOW), where the words are all possible 64 triplets of the four nucleotides denoted as codons in this context [28–30]. Thus all sequences are coded as (normalized) histogram vectors of dimensionality $n = 64$. Mathematically speaking, these vectors are discrete representations of probability densities. Thus, comparison of those vectors can be done using the usual Euclidean distance or, motivated by the latter mentioned density property, by divergences. In the investigations presented later, we applied beside the Euclidean distance the discretized) Kullback-Leibler-divergence

$$D_{KL}(\mathbf{x}, \mathbf{y}) = \sum_{j=1}^{n} x_j \cdot \log(x_j) - \sum_{j=1}^{n} x_j \cdot \log(y_j) \tag{2}$$

for sequence histograms $\mathbf{x}$ and $\mathbf{y}$. Note that the first term in (2) is the negative Shannon-entropy $S(\mathbf{x}) = -\sum_{j=1}^{n} x_j \cdot \log(x_j)$ and $Cr(\mathbf{x}, \mathbf{y}) = \sum_{j=1}^{n} x_j \cdot \log(y_j)$ is the cross-entropy. Yet, other divergences like Rényi-divergences could also be applied [29,31].

Subspace Method. A third approach gaining more and more interest is to represent the nucleotides by unary coding and, hence, to consider a sequence by a respective matrix representation [20]. Here each nucleotide is coded by a unique vector such that the distance between them is always the unit. Thus, a sequence s of length n delivers a matrix $\mathbf{X}$ of rank $k = 3$.

One idea is to take these matrices as representation of greyscale images and to apply deep convolutional networks for data analysis [32]. An interesting alternative is the matrix interpretation as subspaces. The three-dimensional subspace $[\mathbf{X}]$ in the $\mathbb{R}^n$ spanned by $\mathbf{X}$ can be taken as a point in the Grassmannian manifold $\mathscr{G}(k, n)$. The Grassmann manifold $\mathscr{G}(k, n)$ equipped with the Riemann geometry is the space of all k-dimensional linear subspaces [33], i.e. a matrix $\mathbf{X}$ determines via $[\mathbf{X}]$ a certain point in the Grassmann manifold $\mathscr{G}(k, n)$. The Riemann geometry determines the distance measure (metric)

$$d_g([\mathbf{X}], [\mathbf{Y}]) = \sqrt{\sum_{j=1}^{k} \sigma_j^2} \tag{3}$$

denoted as the *geodesic path distance* (GPD) in the Grassmann manifold $\mathscr{G}(k, n)$, where $\sigma_1, \ldots, \sigma_k$ are the subspace angles (SSA) between the subspaces $[\mathbf{X}], [\mathbf{Y}] \in \mathscr{G}(k, n)$ [34,35].[1] The geodesic path between $[\mathbf{X}]$ and $[\mathbf{Y}]$ is given by the parametric equation

$$\mathbf{G}(\tau, \mathbf{X}, \mathbf{Y}) = \mathbf{X} \cdot \mathbf{V} \cdot \cos(\boldsymbol{\Theta} \cdot t) + \mathbf{U} \sin(\boldsymbol{\Theta} \cdot t) \tag{4}$$

with $\mathbf{G}(0, \mathbf{X}, \mathbf{Y}) = [\mathbf{X}]$ and $\mathbf{G}(1, \mathbf{X}, \mathbf{Y}) = [\mathbf{Y}]$ are valid. Here, the quantities $\mathbf{U}$, $\boldsymbol{\Theta}$, and $\mathbf{V}$ are obtained from the singular value decomposition

$$\mathbf{U}\boldsymbol{\Sigma}\mathbf{V} = (\mathbf{I} - \mathbf{X} \cdot \mathbf{X}^T)\mathbf{Y}(\mathbf{X}^T\mathbf{Y})^{-1} \tag{5}$$

---

[1] For computational convenience it is usually assumed that both matrices $\mathbf{X}$ and $\mathbf{Y}$ are orthonormal, which can always be obtained by Gram-Schmidt-orthonormalization. We will take this assumption here, too. If this assumption is dropped the procedure is still valid but more complicated. We refer to [34].

together with $\boldsymbol{\Theta} = \tan(\boldsymbol{\Sigma})$. If $\mathbf{X}^T\mathbf{Y}$ is not invertible, the pseudo-inverse can be applied in (5) for approximation. Further, $P = (\mathbf{I} - \mathbf{X} \cdot \mathbf{X}^T)$ is an orthoprojector matrix and $\boldsymbol{\Sigma}$ is diagonal with the singular values $\sigma_1, \ldots, \sigma_n$ where $\sigma_{k+1} = \ldots = \sigma_n = 0$ and $\sigma_1, \ldots, \sigma_k$ are the subspace angles from (3).

4   Generalized Learning Vector Quantization for Classification Learning

Learning vector quantization (LVQ), as introduced by KOHONEN in [10], assumes data classes $1, \ldots, C$ and data $\mathbf{x} \in X \subseteq \mathbb{R}^n$. The task is to distribute a set $W = \{\mathbf{w}_1, \ldots, \mathbf{w}_M\}$ of prototype vectors $\mathbf{w}_j \in \mathbb{R}^n$ to approximate the class distribution of the data. For this purpose, each prototype $\mathbf{w}_j$ is equipped with a class labels $c(\mathbf{w}_j)$ such that at least one prototype is responsible for each class. The prototypes are adapted according to an attraction-repelling-scheme (ARS) in dependence whether the class label of a presented training sample coincides with that of the best matching prototype $\mathbf{w}_{s(\mathbf{x})}$ determined by a winner-take-all competition (WTAC)

$$s(\mathbf{x}) = \operatorname{argmin}_{j=1\ldots M}(d(\mathbf{x}, \mathbf{w}_j)) \tag{6}$$

where d is a given dissimilarity measure [36]. We denote $\mathbf{w}_{s(\mathbf{x})}$ also as the winner prototype of the competition. After training a data sample is classified according to $c(\mathbf{x}) = c(\mathbf{w}_{s(\mathbf{x})})$.

Generalized LVQ is a cost function based variant of LVQ with

$$E_{GLVQ}(W, X) = \sum_{k=1}^{N} \varphi(\mu(\mathbf{x}_k, W)) \tag{7}$$

as cost function to be minimized during learning of labeled training data $\mathbf{x}_k \in X$ [11]. The function $\varphi(z)$ is monotonically increasing frequently chosen as the identity function $\operatorname{id}(z) = z$ or the sigmoid function $\phi(z, \theta) = \frac{1}{1+\exp\left(\frac{z}{\theta}\right)}$. The function

$$\mu(\mathbf{x}, W) = \frac{d(\mathbf{x}, \mathbf{w}^+) - d(\mathbf{x}, \mathbf{w}^-)}{d(\mathbf{x}, \mathbf{w}^+) + d(\mathbf{x}, \mathbf{w}^-)} \tag{8}$$

is the so-called classifier function. Here, $\mathbf{w}^+$ is the best matching prototype regarding a training vector $\mathbf{x}$ with label $c(\mathbf{x})$ with the same class label whereas $\mathbf{w}^-$ denotes the best matching prototype of all prototypes of the other classes. Hence, $\mu(\mathbf{x}, W) \in [-1, 1]$ takes negative values if $\mathbf{x}$ is correctly classified. Thus, $E_{GLVQ}(W, X)$ approximates the overall classification error [37].

4.1   Stochastic Gradient Descent Learning in GLVQ

Stochastic gradient descent learning (SGDL) with respect to E_{GLVQ} gives

$$\Delta\mathbf{w}^{\pm} \propto -\xi(\mathbf{x}, \mathbf{w}^{\pm}) \cdot \frac{\partial\mu}{\partial d^{\pm}(\mathbf{x})} \frac{\partial d^{\pm}(\mathbf{x})}{\partial\mathbf{w}^{\pm}} \tag{9}$$

requiring the differentiability of the dissimilarity measure d. The scaling factor

$$\xi\left(\mathbf{x}, \mathbf{w}^{\pm}\right) = \frac{\partial E\left(\mathbf{x}_k\right)}{\partial \varphi} \cdot \frac{\partial \varphi}{\partial \mu} \qquad (10)$$

realizes the ARS with the local error $E\left(\mathbf{x}_k\right) = \varphi\left(\mu\left(\mathbf{x}_k, W\right)\right)$ and the short hand notation $d^{\pm}\left(\mathbf{x}\right) = d\left(\mathbf{x}, \mathbf{w}^{\pm}\right)$.

For vector quantization learning in Grassmannian manifolds the geodesic path (4) can be used, as already suggest for self-organizing maps by KIRBY and PETERSON [38]. The adaptation becomes

$$\Delta\mathbf{W}^{\pm} \propto \mathbf{W}^{\pm} + \xi\left(\mathbf{X}, \mathbf{W}^{\pm}\right) \cdot \left(\mathbf{G}\left(1 - \varepsilon, \mathbf{X}, \mathbf{W}\right) - \mathbf{W}^{\pm}\right) \qquad (11)$$

paying attention to the situation that the data $\mathbf{X}$ as well as the prototypes $\mathbf{W}$ are matrices in this case. Again, the factor $\xi\left(\mathbf{X}, \mathbf{W}^{\pm}\right)$ realizes the ARS [39]. The positive quantity $\varepsilon \ll 1$ realizes the learning rate.

4.2   Median and Relational GLVQ

Median and relational GLVQ only require dissimilarity values $d_{ij} = d\left(\mathbf{x}_i, \mathbf{x}_j\right)$ collected into the matrix $\mathbf{D}$. The matrix $\mathbf{D}$ is said to be Euclidean embeddable if there exist a mapping $\tilde{\mathbf{x}} = \psi\left(\mathbf{x}\right)$ such that $d_{ij} = d_E\left(\tilde{\mathbf{x}}_i, \tilde{\mathbf{x}}_j\right)$ is valid, where $d_E\left(\tilde{\mathbf{x}}_i, \tilde{\mathbf{x}}_j\right)$ is the Euclidean distance between $\tilde{\mathbf{x}}_i$ and $\tilde{\mathbf{x}}_j$. For those data the prototypes can be supposed to be linear combination of the data, i.e. we have $\mathbf{w}_l = \sum_j \alpha_{lj}\mathbf{x}_j$ such that the prototype update is realized as SGDL with respect to the coefficients α_{lj} [40]. A sufficient condition for the Euclidicity of $\mathbf{D}$ is that the corresponding similarity matrix $\mathbf{S}$ is positive semi-definite [41].

Median-GLVQ optimizes the GLVQ cost function (7) by an Expectation-Maximization approach [12]. Here the prototypes are restricted to be data objects. Yet, the optimization still works under weaker assumptions regarding the matrix $\mathbf{D}$, particularly the Euclidicity may be violated compared to relational methods.

5   Results

We performed various experiments to investigate the distinguishability between artificial and biological RNA nucleotide sequences based on the previously described data sets S_a, T_a and S_b, T_b for artificial and biological sequences, respectively.

First we calculated all dissimilarity values between the data. We used the NCD, BoW equipped with the Euclidean distance and Kullback-Leibler divergence as well as the subspace geodesic path distance. Additionally, we applied for the latter case the Frobnius-norm as distance measure.

Training is performed on $S_a \cup S_b$ whereas the test data set is $T_a \cup T_b$. For all dissimilarity measures we applied several Median GLVQ runs with different size of prototype sets. Additionally, for the differentiable dissimilarities, trained also a standard (continuous) GLVQ. For the subspace approach with the geodesic path distance we performed the continuous GLVQ using the adaptation rule (11). Additionally, for comparison, a standard kNN is applied for the test cae, where the training set is taken as reference. All results are depicted in Table 1.

Table 1. Classification results for detection of artificial and biological RNA sequences using various dissimilarity measures and accordingly selected GLVQ variants. Both, training and test results are depicted. Additionally, kNN results for several k-values are presented for comparison.

$k/|W|$	kNN			Median GLVQ			GLVQ		
	1	3	5	2	10	20	2	10	20
Training									
NCD				66.6	63.4	63.5			
BoW Euclid				65.8	66.1	66.1	77.8	87.9	91.1
BoW KL				83.2	95.2	95.9	77.9	87.0	93.2
Subspace Frob				97.8	87.6	82.3			
Subspace Geodesic				70.5	90.3	70.0	59.5	68.5	70.6
Test									
NCD	60.1	64.0	63.8	63.1	62.1	61.6			
BoW Euclid	65.4	69.1	69.8	60.4	68.7	72.2	74.3	78.5	76.1
BoW KL	58.4	58.5	60.3	63.5	68.6	70.5	72.1	71.1	72.8
Subspace Frob	60.6	61.6	62.3	51.5	49.7	50.0			
Subspace Geodesic	51.2	51.3	52.3	50.4	52.7	52.3	51.0	51.5	53.5

First, a big difference between training and test can be detected. This observation leads to the conclusion that the methods have the tendency to overadapt to the training data. Thus, carefull training data generation is demanded.

Second, the most successful dissimilarity measure in [5] was the Levenstein-distance [42] yielding an accuracy of 68.4%, which, however, is not alignment free. Our best classification results delivers GLVQ for BoW with the Euclidean distance followed by BoW with Kullback-Leibler-divergence, both outperforming the reported result as well as better than a simple kNN.

Otherwise, the subspace methods and the NCD showed only a weak performance in test. The letter one, however, is still better than the NCD-result of 51.6% reported in [5]. Yet, the generally weak results for these dissimilarities are surprising, because these dissimilarities/methods delivered promising results in other sequence analysis considerations [20].

One idea to reduce the discrepancy between the good training results and the weak test accuracy is to investigate adequate regularization techniques.

6   Conclusion

In this study we considered several dissimilarity measures and LVQ methods for classification of biological and artificial RNA sequences. The discrimination of those sequences is important for detection of life signatures in RNA to better understand the origin of life.

Depending on the dissimilarity measure in use an appropriate LVQ variant has to be applied. In case of non-differentiable dissimilarity measures we used Median GLVQ. For the subspace geodesic path distance the adaptation of the prototypes in GLVQ were done along the geodesic path.

The best classification results are delivered for BoW-coding of the sequences with the Euclidean distance as dissimilarity measure. The accuracy clearly outperforms best results obtained so far using rank statistics instead of classification schemes, which shows that GLVQ delivers better results than the rank statistics.

Future work has to pay attention to the large discrepancy between training and test results signaling the tendency of overfitting. This is particular valid for the subspace methods. Here, regularization techniques should be incorporated. Furthermore, additional dissimilarity measure should be considered including approaches based on Fourier analysis of sequence matrices [43], chaos game representation of sequences [44] and, natural vector representation [45–47].

Acknowledgement. M.K. was supported by grants of the European Social Fond (ESF) for a Young Researcher Group 'MACS' in cooperation with the TU Bergakademie Freiberg (Germany) and for the project titled 'Digitale Produkt- und Prozessinnovationen' at the UAS Mittweida.

References

1. Gilbert W (1986) Origin of life: the RNA world. Nature 319(6055):618
2. Neveu M, Kim H-J, Benner SA (2013) The "Strong" RNA world hypothesis: fifty years old. Astrobiology 13(4):391–403
3. Rich A (1962) On the problems of evolution and biochemical information transfer. In: Kasha M, Pullman B (eds) Horizons in biochemistry. Academic Press, pp 103–126
4. Cech TR (2011) The RNA worlds in context. Cold Spring Harb Perspect Biol 4(7):a006742
5. Wasik S, Szostak N, Kudla M, Wachowiak M, Krawiec K, Blazewicz J (2019) Detecting life signatures with RNA sequence similarity measure. J Theor Biol 463:110–120
6. Szostak N, Synak J, Borowski M, Wasik S, Blazewicz J (2017) Simulating the origins of life: the dual role of RNA replicases as an obstacle to evolution. PLOS ONE 12(7):1–28
7. Eigen M (1971) Selforganization of matter and the evolution of biological macromolecules. Die Naturwiss 58(10):465–523
8. Quastler H (1953) Essays on the use of information theory in biology. University of Illinois Press, Urbana
9. Szostak N, Wasik S, Blazewicz J (2017) Understanding life: a bioinformatics perspective. Eur Rev 25(2):231245

10. Kohonen T (1988) Learning vector quantization. Neural Netw 1(Suppl. 1):303
11. Sato A, Yamada K (1996) Generalized learning vector quantization. In: Touretzky DS, Mozer MC, Hasselmo ME (eds) Advances in neural information processing systems, vol 8. Proceedings of the 1995 Conference. MIT Press, Cambridge, pp 423–429
12. Nebel D, Hammer B, Frohberg K, Villmann T (2015) Median variants of learning vector quantization for learning of dissimilarity data. Neurocomputing 169:295–305
13. Wasik S, Prejzendanc T, Blazewicz J (2013) ModeLang - a new approach for experts-friendly viral infections modeling. Comput Math Methods Med 2013:8
14. Wasik S (2018) Modeling biological systems using crowdsourcing. Found Comput Decis Sci 43(3):219–243
15. Guogas L, Hogle J, Gehrke L (2004) Origins of life and the RNA world: evolution of RNA-replicase recognition. In: Norris R, Stootman F (eds) Bioastronomy 2002: life among the stars. IAU Symposium, vol 213, p 321, June 2004
16. Brister JR, Ako-adjei D, Bao Y, Blinkova O (2014) NCBI viral genomes resource. Nucleic Acids Res 43(D1):D571–D577
17. Eigen M, Schuster P (1982) Stages of emerging life—five principles of early organization. J Mol Evol 19(1):47–61
18. Sharp SJ, Schaack J, Cooley L, Burke DJ, Söll D (1985) Structure and transcription of eukaryotic tRNA genes. CRC Crit Rev Biochem 19(2):107–144
19. Azad RK, Li J (2013) Interpreting genomic data via entropic dissection. Nucleic Acids Res 41(1):e23
20. Mohammadi M, Biehl M, Villmann A, Villmann T (2017) Sequence learning in unsupervised and supervised vector quantization using Hankel matrices. In: Rutkowski L, Korytkowski M, Scherer R, Tadeusiewicz R, Zadeh LA, Zurada JM (eds) Proceedings of the 16th international conference on artificial intelligence and soft computing - ICAISC. LNAI, Zakopane. Springer, Cham, pp 131–142
21. Blaisdell BE (1986) A measure of the similarity of sets of sequences not requiring sequence alignment. Proc Natl Acad Sci USA 83:5155–5159
22. Vinga S, Almeida JS (2004) Alignment-free sequence comparison – a review. Bioinformatics 20(2):206–215
23. Cilibrasi R, Vitányi PMB (2005) Clustering by compression. IEEE Trans Inf Theory 51(4):1523–1545
24. Li M, Chen X, Li X, Ma B, Vitanyi PMB (2004) The similarity metric. IEEE Trans Inf Theory 50(12):3250–3264
25. Kolmogorov AN (1965) Three approaches to the quantitative definition of information. Probl Inf Transm 1(1):1–7
26. Ziv J, Lempel A (1977) A universal algorithm for sequential data compression. IEEE Trans Inf Theory 23(3):337–343
27. Huffman D (1952) A method for the construction of minimum-redundancy codes. Proc IRE 40(9):1098–1101
28. Vinga S (2004) Information theory applictions for biological sequence analysis. Bioinformatics 15(3):376–389
29. Vinga S, Almeida JS (2004) Rényi continuous entropy of DNA sequences. J Theor Biol 231:377–388
30. Fianacca A, LaPaglia L, LaRosa M, LoBosco G, Renda G, Rizzo R, Galio S, Urso A (2018) Deep learning models for bacteria taxonomic classification of metagenomic data. BMC Bioinform 19(Suppl. 7):198
31. Rényi A (1961) On measures of entropy and information. In: Proceedings of the fourth Berkeley symposium on mathematical statistics and probability. University of California Press, Berkeley

32. Nguyen NG, Tran VA, Ngo DL, Phan D, Lumbanraja FR, Faisal MR, Abapihi B, Kubo M, Satou K (2016) DNA sequence classification by convolutional neural network. J Biomed Sci Eng 9:280–286

33. Hamm J, Lee DD (2008) Grassmann discriminant analysis: a unifying view on subspace-based learning. In: Proceedings of the 25th international conference on machine learning, pp 376–388

34. Absil P-A, Mahony R, Sepulchre R (2004) Riemannian geometry of Grassmann manifolds with a view on algorithmic computation. Acta Appl Math 80:199–220

35. Wedin PA (1983) On angles between subspaces of a finite dimensional inner product space. Lecture notes in mathematics, vol 973. Springer, Heidelberg, pp 263–285

36. Nebel D, Kaden M, Villmann A, Villmann T (2017) Types of (dis−)similarities and adaptive mixtures thereof for improved classification learning. Neurocomputing 268:42–54

37. Kaden M, Riedel M, Hermann W, Villmann T (2015) Border-sensitive learning in generalized learning vector quantization: an alternative to support vector machines. Soft Comput 19(9):2423–2434

38. Kirby M, Peterson C (2017) Visualizing data sets on the Grassmannian using self-organizing maps. In: Proceedings of the 12th workshop on self-organizing maps and learning vector quantization (WSOM 2017), Nancy, France. IEEE Press, Los Alamitos, pp 32–37

39. Villmann T (2017) Grassmann manifolds, Hankel matrices and tangent metric models in classification learning. Mach Learn Rep 11(MLR-02-2017):22–25 http://www.techfak.uni-bielefeld.de/~fschleif/mlr/mlr_0_2017.pdf, ISSN:1865-3960

40. Hammer B, Hofmann D, Schleif F-M, Zhu X (2014) Learning vector quantization for (dis-)similarities. Neurocomputing 131:43–51

41. Pekalska E, Duin RPW (2006) The dissimilarity representation for pattern recognition: foundations and applications. World Scientific, Singapore

42. Levenshtein VI (1966) Binary codes capable of correcting deletions, insertions, and reversals. Sov Phys Dokl 10:707–710

43. Yin C, Chen Y, Yau SS-T (2014) A measure of DNA sequence similarity by fourier transform with applications on hierarchical clustering. J Theor Biol 359:18–28

44. Almeida JS, Carrico JA, Maretzek A, Noble PA, Fletcher M (2001) Analysis of genomic sequences by chaos game representation. Bioinformatics 17(5):429–437

45. Deng M, Yu C, Liang Q, He RL, Yau SS-T (2011) A novel method of characterizing sequences: genome space with biological distance and applications. PLoS ONE 6(3):e17293

46. Li Y, He L, He RL, Yau SS-T (2017) A novel fast vector method for genetic sequence comparison. Nat Sci Rep 7(12226):1–11

47. Li Y, Tian K, Yin C, He RL, Yau SS-T (2016) Virus classification in 60-dimensional protein space. Mol Phylogenet Evol 99:53–62

Simultaneous Display of Front and Back Sides of Spherical SOM for Health Data Analysis

Niina Gen[1], Tokutaka Heizo[2(✉)], Ohkita Masaaki[2],
and Kasezawa Nobuhiko[3]

[1] GAUSS, Inc., Tokyo, Japan
[2] Tottori University, Tottori, Japan
tokuhema@hal.ne.jp
[3] Fuji-Ikiiki Hospital, Fuji, Japan

Abstract. We propose to simultaneously display the front and back sides of the spherical SOM so that cluster locations can be expressed in terms of phase relations even if they are on opposite sides of the spherical SOM. The technique is showcased on the medical health care-data. Furthermore, the component map was converted numerically (DIM (Dimensional Interaction Map) mode) for the medical data case and the result compared with that obtained with the front and back sides of the map.

Keywords: Spherical SOM · Simultaneous front and back display · Health care data · DIM mode (Component numerical analysis)

1 Introduction

A self-organizing Map (SOM) develops a low-dimensional discrete representation, usually in terms of a flat lattice of neurons, of the originally high-dimensional data manifold. In order to curb distortions due to the finite nature of the lattice (border effect), Helge Ritter proposed the spherical self-organizing map (SOM), in which case neurons are arranged on a spherical surface [1]. The spherical SOM has been shown to be beneficial for clustering analysis of various data sets [2, 3] as clusters are represented onto the spherical surface thereby preserving possible phase relationships, a clear advantage over the flat SOM (cf., border effect). A drawback of the spherical SOM, however, is the lack of overview of the whole lattice. In this contribution, a method is proposed that enables one to simultaneously view the front- and the back sides of the spherical lattice. In this way, the spherical surface SOM can enjoy the benefit of the planar SOM in providing an overview of the whole map [4, 5]. We demonstrate our technique on clustering the data sets of the prefectural medical health data (Sect. 3). A spherical SOM on 642 nodes was trained on the data using 500 iterations. Finally, for the data set, we also analyzed the results using the newly developed DIM mode (numerical conversion of component map). The algorithm for the DIM mode is described in the following section.

T. Heizo and O. Masaaki—Emeritus Professor.

A. Vellido et al. (Eds.): WSOM 2019, AISC 976, pp. 334–339, 2020.
https://doi.org/10.1007/978-3-030-19642-4_33

2 Analysis for Relationship Between Components

In the DIM mode of our tool [5], one is able to assess the strength (intensity) of each component in the map, roughly divided into red and blue intensities. Due to the characteristics of the DIM mode, the MAX value of each component does not necessarily correspond to red. Let us denote the red intensity at each position i on the map for component k as $R_k[i]$ and the intensity of blue as $B_k[i]$. The similarity of the distribution of the components n and m on the map is defined by the following equations:

$$f(R_n, R_m) = \frac{\sum \left(R_n[i] - \overline{R_n} \right) \left(R_m[i] - \overline{R_m} \right)}{\sqrt{\sum \left(R_n[i] - \overline{R_n} \right)^2} \sqrt{\sum \left(R_m[i] - \overline{R_m} \right)^2}} \tag{1}$$

$$f(B_n, B_m) = \frac{\sum \left(B_n[i] - \overline{B_n} \right) \left(B_m[i] - \overline{B_m} \right)}{\sqrt{\sum \left(B_n[i] - \overline{B_n} \right)^2} \sqrt{\sum \left(B_m[i] - \overline{B_m} \right)^2}} \tag{2}$$

where, the value $\overline{R_k}$ or $\overline{B_k}$ represents mean value.

$$\text{component similarity} = \begin{cases} f(R_n, R_m) & \text{if } |f(R_n, R_m)| \geq |f(B_n, B_m)| \\ f(B_n, B_m) & \text{if } |f(R_n, R_m)| < |f(B_n, B_m)| \end{cases} \tag{3}$$

Equation (1) shows the similarity of the distribution on the map with respect to the intensity of red, Eq. (2) the similarity of the distribution on the map with respect to the intensity of blue. When the component similarity equals to 0.35 or more, then it is decided that there exists a positive relation, when it is 0.65 or more, it is decided that the relation is strong. When the component similarity is between −0.35 and −0.65, then it is deemed that there is a negative relation, when it is less than −0.65, it is decided that the relation is strong.

3 Analysis of Health Data at Prefectural Level

We have considered health data released by 47 prefectures in Japan. Table 1 shows part of the data with 11 prefectures shown row-wise and the 15 inspected components listed column wise. Data are normalized row-wise.

In the dendrogram of Fig. 1, at first glance, the relationship between the 15 components cannot be properly understood. When training the spherical SOM on the other hand, possible relations between these components can be uncovered by inspecting the front and back views simultaneously (Fig. 2).

From Fig. 2, we observe that health checks are successful in locations where the number of public health nurses is large, resulting in low cancer death rate.

From here onwards, we report our results with the DIM (Dimensional Interaction Map) mode, the component map of the spherical SOM organized by the weights of each node[x]. When components are similar, we express this as a positive similarity,

Table 1. Excerpt of prefectural health data.

	Hokkaido	Aomori	Iwate	Miyagi	Akita	Yamagata	Fukushima	Ibaraki	Tochigi
DockLifestyleRelatedDiseaseItemAbnormality	0.6816	0.4358	0.4358	0.5866	0.492	0.4916	0.5698	0.3464	0.486
MetabolicMedicalExpenses	0.2334	0.3794	0.5356	0.2458	0.454	0.2014	0.3539	0.718	0.1107
ExerciseHabits	0.4535	0.4234	0.4264	0.6667	0.541	0.3994	0.5195	0	0.7387
RatioOfHigh-doseDrinkers	0.4444	0	0.4242	0.3434	0.081	0.2727	0.3939	0.3939	0.1818
SmokingHabits	0	0	0.4009	0.1582	0.474	0.5581	0.2525	0.3612	0.2249
MeatFishRatio	0.5834	0.8377	0.8232	0.6721	0.736	0.5238	0.8349	0.6326	0.5635
Lifestyle-relatedDiseaseMortalityRate	0.4984	0.2302	0.2159	0.6416	0.064	0.1533	0.3248	0.559	0.5711
CancerMortalityRate	0.373	0.1543	0.403	0.6607	0	0.1566	0.4764	0.6135	0.6764
NumberOfPublicHealthNurses	0.6895	0.5428	0.6112	0.4059	0.619	0.6186	0.5086	0.2665	0.3545
65YearsOldLifeExpectancy	0.5068	0	0.2635	0.4932	0.108	0.4392	0.223	0.2838	0.1081
MedicalExpenseForTheElderly	0.0036	0.6997	0.851	0.6616	0.675	0.8415	0.6645	0.7863	0.7928
PrefecturalIncome	0.3251	0.312	0.2454	0.3799	0.265	0.2846	0.5561	0.6867	0.6697
LaborForcePopulationratio	0.1905	0.5833	0.7024	0.381	0.417	0.5952	0.5119	0.7024	0.7857
MealTime	0.0769	0.3077	0.4615	0.3846	0.923	0.3077	0.6154	0.6923	0.4615
MedicalExamination-medicalTreatmentTime	0.6667	0.1667	0.5	0.5	0.167	1	0	0.3333	0.5

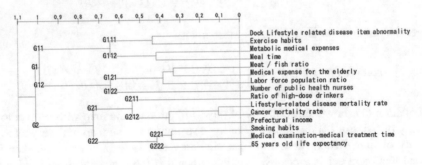

Fig. 1. The dendrogram up to nc 3 (level 3) obtained after clustering data of Table 1 with spherical SOM.

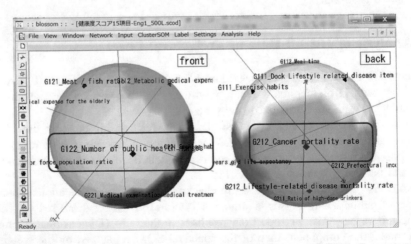

Fig. 2. The example with G122_Number of public health nurses and G212_Cancer mortality rate marked.

in the opposite case as a negative similarity. The original data of Table 1 is displayed in Table 2 that lists 15 components for each one of 47 prefectures were visualized and the positive and negative similarities are shown in Fig. 3.

Table 2. 47 prefectures (column) with their 15 specific health-related components, normalized row-wise. Row labels are difficult to see due to the convenience of space, but the necessary details are described in Fig. 3 and Table 3.

	A	B	C	D	E	F	G	H	I	J	K	L	M	N	O	P
	DockLifestyleRelatedDisease	cMedica	ercise Hal	High-dose	oking Hal	atFishRae	dDiseaser	Mortalit	PublicHe	)ldLifeExpenseFor	ecturalIncce	Popula	MealTim	don-medicalT		
Hokkaido	0.6816	0.2334	0.4535	0.4444	0	0.5834	0.4984	0.373	0.6895	0.5068	0.0036	0.3251	0.1905	0.0769	0.6667
Aomori	0.4358	0.3794	0.4234	0	0	0.8377	0.2302	0.1543	0.5428	0	0.6997	0.312	0.5833	0.3077	0.1667
Iwate	0.4358	0.5356	0.4264	0.4242	0.4009	0.8232	0.2159	0.403	0.6112	0.2635	0.851	0.2454	0.7024	0.4615	0.5
Miyagi	0.5866	0.2458	0.6667	0.3434	0.1582	0.6721	0.6416	0.6607	0.4059	0.4932	0.6616	0.3799	0.381	0.3846	0.5
Akita	0.4916	0.4544	0.5405	0.0808	0.4736	0.7357	0.0635	0	0.6186	0.1081	0.6747	0.265	0.4167	0.9231	0.1667
Yamagata	0.4916	0.2014	0.3994	0.2727	0.5581	0.5238	0.1533	0.1566	0.6186	0.4392	0.8415	0.2846	0.5952	0.3077	1
Fukushima	0.5698	0.3539	0.5195	0.3939	0.2525	0.8349	0.3248	0.4764	0.5086	0.223	0.6645	0.5561	0.5119	0.6154	0
Ibaraki	0.3464	0.718	0	0.3939	0.3612	0.6326	0.559	0.6135	0.2665	0.2838	0.7863	0.6867	0.7024	0.6923	0.3333
Tochigi	0.486	0.1107	0.7387	0.1818	0.2249	0.5635	0.5711	0.6764	0.3545	0.1081	0.7928	0.6697	0.7857	0.4615	0.5
Gunnma	0.486	0.8778	0.7057	0.2626	0.2449	0.8894	0.561	0.6045	0.5306	0.5135	0.6903	0.5235	0.8214	0.6923	0.6667
Saitama	0.648	0.6649	0.7988	0.4545	0.2482	0.6447	0.8978	0.9124	0.0416	0.4662	0.6079	0.6802	0.7738	0.8462	0.3333
Chiba	0.4525	0.6595	0.3123	0.6667	0.2607	0.6568	0.8238	0.8742	0.1834	0.5338	0.781	0.7082	0.5595	0.6923	0.3333
Tokyo	0.6257	0.5647	0.5946	0.2626	0.4262	0.5977	0.8527	0.791	0.0709	0.777	0.4897	1	0	1	0.3333
Kanagawa	0.7654	0.4146	0.8048	0.3939	0.3056	0.5211	0.9654	0.9064	0	0.7432	0.617	0.8531	0.6667	0.3846	0.3333
Niigata	0.3464	0.446	0.5075	0.3333	0.3571	0.8928	0.2898	0.3251	0.5306	0.5608	0.8546	0.4745	0.6429	0.4615	0.5
Toyama	0.2402	0.7941	0	0.9596	0.6212	0.8503	0.3829	0.2734	0.6797	0.5743	0.601	0.6906	0.7738	0.2308	0.6667
Ishikawa	0.486	0.3172	0.5676	0.5152	0.3357	0.7334	0.5683	0.5341	0.4841	0.6351	0.338	0.6051	0.5714	0.2308	0.1667

The first row in the table is discribed as follows: DockLifestyleRelatedDiseaseItemAbnormality, MetabolicMedicalExpenses, ExerciseHabits, RatioOfHigh- doseDrinkers, SmokingHabits, MeatFishRatio, Lifestyle-relatedDiseaseMortalityRate, CancerMortalityRate, NumberOfPublicHealthNurses, 65YearsOldLifeExpectancy, MedicalExpenseForTheElderly, PrefecturalIncome, LaborForcePopulationratio, MealTime, MedicalExamination-medicalTreatmentTime

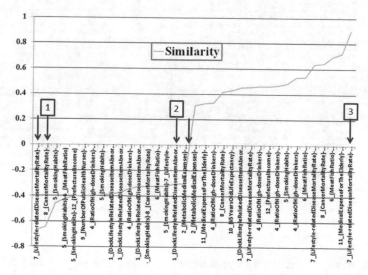

Fig. 3. Part of the DIM-mode results (similarities) calculated from the health data of Table 2.

Figure 3 shows the similarity in the health data for 35 combinations. Originally, there are as many as 105 combinations possible but we omitted 70 with intermediate similarity levels. Since the component names are quite long, for the sake of readability, we have omitted the opposite components (see Table 3 for a complete description). Also, the most strong inverse relationship is the "7_(Lifestyle-related Disease Mortality Rate) 9_(Number of Public Health Nurses)", on the left side. In other words, in prefectures with few public health nurses disease mortality rate is also high.

Numbers 7 and 9 refer to the order in which the components appear in Table 2. Conversely, the right end shows the highest similarity relationship. The relation is "7_(Lifestyle-related Disease Mortality Rate)-8_(Cancer Mortality Rate)". This is to say that the highest "death rate" is the "cancer mortality rate" even if it is seen by prefecture data.

Next, we compare the front/back simultaneous display of the spherical surface SOM with the DIM mode results. In particular, we focus on No. 1, indicated by the arrows in Fig. 3. Note that No. 1 is shown in Fig. 2. Finally, the example at the far right of No. 3 is shown in Fig. 3. Note that the discussion of No. 2 in Fig. 3 is omitted for the sake of space.

Table 3. Excerpt of Fig. 3. Blue and red arrows indicate continuation of the table. Marked in yellow are the examples that are further discussed.

Component relation	Similarity		Description
7_(Lifestyle-relatedDiseaseMortalityRate) 9_(NumberOfPublicHealthNurses)	-0.6673	●	Strong reverse relation
8_(CancerMortalityRate) 9_(NumberOfPublicHealthNurses)	-0.64676	○	reverse relation
5_(SmokingHabits) 11_(MedicalExpenseForTheElderly)	-0.46017	○	reverse relation
5_(SmokingHabits) 6_(MeatFishRatio)	-0.42521	○	reverse relation
5_(SmokingHabits) 12_(PrefecturalIncome)	-0.40125	○	reverse relation
9_(NumberOfPublicHealthNurses) 12_(PrefecturalIncome)	-0.4006	○	reverse relation
4_(RatioOfHigh-doseDrinkers) 6_(MeatFishRatio)	-0.39816	○	reverse relation
5_(SmokingHabits) 13_(LaborForcePopulationratio)	-0.3963	○	reverse relation
1_(DockLifestyleRelatedDiseaseItemAbnormality) 6_(MeatFishRatio)	-0.38213	○	reverse, intertwined with multiple relationships or nonlinear
1_(DockLifestyleRelatedDiseaseItemAbnormality) 15_(MedicalExamination-medicalTreatmentTime)	-0.37751	○	reverse, intertwined with multiple relationships or nonlinear
4_(RatioOfHigh-doseDrinkers) 11_(MedicalExpenseForTheElderly)	-0.3401	△	weak, but a little reverse relationship.
1_(DockLifestyleRelatedDiseaseItemAbnormality)-5_(SmokingHabits)	-0.326	△	weak, but a little reverse relationship.
5_(SmokingHabits)-8_(CancerMortalityRate)	-0.32536	△	weak, but a little reverse relationship.
6_(MeatFishRatio)-15_(MedicalExamination-medicalTreatmentTime)	-0.3248	△	weak, but a little reverse relationship.
5_(SmokingHabits)-7_(Lifestyle-relatedDiseaseMortalityRate)	-0.31996	△	weak, but a little reverse relationship.
1_(DockLifestyleRelatedDiseaseItemAbnormality)-2_(MetabolicMedicalExpenses)	-0.30379	△	weak, but a little reverse relationship.
2_(MetabolicMedicalExpenses)-12_(PrefecturalIncome)	-0.30217	△	weak, but a little reverse relationship.
5_(SmokingHabits)-14_(MeatTime)	-0.25027	×	very weak reverse relation
10_(65YearsOldLifeExpectancy)-11_(MedicalExpenseForTheElderly)	-0.15231	×	very weak reverse relation
2_(MetabolicMedicalExpenses)-5_(SmokingHabits)	0.309209	△	weak, but a little positive relationship.

Component relation	Similarity		Description
4_(RatioOfHigh-doseDrinkers)-12_(PrefecturalIncome)	0.333443	△	weak, but a little positive relationship.
8_(CancerMortalityRate)-10_(65YearsOldLifeExpectancy)	0.415392	○	positive, intertwined with multiple relationships or nonlinear
10_(65YearsOldLifeExpectancy)-15_(MedicalExamination-medicalTreatmentTime)	0.430149	○	positive relation
1_(DockLifestyleRelatedDiseaseItemAbnormality)-3_(ExerciseHabits)	0.452903	○	positive relation
7_(Lifestyle-relatedDiseaseMortalityRate)-10_(65YearsOldLifeExpectancy)	0.455943	○	positive, intertwined with multiple relationships or nonlinear
4_(RatioOfHigh-doseDrinkers)-15_(MedicalExamination-medicalTreatmentTime)	0.461062	○	positive relation
12_(PrefecturalIncome)-13_(LaborForcePopulationratio)	0.462202	○	positive relation
4_(RatioOfHigh-doseDrinkers)-5_(SmokingHabits)	0.468077	○	positive relation
5_(SmokingHabits)-15_(MedicalExamination-medicalTreatmentTime)	0.483858	○	positive relation
4_(RatioOfHigh-doseDrinkers)-10_(65YearsOldLifeExpectancy)	0.533077	○	positive, intertwined with multiple relationships or nonlinear
6_(MeatFishRatio)-13_(LaborForcePopulationratio)	0.535449	○	positive relation
7_(Lifestyle-relatedDiseaseMortalityRate)-12_(PrefecturalIncome)	0.636102	○	positive relation
8_(CancerMortalityRate)-12_(PrefecturalIncome)	0.644258	○	positive relation
6_(MeatFishRatio)-11_(MedicalExpenseForTheElderly)	0.694706	●	strong positive relation
11_(MedicalExpenseForTheElderly)-13_(LaborForcePopulationratio)	0.721769	●	strong positive relation
7_(Lifestyle-relatedDiseaseMortalityRate)-8_(CancerMortalityRate)	0.900828	●	strong positive relation

However, since there are a large number of entries in the range from −0.30 to 0.30 we just consider two of them (marked in yellow) in our discussion. In the table, the up and down red arrows indicate the continuation of the table.

The upper part that is labeled in the front view in Fig. 4 is the Cancer Mortality Rate, the lower part the Life style-related Disease Mortality Rate.

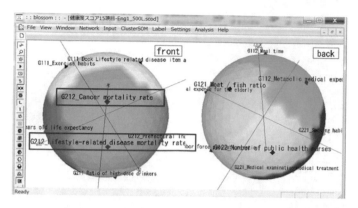

Fig. 4. Example at the right end of 7_(Lifestyle-related Disease Mortality Rate)-8_(Cancer Mortality Rate) in Fig. 3. The DIM relation is shown in the arrow No. 3 of Fig. 3.

4 Summary

We proposed to simultaneously inspect the front and back views of a spherical SOM so as to parallel the overview offered by the flat SOM. The spherical SOM has the additional advantage in supporting the mutual locations of the clusters to be expressed in terms of phase relations and we showcased this on the data sets of the prefectural health data. By simultaneously inspecting the front and back views of the sphere, we could unveil and draw conclusions about clusters on opposite sides. Furthermore, analysis in DIM mode of the health data map was performed. Using the method, the relationship between the front and back views was discussed more quantitatively.

Acknowledgement. We sincerely thankful to Prof. M. V. Hulle of KU Leuven for correcting our English.

References

1. Ritter H (1999) Self-organizing maps on non-Eucledian spaces. In: Oja E, Kaski S (eds) Kohonen maps. Elsevier, pp 95–110
2. Oyabu M (2002) Development of spherical SOM. In: Proceedings of SCI, Orland, Florida, USA, XII, pp 384–387
3. Schmidt CR, Rey SJ, Skupin A (2010) Effects of irregular topology in spherical self-organizing maps. In: International regional science review, pp 215–229, 30 December 2010
4. Tokutaka H, Ohkita M, Kasezawa M, Niina G (2018) Simultaneous display of front and back sides by spherical SOM method for health data analysis. In: SOM study meeting, Hiroshima, 21 March 2018
5. Tokutaka H, Ohkita M, Kasezawa N, Niina G (2018) Simultaneous display of front and back sides by spherical SOM method for the analysis of the health data and other. In: BMFSA 2018 conference, Kanazawa, 3–4 November 2018

Author Index

© Springer Nature Switzerland AG 2020
A. Vellido et al. (Eds.): WSOM 2019, AISC 976, pp. 341–342, 2020.
https://doi.org/10.1007/978-3-030-19642-4

Printed in the United States
By Bookmasters